Developments in Geotectonics 11

STYLES OF FOLDING

FURTHER TITLES IN THIS SERIES

1. J. AUBOUIN
GEOSYNCLINES

2. R.W. VAN BEMMELEN
GEODYNAMIC MODELS

3. A. SUGIMURA AND S. UYEDA
ISLAND ARCS, JAPAN AND ITS ENVIRONS

4. A.R. RITSEMA (Editor)
THE UPPER MANTLE

5. C. LOMNITZ
GLOBAL TECTONICS AND EARTHQUAKE RISK

6. X. LE PICHON, J. FRANCHETEAU AND J. BONNIN
PLATE TECTONICS

7. R.W. GIRDLER (Editor)
EAST AFRICAN RIFTS

8. S. MUELLER (Editor)
THE STRUCTURE OF THE EARTH'S CRUST

9. N. PAVONI AND R. GREEN (Editors)
RECENT CRUSTAL MOVEMENTS

10. S.W. CAREY
THE EXPANDING EARTH

Developments in Geotectonics 11

STYLES OF FOLDING

MECHANICS AND MECHANISMS OF FOLDING OF NATURAL ELASTIC MATERIALS

by
ARVID M. JOHNSON
Stanford University, Stanford, Calif., U.S.A.

with shorter contributions by
Elmont Honea
Ze'ev Reches
Benjamin M. Page
Stephenson D. Ellen
and
Ivar B. Ramberg

ELSEVIER SCIENTIFIC PUBLISHING COMPANY
Amsterdam - Oxford - New York 1977

ELSEVIER SCIENTIFIC PUBLISHING COMPANY
335 Jan van Galenstraat
P.O. Box 211, Amsterdam, The Netherlands

Distributors for the United States and Canada:

ELSEVIER NORTH-HOLLAND INC.
52, Vanderbilt Avenue
New York, N.Y. 10017

Library of Congress Cataloging in Publication Data

Johnson, Arvid M
Styles of folding.

(Developments in geotectonics ; 11)
Includes bibliographies and indexes.
1. Folds (Geology) 2. Geology--Utah--Arches National Park. 3. Geology--California--Coast Range.
I. Title. II. Series.
QE606.J63 551.8'7 76-45457
ISBN 0-444-41496-7

Printed in The Netherlands

This book is dedicated to my wife,

DEANNA DUNN JOHNSON

for patience, love, and tolerance during the long periods of comatose behavior while I was doing this research.

PREFACE

The research reported in the eight papers comprising Chapters 2 through 9 was done over a nine year period, starting in 1966, but most of it was done between September of 1972 and September of 1975. I first became interested in folding theory as a graduate student at The Pennsylvania State University, in 1962. At that time Professor Laurence Lattman, currently Dean of Earth Sciences at the University of Utah, led a departmental field trip, in part to see folds in West Virginia which had been studied by Currie, Patnode and Trump of Gulf Research in Pittsburg. I later read the paper by these investigators and was most impressed by the combination of field work, theoretical analysis and experimentation which the paper displayed.

After I joined the faculty at Stanford University, Stephenson Ellen, a former student, and I began a study of folds, as described in the introduction of Chapter 3. Stephenson examined folds in many parts of the United States. We were disappointed that none of the folds appeared to be sinusoidal in form, as we believed to be indicated by theories by Currie et al., Biot and Ramberg. Indeed, most of the folds appeared to be kink-like, chevron-like, or a combination of arc and cusp structures resembling parts of concentric folds. In 1967, as I remember, Maurice Biot came to Stanford to give a few lectures in the departments of Applied Mechanics and Geology. Stephenson Ellen and I told him about the folds we had seen, including those in the Franciscan Complex near San Francisco (described in Chapter 7), and asked him how we might proceed to understand the folds in the Franciscan. He indicated that he already had developed the theory for such folds and had presented it in his book (Biot, 1965a) *, and recommended that we study the book. We organized a seminar based on Biot's book, and the struggle to understand Biot's theories was thus begun and is still incomplete.

During this time I took my family on a vacation through parts of Utah and Arizona and one of our stops was at Arches National Monument, near Moab, Utah. The paper in Chapter 2, published in 1969, is a result of studying the folds within the monument and attempting to understand part of Biot's theory of folding. The theory presented in Chapter 2 is based on Biot's theory for viscous materials but was completely rederived. The wavelengths are equivalent to those derived earlier by Biot, although details of the theory are somewhat different. The content of that paper was rewritten and formed one chapter of my textbook (Johnson, 1970, Chapter 6).

Many of the folds in the Carmel Formation at Arches National Monument are more complex than those described in the paper. In the paper I indicated that the folds probably were a result of regional shortening. However, the folds are widespread in the Carmel even where there is no evidence of regional shortening, as in the area of the Henry Mountains and San Rafael Swell. The Carmel typically is rich in gypsum so it is possibly that the folds are a result of volume changes associated with the conversion of anhydrite to gypsum as well as of regional shortening. However, the source of the layer-parallel normal stresses that produced the folds does not enter the theoretical analysis. Certainly the most important contribution of the analysis is the application of sinusoidal folding theory of multilayers to gain some understanding of a field example. That example remains as one of the few that seems to correlate well with a theoretical fold pattern that Biot called "internal instability".

Chapter 3, which is the introduction to a series of papers begun in 1972, indicated an early plan of our research into folding theory. The plan was to develop theory to explain conditions under which a variety of fold types might develop. At that time we believed that a key to understanding of fold forms rested in the theory of characteristics. The idea was that different fold types involved either no discontinuities (similar folds) or discontinuities in slope (chevron and kink folds) or discontinuities in curvature (concentric). Elmont Honea and I tried many times, over a period of more than a year, to derive a set of hyperbolic equations for elastic material which would allow the various discontinuities to exist, according to the plan set out in Chapter 3. Thanks in part to criticisms by Hans Ramberg, Maurice Biot, Peter Cobbold and David Pollard, we were finally dissuaded from this approach. Further, we could not derive four real characteristic directions as required for there to be discontinuities. Thus, we were forced to take another approach, as described in Chapter 5. I consider the approach to understand different fold forms reported in Chapter 3 to be interesting, but not partic-

---

* See references at end of Chapter 1.

ularly fruitful. I have since derived hyperbolic equations with four real characteristics by considering a multilayer with contact strength, but the analysis will not be presented because I fail to recognize its value.

Chapter 4 is a brief, but I hope clear, presentation of linearized equations describing the incremental deformation of materials with memory. The analysis appears to be exact through eqs. 20 of that chapter, and the resulting equations are special cases of equations of three-dimensional elasticity derived by Novozhilov (1953). The remainder of the equations are different from those derived by Novozhilov, and they incorporate two special assumptions. One of the goals of Chapter 4 was to derive equilibrium equations and boundary conditions that are simpler than those derived by Biot (e.g., 1965a), and to eliminate several special assumptions in Biot's analysis. Unfortunately, the resulting equations are nearly as complex as Biot's. In Chapter 10 I have completely rederived the equations necessary for solving folding problems involving elastic materials and have removed the two special assumptions made in Chapter 4. The analysis is relatively easy to follow, yet is exact to first order. As indicated in Chapter 10, predictions of the two theories appear to be equivalent, at least for several folding problems solved by the two different methods.

One of the most important results of the derivations presented in Chapters 4 and 10 is that initial normal stresses change the resistance of the materials to incremental shear; the larger the compressive normal stresses, the lower the shear resistance of the material. This is discussed in some detail at the end of Chapter 4.

Chapter 5 uses the equations derived in Chapter 4 to solve for wavelengths of sinusoidal folds in single layers and multilayers. The theoretical predictions agree quite closely with experimental measurements using multilayers of gelatin and rubber. The most significant results are the explanations of "pinches" at the edges of deep beams of rubber and gelatin, the explanation of "elastic yielding" of crests and troughs of folds to produce chevron forms, and the explanation of the transition from sinusoidal to concentric-like to chevron forms. All these explanations are based on higher-order effects and are therefore unexplained in terms of the linearized equations alone. We arrived at these conclusions by means of a combination of theoretical analysis and measurements of experimental folds produced in single layers and multilayers. The conclusions should be valid for materials with "memory". Thus, I see no reason to change the explanations of chevron folding if yielding at the hinge is irrecoverable rather than recoverable or if the yielding is an expression of nonlinear elastic behavior as discussed elsewhere (Johnson, 1970, p. 310). The explanation for the transformation from sinusoidal to concentric-like to chevron forms is consistent with experimental observations, but I would be more nearly satisfied if it were based on a complete, higher-order analysis. Thus the explanations represent extrapolations of the theory we have developed and are therefore in need of verification or rejection in terms of exact analyses of higher-order effects.

Chapter 6 introduces a theory of kink folding in elastic multilayers. The theory is based on the mechanism of plastic yielding between layers, allowing contact strength to be overcome locally. Bending resistance of layers is required, apparently, to force widths of kink bands to be finite. The fundamental concept, however, is *yielding*. The concepts of yielding of contacts between layers and of yielding instability are more nearly completely developed in Chapter 8, where we are primarily concerned with asymmetric folding. The concept that the fundamental mechanism of kink folding is yielding of contacts was deduced theoretically by deriving the "kink form", described in Chapter 6. We deduced that folds would be kink-like in multilayers of nonlubricated strips of rubber and be sinusoidal in the same strips lubricated. Experiments verified this deduction, as shown in Chapter 6, so we proceeded to develop the theories explained there and in Chapter 8. This is one example where experiments were used truly as a verification as well as a guide for our analysis. The suspicion that a combination of initial layer-parallel shear stress and compression produce monoclinal kink bands derived from observations of experiments made by Stephenson Ellen (Johnson, 1970, p. 319). Although most of the theory and the experiments are restricted to elastic layers with contact strength, the fundamental mechanisms are independent of the rheological properties of the layers.

Chapter 7 deals with asymmetric folds in interbedded cherts and shales of the Franciscan Complex. I doubt that the Franciscan rocks behaved as specified by the theory of deformation of materials with "memory", as we assumed in applying some of the theory to understand the folds in these rocks. I also doubt that the theoretical and experimental observations on which our interpretation is based would be changed appreciably if we analyzed the rocks as viscous-plastic rather than as elastic-plastic materials. Nevertheless, the viscous-plastic or nonlinear viscous models of folding of multilayers with contact strength should be analyzed in detail because they may provide some surprises.

Chapter 9 is an analysis of some folds in Tertiary rocks in the Coast Ranges of California. The most important theoretical contributions are general solutions for folding of multilayers in confining media with a wide range of stiffnesses, a simplified method of estimating wavelengths of folds in multilayers containing layers with different thicknesses or properties, and a method of estimating ratios of principal horizontal stresses by measuring axial ratios of doubly plunging folds. Thus the field study of a specific area provided motivation for performing theoretical analyses which appear to have general value.

Chapter 10 is primarily concerned with some of the loose ends left in other chapters, as mentioned in earlier paragraphs, but it also extends the theory to include effects of initial irregularities on the growth of fold forms and to include specifically a nonlinear material of the power-law form, both for single-layer and multilayer folds. Also, dominant wavelengths in viscous materials and wavelengths requiring minimal compression in elastic materials, according to theories of single-layer folding by Biot, Ramberg, Fletcher, and the linearized theory presented here, are compared in Chapter 10.

The most important mechanisms of folding recognized thus far are summarized in Chapter 11 for multilayered materials with a wide range of properties. In addition, Chapter 11 summarizes the mechanical analyses of folding presented in Chapter 2 through 10, concerning folding of materials with memory. Thus I have attempted to gather together what I consider to be the most important principles developed during the writing of the papers, principles which are scattered almost at random throughout the papers, in part because the papers were written over a period of several years and in part because concepts were clarified or rejected as the work progressed.

We begin in Chapter 1 with a brief critical summary of most mechanisms and mechanical analyses of folding presented in English language literature up to the time of preparing this manuscript for publication. The chapter should help place the series of papers in perspective.

ACKNOWLEDGEMENTS

The development of the analyses of folding presented in this series of papers has been a group effort. Some members of the group appear as co-authors of the papers. Some others, however, have been recognized only through brief acknowledgements and others have not been mentioned at all heretofore. I wish to thank these people, while absolving them from responsibility for errors the book contains.

Dr. David Pollard, U.S. Geological Survey, reviewed every paper, usually in two drafts, suggested innumerable improvements, raised questions and made suggestions which, in several cases, led to complete revisions and new insights. Dr. Raymond Fletcher of Stanford University contributed similarly to improvements of the first chapter and the last four chapters. One of his excellent suggestions led to the derivation of the plan shapes of doubly plunging folds, reported in Chapter 9. Professor Hans Ramberg, Dr. Ove Stephansson, Dr. Peter Cobbold and Dr. Maurice Biot were largely responsible for the marked change in direction of the research between the papers in Chapters 3 and 4. Their criticisms and questions concerning methods and conclusions are gratefully acknowledged. Discussion of parts of the research following lectures at several universities were helpful in formulating and correcting ideas. In particular, brief discussions with Dr. Gerhard Oertel and Dr. Ronald Shreve of U.C. Los Angeles, with Dr. Lionnel Weiss of U.C. Berkeley, and with Dr. Ronald Parker of the University of Wyoming were most fruitful.

I am indebted to Perfecto Mari for preparing most of the many figures and photographs for this series of papers.

The plan of the series of papers was conceived while I was in Sweden as a Fellow of the John Simon Guggenheim Memorial Foundation, during 1971—72, and experimentation and writing were begun at that time. I am grateful to the Foundation for financial support and to Professor Hans Ramberg for providing office space and facilities during that wonderful, productive year in Sweden.

Earlier, joint research with Stephenson Ellen was supported by the Shell Oil Company Fund of the Department of Geology, Stanford, by the Penrose Fund of the Geological Society of America, and by a grant from the National Science Foundation to study intrusions, Grant no. GA-14899. Research during 1972—74 was supported by the National Science Foundation, Grant no. GA-36917, to study folding.

I would like to thank students who took my course on folding theory during 1974 for helping to formulate some of the problems studied in this series of papers. Attila Aydin, Tom Crough, Dave Scofield, Elizabeth Hill, Jon Fink and Ze'ev Reches were particularly helpful.

I found persons associated with *Tectonophysics* to be especially helpful with the manuscripts for the papers. I would like to be able to thank many unknown reviewers for suggestions and criticisms of the papers. Also, A.V. Stunt of the Editorial Office and J. Tuzo Wilson, Editor, were most courteous and helpful. I am grateful to *Tectonophysics* for being willing to accept rather long papers, which seem to be my style.

Finally, I wish to thank the co-authors, Dr. Elmont Honea of Humboldt State College, Arcata, California, Dr. Stephenson Ellen of the U.S. Geological Survey, Ze'ev Reches, graduate student at Stanford, Dr. Benjamin Page of Stanford, and Dr. Ivar Ramberg, University of Oslo, Norway, for many hours and days of exciting research and for their continued enthusiasm. Many of the key ideas were developed jointly with Elmont.

Stanford University,
Spring 1976

ARVID M. JOHNSON

CONTENTS

Chapter 1

MECHANISMS AND MECHANICS OF FOLDING

INTRODUCTION

Much of our current understanding of folds in rock has derived from field-based studies of folds, primarily before 1900. The field studies were devoted to developing understanding of *mechanisms* of folding, such as slip between layers, bending of layers, and yielding of hinges. Research into the *mechanics* of folding started early also, in the early 1900's, and has gathered considerable momentum within the last two decades. In general the researchers of mechanisms of folding have utilized no theory, whereas researchers of mechanics of folding have been concerned primarily with the theory of growth of sinusoidal forms, which are uncommon or difficult to recognize in the field.

In this chapter I shall briefly review contributions which I think have been most significant to the development of current understanding of folds. The review emphasizes folds which apparently developed under conditions of layer-parallel shortening, a condition deduced long ago by Sir James Hall (1851). It nearly totally ignores effects of gravity on folding. This bias in the review does not reflect a feeling that gravity is unimportant in folding. Indeed, many investigators have shown that gravity instability is a common mechanism of folding. Biot and Odé (1965), Odé (1966), Ramberg, (1967), and Stephansson (1971, 1972) considered two-dimensional gravity instabilities, and Biot (1966b) and Fletcher (1967, 1972) considered gravity instability in three dimensions. Ramberg (1970) treated a combination of gravity and buckling instability. Nevertheless, these and other studies of effects of gravity will not be reviewed because we ignore such effects in our analyses of folding reported in following chapters. In this important respect the review is known to be incomplete.

MECHANISMS OF FOLDING

Willis

One of the earliest contributions to our understanding of mechanisms of folding was by Bailey Willis (1893), who studied folds in the Appalachian Mountains and who folded a variety of experimental materials in the laboratory. Willis imagined that there are *competent* folds and *incompetent* folds. Incompetent folds form in materials that offer no appreciable resistance to bending, and the strata are simply flexed by differential vertical movements. Competent folds form in strata which are competent to resist layer-parallel stresses and are capable of buckling. Thus, he apparently introduced the concepts of passive and active roles of layering in folding, at least in the English language literature.

Willis also introduced the concept that initial dips play a significant role in determining where folds will initiate. This concept is discussed at some length in Chapter 2 and elsewhere (Johnson, 1970, chapters 3 and 4), and used there to show that initial amplitudes of layers may control which wavelengths of folds will predominate in elastic layers. The effect of initial perturbations on subsequent folding was studied by William Hobbs (1915), by Biot et al. (1961); and more recently by Peter Cobbold (1975) and by Fletcher and Sherwin (in preparation).

Van Hise

Three years after Willis published his study of the folds in the Appalachians, C.R. van Hise (1896a) presented generalizations of his experiences with folded rocks. Van Hise introduced the concepts of *concentric* (Fig. 1A) and *similar* (Fig. 1B) folds. He indicated (op. cit., p. 209) that concentric folding involves slippage between layering. In similar folding, however, he indicated that there must be flowage of material from limbs toward crests and troughs of the folds, and there may also be slippage between layers (op. cit., p. 210 and 211). In fact, the similar pattern could be produced by homogeneous shortening and thickening of an initial wave form. However, Van Hise recognized two important mechanisms of folding which still dominate folding theory today.

Van Hise also was concerned with the asymmetry of folds. He deduced that asymmetry can be a result of interactions of folds with different wavelengths, forming an anticlinorium or a synclinorium (1896b). If the small

Fig. 1. Idealized fold forms conceived by Van Hise (1896a, figs. 4 and 5, p. 210). A. Concentric or parallel folds. B. Similar folds.

folds developed by differential displacements normal to layering in larger folds, or if the small folds developed before the larger folds, the traces of axial planes would fan outwards in anticlinoria (Fig. 2A) and inwards in synclinoria. Further, he followed James Dana (1847, p. 33) in suggesting that the outward fanning of axial traces of small folds in an anticlinorium would be enhanced by flowage due to gravity (Fig. 2B). The outward fanning shown in Figs. 2A and B he called "normal". "Abnormal" composite folds are those where the axial traces of small folds in anticlinoria dip outward (Fig. 2C) and in synclinoria dip inward. The "abnormal" folds are a result of layer-parallel slip associated with the larger fold (Figs. 3A, B and C). These latter are now called "drag folds". Some geologists have the misunderstanding that drag folds are produced by layer-parallel shear, without layer-parallel shortening (e.g., Billings, 1972, p. 54 and 55).

In any case, Van Hise's explanation of drag folds is consistent with Ramberg's (1963b), who has shown that the layers that develop drag folds are folded as a result of layer-parallel shortening and that layer-parallel shear eventually makes the folds asymmetric (Figs. 3D and E).

De Sitter

De Sitter (1937) decreed that the typical fold due to layer-parallel shortening is concentric. Kuenen and De Sitter (1938) produced concentric-like experimental folds in rubber, stacks of paper, clay and paraffin. They noted that compression occurs on the inside of arcs and extension on the outside of arcs of the experimental folds in nonlayered paraffin, clay and rubber, as

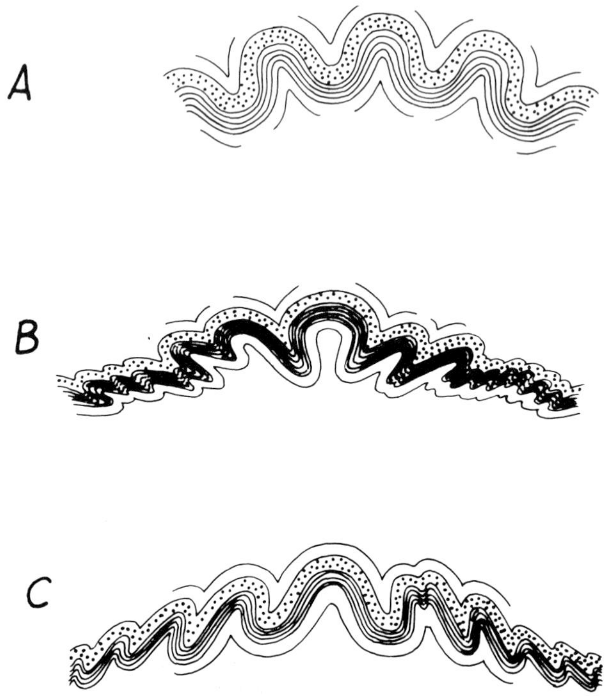

Fig. 2. Patterns of superposed folds. A. Pattern of folds where smaller folds formed before larger folds. Axial traces fan outward on anticlines (after Van Hise, 1896b, fig. 7, p. 320). B. Outward fanning of axial traces of small folds enhanced by flowage due to gravity (op. cit., fig. 8, p. 321). C. "Drag folds", in which axial traces of small folds fan inward on anticlines, as a result of layer-parallel slip associated with growth of large anticline (op. cit., fig. 12, p. 324).

did Van Hise (1896a) in natural folds. They recognized that concentric-like forms developed in highly layered materials such as paper display quite different internal deformations from concentric-like forms developed in non-layered materials. Thus their experiments emphasized once more the importance of strain on the one hand and layer-parallel slip on the other, as recognized by Van Hise (1896a).

Carey

Mechanisms of folding became of special interest to structural geologists again during the early and middle 1960's. Warren Carey (1962, p. 98 and 116) indicated that similar folding occurs in isotropic materials in which layering is passive, whereas concentric folds can develop only in anisotropic or layered materials in which the layers have markedly different physical properties. Thus his basic notions of mechanisms are the same as Van Hise's. He imagined that similar folds develop, however, by differential velocity gra-

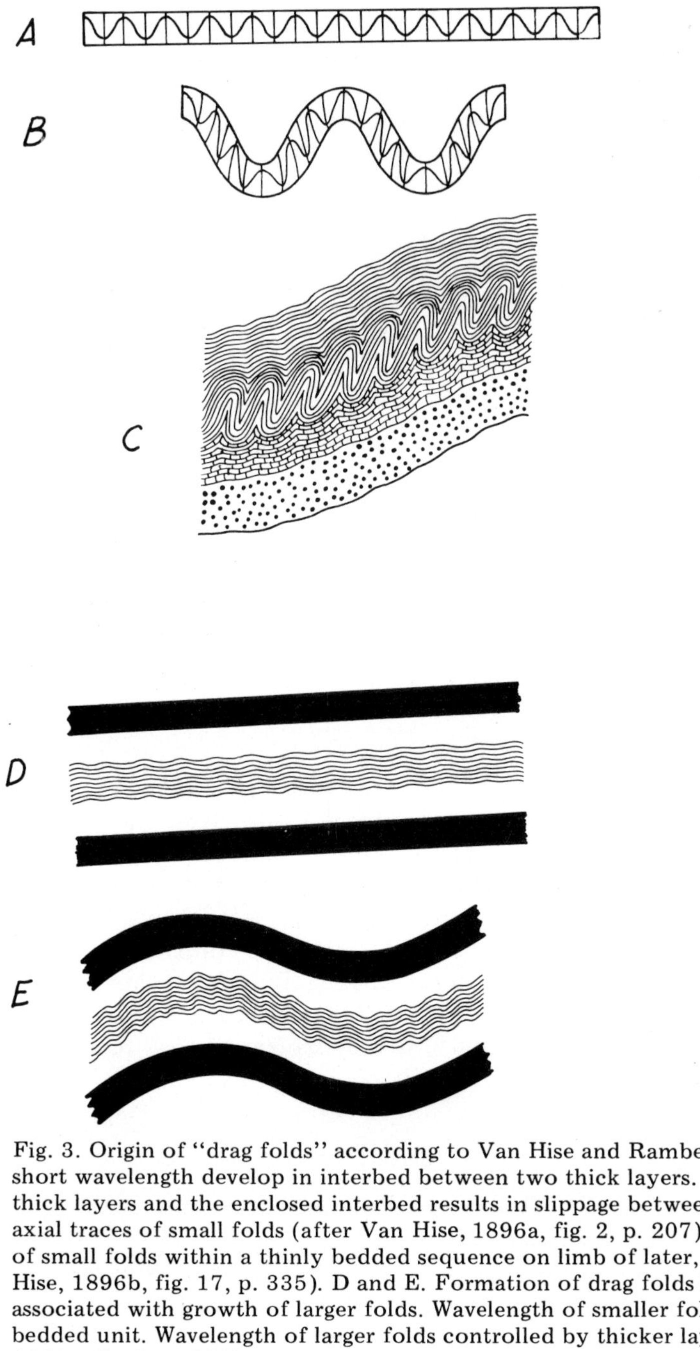

Fig. 4

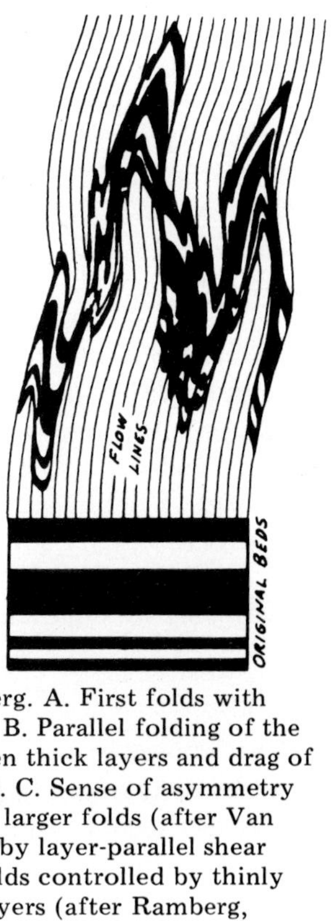

Fig. 3. Origin of "drag folds" according to Van Hise and Ramberg. A. First folds with short wavelength develop in interbed between two thick layers. B. Parallel folding of the thick layers and the enclosed interbed results in slippage between thick layers and drag of axial traces of small folds (after Van Hise, 1896a, fig. 2, p. 207). C. Sense of asymmetry of small folds within a thinly bedded sequence on limb of later, larger folds (after Van Hise, 1896b, fig. 17, p. 335). D and E. Formation of drag folds by layer-parallel shear associated with growth of larger folds. Wavelength of smaller folds controlled by thinly bedded unit. Wavelength of larger folds controlled by thicker layers (after Ramberg, 1964a, fig. 9. p. 322).

Fig. 4. Formation of similar folds according to Carey. Original layers shown at bottom of figure. Folds develop by simple shear along "flow lines", and involve no overall shortening in direction parallel to original layering (after Carey, 1962, fig. 1, p. 96). No mechanism has been proposed to produce folds in the way visualized by Carey.

dients along layers, and that axial traces are flow lines in the material, and that there need be no layer-parallel shortening (Fig. 4). He did not explain possible mechanisms that would produce the differential movements required to produce a similar fold pattern, and his proposed mechanism must be considered to be unsound. Nevertheless, it persists in so-called card-deck analogies of folding (e.g., Ragan, 1973).

Ramberg

Hans Ramberg (1963c) suggested that folds can be of the buckling type, which are a result of layer-parallel shortening, and of the bending type, which are a result of passive deformation of layers, for example, draping around stiff inclusions. He suggested (op. cit., p. 9) that flow folds are either of the buckling or the bending type and do not form by distinctly different mechanisms. I believe that this is the first suggestion that concentric and similar folds can form by the same fundamental mechanisms. Ramberg showed experimentally that, where a multilayer comprised of layers of uniform thickness was buckled in a softer medium, similar chevron folds developed near midheight in the multilayer whereas concentric-like folds developed near the boundaries. He indicated that concentric-like folds and chevron folds are types of buckling folds which are different responses of the same material to different boundary conditions.

Paterson and Weiss

Paterson and Weiss (1966) are largely responsible for our current understanding of the geometry of kink folds. They produced folds in samples of phyllite in which foliation was inclined at various angles to the direction of maximum compression. They produced monoclinal and conjugate kink bands as well as chevron folds (Fig. 5). The most important contributions of their work, I believe, are the geometric analysis of ideal kink bands and the demonstration that some chevron folds can result from interaction of conjugate kink folds. They did not suggest a plausible mechanism of kink folding, nor did they mechanically analyze kink folding.

Ghosh

Subir Ghosh (1968), one of Hans Ramberg's students, experimented with multilayers of modelling clay with thin veneers of lubricant and showed that fold forms can be changed in these materials from rounded, sinusoidal or concentric-like forms, or sharp chevron forms, to kink forms merely by changing the lubrication between layers. Kink bands were developed in multilayers which were imperfectly lubricated whereas rounded forms were developed where lubrication was high. He also noted that kink folds can transform into chevron folds with increasing deformation.

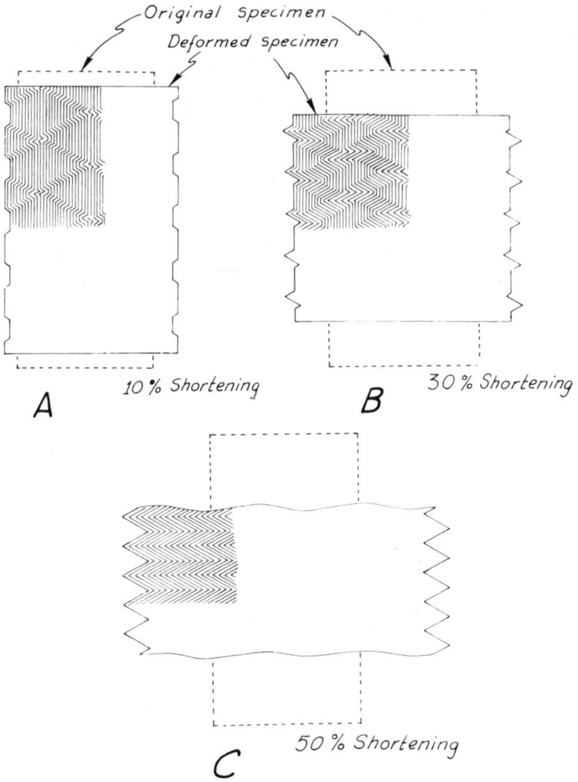

Fig. 5. Geometry and evolution of kink bands in foliated materials according to Paterson and Weiss. A. Pattern of uniform, conjugate kink bands after specimen has been shortened 10%. B. Widened kink bands, after 30% shortening of specimen. Final pattern of chevron folds, developed after 50% shortening (after Paterson and Weiss, 1966, fig. 18, p. 369).

Thus, Ramberg recognized that concentric folds and chevron folds can be genetically related. Paterson, Weiss and Ghosh recognized that kink folds and chevron folds can be genetically related. Ghosh more clearly than other investigators showed that the forms of folds can be changed from rounded to kink-like in the same material merely by changing the properties of the lubricant between layers.

Biot

Maurice Biot, who has made many of the key contributions to sinusoidal folding theory, also has presented some ideas about mechanisms by means of which various fold forms might develop. The most important mechanism, introduced independently by Biot and Ramberg, is that of *instability.* The basic fold form, according to Biot, is sinusoidal for low-amplitude folds (e.g., 1957, 1959, 1961, 1965a). He indicated that chevron shapes could develop where hinges of sinusoidal folds yielded plastically (Biot, 1961). Chapple

(1969) studied the development of sharp hinges in isolated layers of plastico-viscous material. Biot provided one explanation of concentric folds. He indicated that they are results of *bending* of a homogeneous layer or a multilayer containing layers with nearly the same properties (1965d). This explanation does not apply to concentric-like forms produced in multilayers allowing easy slip, as a stack of paper (e.g., Kuenen and De Sitter, 1938). Kink folds, he hypothesized, develop where layering is inclined at about 45 degrees to the direction of maximum compression (1965a). This hypothesis is inconsistent with observation.

Chapple and Spang

William Chapple and John Spang (1974) studied a small, asymmetric, concentric-like fold in limestone. They determined strains within the fold by measuring the degree of twinning of calcite grains. The measurements indicated that the strains are much too low to account for the amount of deformation produced by folding according to any model of deformation. Further, they found abundant field evidence for bedding-plane slip. Theoretical analysis of folding of an anisotropic viscous material indicated that the amount of bedding-plane slip was equivalent to about 24 cm of relative displacement between bedding planes separated about 30 cm. Thus, this is an example where large magnitudes of bedding-plane slip have been documented in the field.

One of the most important contributions of their study, however, is the proposal of a new mechanism by means of which asymmetric folds can develop. They suggest that the asymmetry of the small fold in the limestone was *inherited* from early stages of folding. In order for this to be possible, they indicate, the rock must have behaved much as anisotropic material. If it had been isotropic, the symmetric components of the fold would have grown much faster than the asymmetric components. Thus, the mechanism they propose is in contrast to that visualized by Van Hise (Fig. 2C), where the asymmetry is due to layer-parallel shear within a larger fold.

MECHANICS OF FOLDING

Single-layer folds — low amplitude

Most of the theoretical research into mechanics of folding has been devoted to understanding of folding of single layers, normally confined by a medium. The theory of single-layer folding is quite highly developed. Much of the valid theoretical research with which I am familiar has been focused on the initiation of folding in which forms are sinusoidal.

Smoluchowski and Bijlaard

According to Tromp (1937), the first mechanical analysis of folding as a buckling problem was by Smoluchowski (1909), who computed the shape of waves in an elastic layer on a dense, inviscid, fluid substratum. The geological problem that interested Smoluchowski was the deformation of the earth's crust on a soft mantle. Bijlaard (1946) used Smoluchowski's analysis to estimate the strength of crustal rocks and to show that the crust should not be folded as a unit, rather it should fail before the buckling stress is reached. According to Smoluchowski, the wavelength, L, of folds in an elastic layer requiring the least layer-parallel stress, is:

$$L = 2\pi \sqrt[4]{BI'/\gamma} \tag{1a}$$

where B is an elastic modulus for a plate,

$$B = E/(1-\nu^2) \tag{1b}$$

$$I' = T^3/12 \tag{1c}$$

where E is Young's modulus, ν is Poisson's ratio, γ is the unit weight of the substratum and T is the thickness of the layer.

Bijlaard (1946) also made an important analysis of the effect of layering on the wavelength. He showed that, if the singler layer of thickness T is replaced with a stack of n layers of individual thickness t which slip effortlessly relative to each other, the wavelength is expressed by eq. 1a if we replace I' in eq. 1c by $nt^3/12$. Thus the wavelength to total thickness and the critical buckling stress would be greatly reduced by the layering. A similar result was derived for folding of a multilayer between soft elastic media by Currie et al. (1962) and by Johnson (1970).

Hans Ramberg and Ove Stephansson (1964) rederived eq. 1a and checked it experimentally with elastic plates of rubber with various thicknesses and elastic properties, on substrata of mercury or solutions of potassium iodide, and found excellent agreement between the theoretical predictions and the experimental results (op. cit., fig. 6). They also derived a theory of buckling of a viscous plate on a dense substratum and checked the predictions theoretically. The paper is an excellent illustration of Ramberg's general approach of interweaving theoretical analysis and experimental observation and measurement.

Gunn

In 1937 Ross Gunn extended Smoluchowski's analysis to include the effect of a fault on the buckling of the earth's crust, in order to explain how geosynclines might form. The form of the deflection is the same as that shown by Johnson (1970, fig. 3.15). Further, Gunn showed that the effect of sedimentation on the synclinal side of the fault and erosion on the anticlinal side of the fault is to enhance the instability produced by the buckling and shear.

Gough, Elan and De Bruyne

Thus, by 1937 several mechanical analyses had been performed concerning the buckling of crustal rocks on a dense substratum. This is the same year that Maurice Biot (1937) derived the solution for the reaction of an elastic medium to the deflection of one surface into sinusoidal waves, a solution which was required in order to derive wavelengths for folding of a single elastic or viscous layer within softer elastic or viscous media.

Several engineers, Gough, Elan and De Bruyne (1940), Goodier (1946) and Bijlaard (1946), derived the expression for the wavelength of a buckled stiff elastic plate in a soft elastic medium:

$$L/T = 2\pi\sqrt[3]{B/6B_0} \tag{2}$$

where B is the modulus of the layer and B_0 is the modulus of the medium, defined as in eq. 1b. The analyses were based on plate theory.

Biot

Maurice Biot, under the instigation of M. King Hubbert of Shell Research and Development in Houston, Texas, entered the field of geological folding theory with a paper in 1957. He derived the expression for the dominant wavelength, L_d, for a viscous layer in a viscous medium:

$$L_d/T = 2\pi\sqrt[3]{\mu/6\mu_0} \tag{3}$$

where μ is the coefficient of viscosity of the stiff layer and μ_0 is the coefficient of viscosity of the soft media. Biot's derivation, based on plate theory, is a general solution for viscoelastic materials (e.g., Biot, 1957, eq. 7.5, p. 454). This paper represents one of the most important contributions Biot has made to folding theory. However, he has made many others. For example, he introduced the concept of wavelength selection into folding theory (Biot, 1961), a concept further developed and improved by Sherwin and Chapple (1968) and by Fletcher and Sherwin (in preparation). He showed that unstable waves with a definite wavelength can develop at the surface of an homogeneous, infinite half-space of a material in which the viscosity of shear modulus decreases exponentially with depth (Biot, 1965a, p. 204). He derived one of the early, first-order solutions for the buckling of a deep beam (Biot, 1965a).

Ramberg

Independently, Hans Ramberg (1959) began a series of analyses of folding of a single viscous layer in a viscous medium, using equations of fluid mechanics for deriving the reaction of the medium, but using plate theory to describe the deformation of the stiff layer. In 1959 he presented an equation equivalent to eq. 3 except that, instead of a six under the cube root sign, he has an undetermined constant, because he was unsure how to evaluate the effect of the medium. In subsequent papers he made attempts to derive expressions for the medium and in 1961 he succeeded, using an infinite-series solution (Ramberg, 1962). He showed that eq. 3 is a first approximation to his solution.

Biot and Ramberg have both derived "exact", thick-plate solutions for the stiff layer (Biot, 1965a; Ramberg, 1970), which will be discussed in following paragraphs.

The subsequent mechanical analyses of folding by Biot, by Ramberg, and by Currie, Patnode and Trump (1962) stimulated considerable interest in folding theory, which has virtually blossomed today.

Fletcher

In the early 1960's Raymond Fletcher began to study the mechanics of gravity instability and the emplacement of gneiss and salt domes (Fletcher, 1967, 1972). The methods used to formulate the boundary conditions for this problem were applied to viscous folding, in which growth of folds is induced by interactions between stresses and perturbations of interfaces between materials with different viscosities. One analysis of folding, for a nonlinear viscous substance, was published in 1974. Subsequently, Fletcher has published an exact analysis of folding of linear viscous materials based on the theory cryptically presented in the 1974 paper (Fletcher, 1977). Fletcher avoided all the special assumptions used by Ramberg and Biot and other investigators to mechanically analyze folding of thick plates, so his analysis is *exact* to first order for low-amplitude folds. Comparison of dominant wavelengths predicted by his theory and by Ramberg's and Biot's indicates that all three theories predict nearly the same dominant wavelengths where the viscosity ratio between layer and medium is high, greater than about 100 : 1 (Fig. 6). For viscosity ratios smaller than about 10 : 1, however, dominant wavelengths predicted by Biot's and Ramberg's theories are in error (Fig. 6). Indeed, Fletcher shows that Ramberg's first-order approximation of his "exact" solution more closely approximates the correct wavelengths than his "exact" solution itself. Biot's "exact" theory overestimates wavelengths; it predicts the correct wavelengths approximately as closely as does the elementary theory, eq. 3.

Ronald Smith (1975) has derived a theory of viscous folding which is equivalent to Fletcher's first-order theory.

The analysis of folding developed by Fletcher is especially significant to future research in folding theory because the analysis readily lends itself to analyses of higher-order effects, such as initial, layer-parallel shear stress, high slopes of interfaces and nonlinear rheological properties. Fletcher (1974) showed that the theory can be readily extended to a power-law material. A homogeneous material with power-law behavior responds to layer-parallel shortening much as an anisotropic material, and dominant wavelengths are functions of the state of initial stress during folding (Fletcher, 1974).

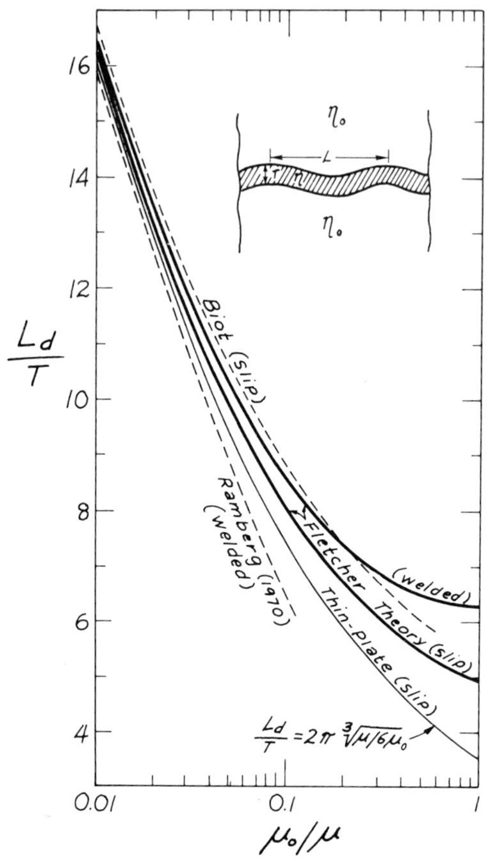

Fig. 6. Relations between dominant wavelength, L_d, and viscosity ratio of media and layer for single layer according to elementary, thin-plate theory of Biot and Ramberg, and to thick-plate solutions of Ramberg, Fletcher, and Biot. Ramberg's theory is for welded contacts and Biot's theory is for free slip at contacts. Fletcher's theory is exact to first order, whereas Ramberg and Biot theories apparently are not (after Fletcher, 1977, fig. 3).

Chapple

Jo-Ann Sherwin and William Chapple (1968) measured ratios of wavelength to thickness for a large number of single-layer folds in quartzite veins confined by slate or phyllite. Using the Biot-Ramberg theory for dominant wavelengths, eq. 3, they computed apparent viscosity ratios of 2 : 1 to 8 : 1, which are much too small for the elementary theory to be applicable. Thus, they derived a theory of wavelength selection which takes account of uniform shortening and thickening of the folded layer, a theory which is valid whether the elementary theory of folding of Biot and Ramberg or the exact theory of Fletcher is used to predict dominant wavelengths (e.g., see Fletcher, 1974 and 1977). Using the elementary theory of Biot and Ramberg, Sherwin and Chapple derived the following equation for the dominant wavelength of the final folds:

$$L_d/T = 2\pi\sqrt[3]{(\mu/6\mu_0[(1+\lambda)/(2\lambda^2)]} \quad (4)$$

where λ is the uniform, finite elongation that occurred during growth of the folds and T is the final thickness of the folded layer. Using eq. 4, Sherwin and Chapple estimated that the apparent viscosity ratios of small quartzite veins in slate or phyllite ranged from about 16 to 21.

Peter Hudleston (1973b) experimentally produced single-layer folds using mixtures of ethyl cellulose and benzyl alcohol. He found that ratios of wavelength to thickness in the folds were best predicted with the theory of Sherwin and Chapple.

Single-layer folds — high amplitude

There have been a few studies of high-amplitude folding. Three basic approaches have been used. Chapple (1968) used elementary plate theory to describe the equilibrium of a stiff viscous layer and the Navier-Stokes equations to describe the deformation of confining media, using finite-difference equations and a computer to perform the analysis. Fletcher (1967, 1972) used an analytical solution and approximate boundary conditions to study high-amplitude growth of a single interface between two viscous materials. Dietrich (1970), Hudleston and Stephansson (1973) and Parrish (1973) used finite-element methods to follow fold development to high amplitudes. All the studies of high-amplitude folding have been for linear viscous materials, except for that of Parrish, who assumed a power-law material without memory. Nobody has studied high-amplitude growth of folds in elastic material, except for a brief study of the "elastica" of an isolated layer by Johnson (1970).

Chapple

The first study of high-amplitude folding, and certainly the most important in terms of new concepts, is that of Chapple (1968). Chapple showed that the infinitesimal-amplitude solutions of Biot and Ramberg do not predict fold-shape development satisfactorily for limb dips greater than five degrees. The fold shapes depend upon the wavelength of the fold relative to the wavelength of the dominant wavelength predicted by the Biot and Ramberg theory, defined in eq. 3. Folds with sharper crests developed where wavelengths are greater than the dominant wavelength (e.g., op. cit., fig. 10, p. 58). Perhaps even more important, Chapple indicated that wavelength selection stops when limb dips reach a magnitude of about 15 degrees, so that the arc-lengths of high-amplitude folds are largely determined by wavelengths of folds at the time the folds had limb dips of about 15 degrees. Hudleston (1973b) performed a series of experiments, developing single-layer folds in ethyl cellulose, and verified these predictions by Chapple.

Dietrich, Hudleston and Stephansson

Dietrich (1970) used the finite-element method to study high-amplitude folds in slightly compressible, linear viscous materials. He found that high-amplitude folds assume different forms depending upon the relative viscosities of the layer and media. In single-layer folds, the fold geometry approaches a concentric-like form or a similar form depending upon whether viscosity contrast is high or low. Hudleston and Stephansson (1973) obtained the same general results using the finite-element method. Further, they noted that the amount of layer-parallel shortening becomes relatively unimportant after limb dips reach angles of 15—25 degrees, a result consistent with that obtained by Chapple. Finally, they found that folds which are nearly parallel for low limb dips can become flattened with growth of amplitude and thereby become transformed into nearly similar folds.

Parrish

David Parrish (1973) used the finite-element method to study high-amplitude folding of power-law materials. He estimated coefficients in the model based on experimental data for quartzite and marble. He found that fold geometry according to the power-law model is similar to that of the linear viscous model used by Dietrich and by Hudleston and Stephansson. The primary difference between the two models is that the folds in power-law materials displayed more flow in the inner hinges of the folds.

Thus, theories of folding of single layers have been highly developed. Theories have been advanced for materials with linear elastic, linear and nonlinear viscous, and viscoelastic properties. Both high- and low-amplitude folds in viscous materials can be described accurately and related to the properties of the layer and medium. Some recent research by Fletcher and Sherwin (in preparation) concerns the determination of theoretical, dominant wavelengths using field data. The problem is that a complex series of inflections is expected to develop in a layer because the layer typically will contain initial dips with a wide range of values and widths and many wavelengths of folds will amplify simultaneously. It is difficult to select the dominant wavelengths by examining the complex series of inflections (e.g., Sherwin and Chapple, 1968, Hudleston, 1973a). The approach of Fletcher and Sherwin is to measure segments of arc of irregular folds and determine the mean of the distribution of ratios of arc length to thickness. They show numerically that the ratio of the dominant wavelength to the thickness is related to the mean ratio of the arc length to thickness and the viscosity ratio of the layer and medium. Thus, by determining the mean of the ratio of arc length to thickness, one can determine the viscosity ratio at the time of folding.

The theory of single-layering folding is highly developed, indeed.

Multilayer folds

Bijlaard

Theories of folding of ideal multilayered materials are nearly as old as theories of folding of single layers. Bijlaard (1946) assumed that a sequence of identical strata, with zero contact strength, buckle with the same form, and showed that the ratio of wavelength to total thickness is greatly reduced by the presence of stratification. Currie, Patnode and Trump (1962) used the same type of analysis for folding of elastic layers confined by very soft media, but allowed layers to have different thicknesses and different moduli. This first attempt by Bijlaard to study folding of multilayers was based on the assumption that the folds have constant amplitudes throughout the multilayer, so similar folding was assumed.

Ramberg

Subsequently, three different approaches to folding of multilayers have evolved. The most accurate and general is that of Ramberg (1960, 1963a, 1964a, 1970). Ramberg uses an exact solution for stresses and displacements within each layer of a multilayer and matches boundary conditions between adjacent layers. The method is most clearly explained in the appendix of his book on folding (Ramberg, 1967) and in his 1970 papers. In various papers Ramberg has used different methods of computing the stresses and displacements within individual layers. The error in his analysis pointed out by Fletcher (1977) in no way detracts from the value of the general method of analysis of multilayer folds developed by Ramberg and, as indicated by Fletcher, Ramberg's approximate theory of stresses and displacements within layers will provide quite accurate estimates of wavelengths if viscosity ratios of stiff to soft layers are greater than about 10 : 1. Ramberg's method of treating multilayers readily lends itself to analysis of multilayers using other theories of folding of viscous or nonlinear viscous substances. For example, the approach of Fletcher and Smith could be used to determine the dominant wavelengths of folds developed within the multilayer.

Ramberg has treated a variety of problems of sinusoidal folding of multilayers, including folding of an infinite stack of identical stiff and soft layers (1961, fig. 12; 1963a, figs. 5 and 6), and folding of a finite multilayer between rigid boundaries (Fig. 7; 1964a, fig. 9). He showed that the thinner layers may buckle before general folding of a complex multilayer. He suggested that his is one way to explain drag folds (Ramberg, 1963b). His concept of the formation of drag folds is consistent with that of Van Hise, Fig. 3. Ramberg and Strömgård (1971) showed that folds with different wavelengths can develop simultaneously in complex multilayers (Fig. 8A), or in multilayers confined by soft media (Ramberg, 1970, p. 225 and 226) above and below (Fig. 8B) or only below (Fig. 8C).

Biot

Maurice Biot has used two different approaches to study folding of multi-

Fig. 7. Pattern of folds in thin layers confined above and below by thick, stiff layers which do not fold (after Ramberg, 1963a, fig. 13, p. 504).

layers comprised of identical soft layers and identical stiff layers, normally tightly bonded together. In one approach (Biot, 1961, 1965b and 1968), he uses elementary plate theory to describe the resistance to bending of individual stiff layers in multilayers. Using this method he has derived solutions for sinusoidal folding of an infinitely long multilayer between rigid boundaries (Fig. 9A) and a finite multilayer containing a discontinuity normal to layering across which bending moments are not transmitted (Fig. 9B). This latter problem was solved for a single layer on a dense substratum by Gunn (1937). Also using the same method, Biot (1965d) solved for the folding of a multilayer confined by soft media (Figs. 9C and D). He showed that where the soft interbeds within the multilayer are nearly as stiff as the stiffer layers, the multilayer buckles much as a homogeneous plate (Fig. 9C). The resulting form is concentric-like. If, however, the soft interbeds are very soft, the multilayer folds with a *similar* form (Fig. 9D).

In quite a different approach, Biot assumed that the multilayer behaves as an anisotropic, thick plate (Biot, 1960, 1964, 1965a, 1966a, 1967). The fold forms according to this theory are sinusoidal and the same as those for a thick plate (1965a). There are two fundamental problems with this theory. It ignores the bending resistance of layers within the multilayer so that, for example, the wavelength of folds in multilayers confined by rigid or stiff boundaries tends to be zero, which is ridiculous, as he indicates himself (1965a). Further, the theory is based on a theory of incremental deformations (Biot, 1939) which appears to describe the behavior of no known material. He indicates (1965a) that the theory describes the behavior of elastic, viscous, and other viscoelastic materials. In fact, however, it does not describe accurately the behavior of linear viscous materials (Fletcher, 1977) and it probably does not describe accurately the behavior of elastic materials, as we

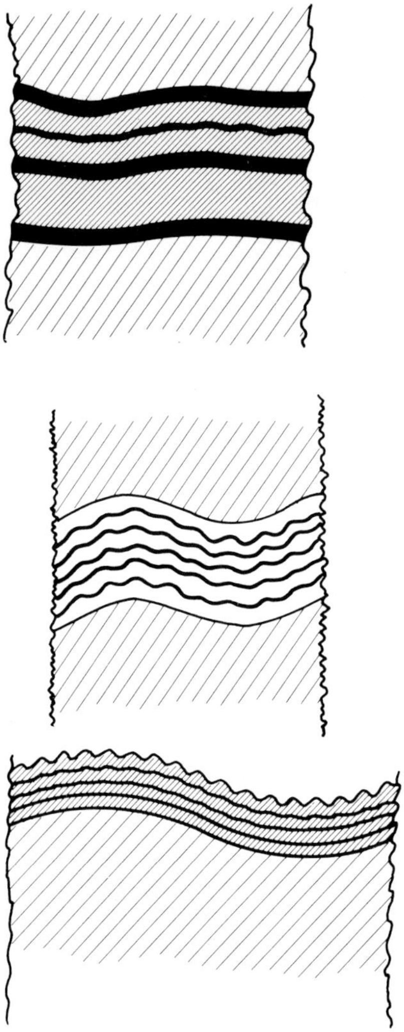

Fig. 8. Simultaneous development of several orders of folds under various conditions. A. Multilayer containing an odd, thin, stiff layer. Thin layer forms short wavelength whereas combination of three thick, stiff layers forms large wavelength (after Ramberg and Strömgård, 1971, fig. 1). B. Thin, stiff layers form short wavelength whereas multilayer as a whole forms long wavelength (after Ramberg, 1970, fig. 9, p. 225). C. Multilayer with free upper surface which forms two wavelengths of folds (op. cit., fig. 26, p. 119).

will show in Chapter 10, so it probably does not describe the behaviors of other viscoelastic materials either.

Biot's theories, nevertheless, are qualitatively valid, and through manipulations of his various theories Biot has contributed more than any other individual to our understanding of folding of simple materials. His work must

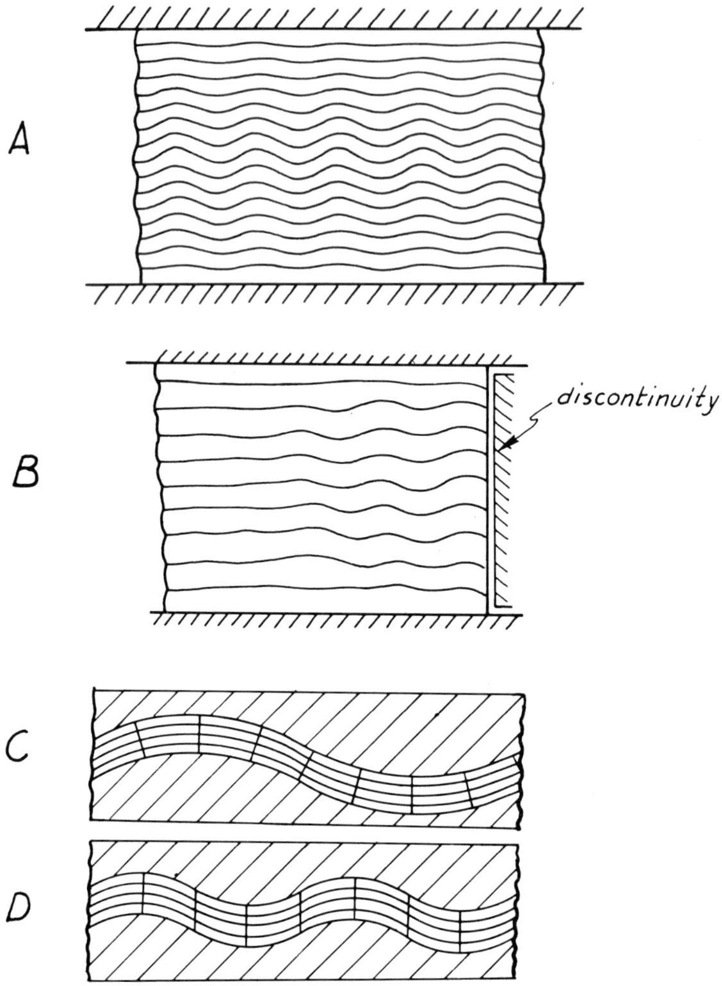

Fig. 9. Some idealized fold patterns according to Biot. A. Pattern of "internal instability", developed where boundaries are rigid (after Biot, 1964, fig. 1, p. 564). B. Pattern of "edge buckling" developed near a vertical discontinuity, such as a fault (after Biot, 1968, fig. 1, p. 125). C and D. Formation of concentric-like pattern (C) and similar pattern (D) of folds (after Biot, 1965b, fig. 3, p. 254).

be considered as pioneer studies, which have provided direction for all other studies to date. For example, he has indicated how one analyzes a large number of ideal fold patterns, including sinusoidal folding of a single layer; formation of sharp hinges in chevron folds; formation of concentric-like, similar and kink folds; development of "internal instability", which is a sinusoidal pattern of folds in a multilayer confined by rigid boundaries; formation of "edge buckling", in which amplitudes of folds die off exponentially from a discontinuity such as a fault; surface instability of an infinite half-space in isotropic, multilayered, or anisotropic materials; folding of deep beams of

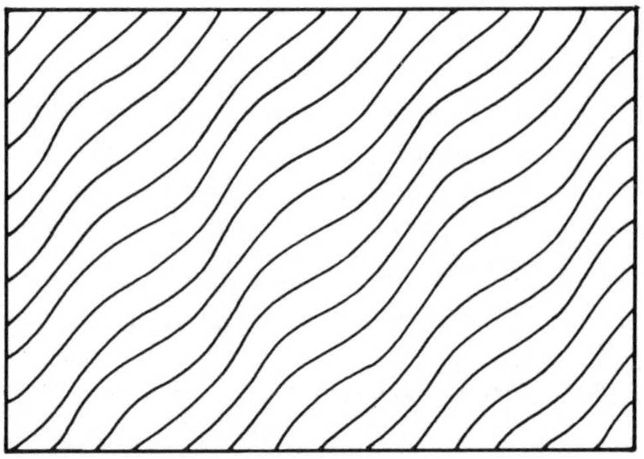

Fig. 10. Pattern of asymmetric folds developed in multilayer in which anisotropy is horizontal but layering is inclined (after Cobbold et al., 1971, fig. 21, p. 52).

isotropic or anisotropic materials; and formation of pinch-and-swell structures. Although some of the analyses are incorrect in detail, or are irrelevant, the analyses are of lasting importance because they point toward ways in which correct and relevant analyses can be made.

Fig. 11. Development of chevron form from sinusoidal form (after Bayly, 1974, fig. 5).

Cobbold, Cosgrove and Summers

Cobbold et al. (1971) used Biot's theory in order to predict the formation of various types of folds within multilayers. Their application of the analysis of kink folding, which is essentially the same as that by myself (Johnson, 1970, chapter 8), is incorrect so it will not be summarized here. However, Cobbold et al. deduce the interesting result that asymmetric folds can form where mechanical layering or other planar anisotropy of a material is oblique to the layering visible in rock, because the sinusoidal folds will be symmetrical about the mechanical layering, not necessarily the visible layering (Fig. 10).

Bayly

Brian Bayly, a former student of Hans Ramberg, has followed a different path from that of most researchers concerned with folding of multilayers. He has primarily studied the development of chevron folds (Bayly, 1964, 1970, 1971, 1974). In his most recent research, he has analyzed relative amounts of energy dissipated in limbs and crests of folds in viscous multilayers. He concludes that angular folds such as chevron folds and kink folds require less energy for high limb dips than rounded folds. Thus, he predicts that a fold will transform from a smooth form at low limb dips to a chevron form for high limb dips (Fig. 11).

REFERENCES CITED IN CHAPTERS 1, 10 AND 11

Bayly, M.B., 1964. A theory of similar folding of viscous materials. Am. J. Sci., 262: 753—766.

Bayly, M.B., 1970. Viscosity and anisotropy estimates from measurements on chevron folds. Tectonophysics, 9: 459—474.

Bayly, M.B., 1971. Similar folds, buckling, and great-circle patterns. J. Geol. 79: 110—118.

Bayly, M.B., 1974. An energy calculation concerning the roundness of folds. Tectonophysics, 24: 291—316.

Biezeno, C.B. and Hencky, H., 1928. On the general theory of elastic stability. Proc. K. Ned. Akad. Wet. Amsterdam, 31: 569—592.

Bijlaard, P.P., 1946. On the elastic stability of thin plates supported by a continuous medium. Proc. K. Ned. Akad. Wet., 49: 1189—1199.

Billings, M.P., 1972. Structural Geology. Prentice-Hall, Englewood Cliffs, N.J. 3rd edit., 606 p.

Biot, M.A., 1937. Bending of an infinite beam on an elastic foundation. Trans. Am. Soc. Mech. Eng., 55: A1—A7.

Biot, M.A., 1939. Non-linear theory of elasticity and the linearized case for a body under initial stress. Philos. Mag., 28: 468—489.

Biot, M.A., 1957. Folding instability of a layered viscoelastic medium under compression. Proc. R. Soc. London, Ser. A, 242: 444—454.

Biot, M.A., 1959. On the instability of folding deformation of a layered viscoelastic medium under compression. J. Appl. Mech., 26: 393—400.

Biot, M.A., 1960. Instability of a continuously inhomogeneous viscoelastic half-space under initial stress. J. Franklin Inst., 270: 190—201.

Biot, M.A., 1961. Theory of folding of stratified viscoelastic media and its implications in

tectonics and orogenesis. Geol. Soc. Am. Bull., 72: 1595—1620.
Biot, M.A., 1963. Internal buckling under initial stress in finite elasticity. Proc. R. Soc. London, Ser. A, 273: 306—328.
Biot, M.A., 1964. Theory of internal buckling of a confined multilayered structure. Geol. Soc. Am. Bull., 75: 563—568.
Biot, M.A., 1965a. Mechanics of Incremental Deformations. Wiley, New York, N.Y., 505 pp.
Biot, M.A., 1965b. Further development of the theory of internal buckling of multilayers. Geol. Soc. Am. Bull., 75: 833—840.
Biot, M.A., 1965c. Theory of viscous buckling and gravity instability of multilayers with large deformation. Geol. Soc. Am. Bull., 76: 371—378.
Biot, M.A., 1965d. Similar folding of the first and second kinds. Geol. Soc. Am. Bull., 75: 251—258.
Biot, M.A., 1966a. Fundamental skin effect in anisotropic solid mechanics. Int. J. Solids Struct., 2: 645—663.
Biot, M.A., 1966b. Three-dimensional gravity instability derived from two-dimensional solutions. Geophysics, 31: 153—166.
Biot, M.A., 1967. Rheological stability with couple stresses and its application to geological folding. Proc. R. Soc. London, Ser. A, 298: 402—423.
Biot, M.A., 1968, Edge buckling of a laminated medium. Int. J. Solids Struct., 4: 125—137.
Biot, M.A. and Odé, H., 1965. Theory of gravity instability with variable overburden and compaction. Geophysics, 30: 213—227.
Biot, M.A., Odé, H. and Roever, W.L., 1961. Experimental verification of the folding of stratified viscoelastic media. Geol. Soc. Am. Bull., 72: 1621—1630.
Blatz, P.J., 1969. Application of large deformation theory to the thermomechanical behavior of rubberlike polymers — porous, filled and unfilled. In: F.R. Eirich (Editor), Rheology. Academic Press, 5: 1—55.
Calladine, C.R. and Drucker, D.C., 1962. Nesting surfaces of constant rate of energy dissipation in creep. J. Appl. Math., 29: 79—84.
Campbell, J.D., 1951. Some aspects of rock folding by shear deformation. Am. J. Sci., 249: 625—639.
Carey, S.W., 1962. Folding. Alberta Soc. Pet. Geol. J., 10: 95—144.
Chapple, W.M., 1968. A mathematical theory of finite-amplitude folding. Geol. Soc. Am. Bull., 79: 47—68.
Chapple, W.M., 1969. Fold shape and rheology: The folding of an isolated viscous-plastic layer. Tectonophysics, 7: 97—116.
Chapple, W.M. and Spang, J.H., 1974, Significance of layer-parallel slip during folding of layered sedimentary rocks. Geol. Soc. Am. Bull., 85: 1523—1534.
Cobbold, P.R., 1975. Fold propagation in a single embedded layer. Tectonophysics, 27: 333—351.
Cobbold, P.R., Cosgrove, J.W. and Summers, J.M., 1971. Development of internal structures in deformed anisotropic rocks. Tectonophysics, 12: 23—53.
Currie, J.B., Patnode, A.W. and Trump, R.P., 1962. Development of folds in sedimentary strata. Geol. Soc. Am. Bull., 73: 461—472.
Dana, J.D., 1847. Geological results of the earth's contraction in consequence of cooling. Am. J. Sci., 2nd ser., 3: 185.
De Sitter, L.U., 1937. Plastic deformation. Leidse Geol. Meded., 9: 1—17.
Dietrich, J.H., 1970. Computer experiments on mechanics of finite amplitude folds. Can. J. Earth Sci., 7: 467—476.
Dietrich, J.H. and Carter, N.L., 1969. Stress-history of folding. Am. J. Sci., 267: 129—154.
Donath, F.A., 1961. Experimental study of shear failure in anisotropic rocks. Geol. Soc. Am. Bull., 72: 985—990.

Donath, F.A., 1971. Some information squeezed out of rock. Am. Sci., 58: 54—72.
Fletcher, R.C., 1967. A Finite-amplitude Model for the Emplacement of Gneiss Domes and Salt Domes. Ph.D. Thesis, Brown University, 221 p.
Fletcher, R.C., 1972. Application of a mathematical model to the emplacement of gneiss domes. Am. J. Sci., 272: 197—216.
Fletcher, R.C., 1974. Wavelength selection in the folding of a single layer with power-law rheology. Am. J. Sci., 271: 1029—1043.
Fletcher, R.C., 1977. Folding of a single viscous layer: exact infinitesimal-amplitude solution. Tectonophysics.
Fletcher, R.C. and Sherwin, J.-A., in preparation. Relation between observed fold arc length and theoretical preferred wavelength in single layer folding.
Fung, Y.C., 1965. Foundations of Solid Mechanics. Prentice-Hall. Englewood Cliffs, N.J., 525 pp.
Ghosh, S.K., 1968. Experiments of buckling of multilayers which permit interlayer gliding. Tectonophysics, 6: 207—249.
Ghosh, S.K., 1970. A theoretical study of intersecting fold patterns. Tectonophysics, 9: 559—569.
Goodier, J.N., 1946. Cylindrical buckling of sandwich plates. J. Appl. Mech., Trans. ASME, 68: 253—260.
Gough, G.S., Elan, C.F. and De Bruyne, N.A., 1940. The stabilization of a thin sheet by a continuous supporting medium. J. R. Aeronaut. Soc., 44: 12—43.
Green, A.E. and Zerna, W., 1968. Theoretical Elasticity. Clarendon Press, Oxford, 457 p.
Green, A.W., Rivlin, R.S. and Shield, R.T., 1952. General theory of small elastic deformation superposed on finite elastic deformations. Proc. R. Soc. London, Ser. A., 211: 128—154.
Gunn, R., 1937. A quantitative study of mountain building on an unsymmetrical earth. J. Franklin Inst., 224: 19—53.
Hall, Sir James, 1851. On the vertical position and convolution of certain strata and their relation with granite. Trans. R. Soc. Edinb., 7: 79—108.
Handin, J. and Hager, R.V., Jr., 1957. Experimental deformation of sedimentary rocks under confining pressure. Bull. Am. Assoc. Pet. Geol., 41: 1—50.
Hobbs, W.H., 1915. Mechanics of formation of arcuate mountains. J. Geol., 22: 166—188.
Hudleston, P.J., 1973a. Fold morphology and some geometrical implications of theories of fold development. Tectonophysics, 16: 1—46.
Hudleston, P.J., 1973b. An analysis of "single-layer" folds developed experimentally in viscous media. Tectonophysics, 16: 189—214.
Hudleston, P.J. and Stephansson, O., 1973. Layer shortening and fold-shape development in the buckling of single layers. Tectonophysics, 17: 299—321.
Johnson, A.M., 1969. Development of folds within the Carmel Formation, Arches National Monument, Utah. Tectonophysics, 8: 31—77.
Johnson, A.M., 1970. Physical Processes in Geology. Freeman Cooper, San Francisco, Calif. 577 p.
Kuenen, Ph.H. and De Sitter, L.U., 1938. Experimental investigation into the mechanism of folding. Leidse Geol. Meded., 9: 217—239.
Mooney, M., 1940. A theory of large elastic deformation. J. Appl. Phys., 11: 582—592.
Novozhilov, V.V., 1953. Foundations of the Nonlinear Theory of Elasticity. Graylock, Rochester, N.Y., 233 p.
Odé, H., 1966. Gravitational Instability of a Multilayered System of High Viscosity. Thesis, Univ. Utrecht, Utrecht, 96 p.
Parrish, D.K., 1973. A nonlinear finite-element fold model. Am. J. Sci., 273: 318—334.
Paterson, M.S. and Weiss, L.E., 1966. Experimental deformation and folding of phyllite. Geol. Soc. Am. Bull., 77: 343—374.
Prager, W., 1961. Introduction to Mechanics of Continua. Ginn, Boston, 230 pp.

Prager, W. and Hodge, P.G., 1951. Perfectly Plastic Solids. Wiley, New York, N.Y., 264 p.
Ragan, D.M., 1973. Structural Geology. Wiley, New York, 208 p.
Ramberg, H., 1955. Natural and experimental boudinage and pinch-and-swell structures. J. Geol., 63: 512—526.
Ramberg, H., 1959. Evolution of ptygmatic folding. Nor. Geol. Tidsskr., 39: 99—151.
Ramberg, H. 1960. Relationship between arc length and thickness of ptygmatically folded veins. Am. J. Sci., 258: 36—46.
Ramberg, H., 1961. Relationship between concentric longitudinal strain and concentric shearing strain during folding of homogeneous sheets of rock. Am. J. Sci., 259: 382—390.
Ramberg, H., 1962. Contact strain and folding instability of a multilayered body under compression. Geol. Rundsch., 51: 405—439.
Ramberg, H., 1963a. Fluid dynamics of viscous buckling applicable to folding of layered rocks. Bull. Am. Assoc. Pet. Geol., 47: 484—505.
Ramberg, H., 1963b. Evolution of drag folds. Geol. Mag., 100: 97—106.
Ramberg, H., 1963c. Strain distribution and the geometry of folds. Bull. Inst. Geol., Univ. Upps., 42: 1—20.
Ramberg, H., 1964a. Selective buckling of composite layers with contrasted rheological properties; a theory for simultaneous formation of several orders of folds. Tectonophysics, 1: 307—341.
Ramberg, H., 1964b. Note on model studies of folding of moraines in piedmont glaciers. J. of Glaciol., 5: 207—218.
Ramberg, H., 1967. Gravity, Deformation and the Earth's Crust. Academic Press, London, 214 p.
Ramberg, H., 1968. Instability of layered systems in the field of gravity. Phys. Earth Planet. Inter., 1: 427—447.
Ramberg, H., 1970. Folding of laterally compressed multilayers in the field of gravity. Phys. Earth Planet. Inter., Part I, 2: 203—232; Part II, 4: 83—120.
Ramberg, H. and Stephansson, O., 1964. Compression of floating elastic and viscous plates affected by gravity; a basis for discussing crustal buckling. Tectonophysics, 1: 101—120.
Ramberg, H. and Strömgård, K., 1971. Experimental tests of modern buckling theory applied on multilayered media. Tectonophysics, 11: 261—272.
Ramsay, J.G., 1967. Folding and Fracturing of Rocks. McGraw-Hill, New York, 568 p.
Rivlin, R.S., 1948. Large elastic deformations of isotropic materials. Philos. Trans. R. Soc. London, Ser. A; Part I, 240: 459, Part II, 240: 491; Part III, 240: 509, Part IV, 241: 379.
Rivlin, R.S., 1949. Large elastic deformations of isotropic materials. Philos. Trans. R. Soc. London, Ser. A; Part VI, 242: 173; Part VII, 243: 34.
Rivlin, R.S., 1956. Large elastic deformations. In: F.R. Eirich (Editor), Rheology, 1: 351—385.
Sherwin, J.-A. and Chapple, W.M., 1968. Wavelengths of single layer folds: A comparison between theory and observation. Am. J. Sci., 266: 167—179.
Smith, R.B., 1975. A unified theory of the onset of folding, boudinage and mullion structure. Geol. Soc. Am. Bull., 86: 1601—1609.
Smoluchowski, M., 1909. Über ein gewisses Stabilitätsproblem der Elastizitätslehre und der Beziehungen zur Entstehung von Faltenbirgen. Anz. Akad. Wiss. Krakau, Math. Naturw., 2: 3 and 727.
Stephansson, O., 1971. Gravity tectonics on Öland. Bull. Geol. Inst. Univ. Upps., New Ser., 3: 37—78.
Stephansson, O., 1972. Theoretical and experimental studies of diapiric structures on Öland. Bull. Geol. Inst. Univ. Upps., New Ser., 3: 163—200.
Timoshenko, S.P. and Gere, J.M., 1961. Theory of Elastic Stability. McGraw-Hill, New York, N.Y., 541 p.

Toupin, R.A. and Bernstein, B., 1961. Sound waves in deformed perfectly elastic materials, acoustical effects. J. Acoust. Soc. Am., 33: 216—225.
Tromp, S.W., 1937. On the Mechanism of the Geological Undulation Phenomena in General and of Folding in Particular and Their Application to the "Roots of Mountains" Problem. Sijthoff, Leiden, 184 p.
Truesdell, C. and Noll, W., 1965. The non-linear field theories of mechanics. In: S. Flugge (Editor), Encyclopedia of Physics, Vol. 3 no. 3. Springer, New York, 602 p.
Van Hise, C.R., 1896a. Studies for students. Deformation of rocks. J. Geol., 4: 195—213.
Van Hise, C.R., 1896b. Studies for students. Deformation of rocks, Part II. J. Geol., 4: 312—353.
Vesić, A.S. and Clough, G.W., 1968. Behavior of granular materials under high stresses. J. Soil. Mech. Found. Div., Am. Assoc. Civ. Eng. Proc., Sm3: 661—668.
Weiss, L.E., 1968. Flexural slip folding of foliated model materials. In: A.J. Baer and D.K. Norris (Editors), Proceedings of the Conference on Research in Tectonics. Geol. Surv. Can. Pap., 68-52: 294—333.
Willis, B., 1893. The mechanics of Appalachian structure. 13th Ann. Rep. U.S. Geol. Surv. (1891-92), p. 211—281.

Chapter 2

FOLDING OF BEDDED SANDSTONES OF THE CARMEL FORMATION, ARCHES NATIONAL MONUMENT, UTAH *

SUMMARY

Crenulations of bedding within the Carmel Formation of southeastern Utah usually have been interpreted to be chaotic structures. In Arches National Monument, however, definite patterns can be recognized and the patterns can be understood in terms of multilayer-folding theory.

In some places, interbedded sandstones and siltstones of the Carmel Formation were confined above and below by relatively massive and rigid Entrada and Navajo sandstones. As a result, the fold pattern within the Carmel is of the internal variety in which the fold amplitudes diminish upwards and downwards from a maximum near the center of the Carmel. The maximum amplitudes usually are not at the center of the Carmel as we would expect if the Carmel had uniform physical properties at the time of folding. Rather the maximum amplitudes are displaced towards the base of the Carmel, apparently indicating that the lower beds were softer than the upper beds.

In other places, the folds within the Carmel Formation extend upwards into the lower part of the massive Entrada Sandstone, where their ampli-

* Originally published as: Johnson, A.M., 1969. Development of folds within Carmel Formation, Arches National Monument, Utah. Tectonophysics, 8: 31—77.

tudes gradually diminish. Two factors seem to control this type of folding and they can be understood in terms of an inequality of the form:

$$\frac{B}{B_0} > \left(\frac{T}{2t_1}\right)^2$$

where B is the modulus of the stiff beds of the multilayer (Carmel), B_0 is the modulus of the confining medium (Entrada), T is the total thickness of the multilayer, and t_1 is the thickness of each of the stiff beds of the multilayer. If this inequality is satisfied, according to theory, folding should be of the general buckling variety in which the folds extend into the overlying Entrada. If this inequality is not satisfied, folding should be of the internal variety. The inequality seems to account for the field observation that, where the Carmel is quite massive, that is where the number of beds, T/t, is quite small, the folds usually deflect the lower part of the Entrada. Where the Carmel is well-bedded, however, that is where T/t, is quite large, the inequality apparently is not satisfied, because the folds are of the internal variety.

The elastic analog of Biot's theory of the multilayer is derived and the assumptions that lead to the theory are emphasized and discussed so that the theory can be modified to account for field observations.

There is field evidence that beds of the Carmel were somewhat irregular prior to folding, rather than being perfectly flat plates as is assumed in theory, so that effects of initial deflections within single layers and within multilayers are analyzed. The analysis indicates that the internal folding of an initially deflected multilayer is of the form:

$$v = \frac{\delta_i}{\left(1 - \frac{P}{P_{\mathrm{mult}}}\right)} \sin\left(\frac{\pi y}{T}\right) \sin\left(\frac{2\pi x}{L}\right)$$

where δ_i is the amplitude and L is the wavelength of the initial deflection, P is the axial load, and P_{mult} is the critical load for buckling of initially flat layers. This solution for the deflection helps us to realize the significance of initial deflections, even small ones, in the process of buckle folding. The solution indicates that as the magnitude of the initial deflection becomes smaller and smaller, the magnitude of the axial load required for significant deflection becomes larger and larger, until, in the limit, when the initial deflection has vanishing amplitude, the axial load equals the critical buckling load, P_{mult}. Thus the critical load is the *maximum* load required for significant folding of real layers, which are never perfectly flat.

Analysis of folding of markedly irregular multilayers indicates that the resultant fold pattern usually will have a dominant wavelength of the *Biot wavelength*:

$$L_B = 2\pi \sqrt{\frac{T}{2\pi}} \sqrt[4]{\frac{B_p I_p}{tbB_n}}$$

which is the dominant wavelength in initially flat beds, unless irregularities with other wavelengths have much greater initial deflections than the irregularities of the Biot wavelength. Here B_p is the modulus parallel to bedding, I_p is the moment of inertia of each stiff bed, and B_n is the modulus normal to bedding.

The Biot wavelength becomes:

$$L_B = 2\sqrt{\frac{T\pi t_1}{\sqrt{12}}} \approx 1.90\sqrt{Tt_1}$$

where t_1 is the thickness of each stiff bed, if the modulus or viscosity of the soft beds is much smaller than that of the stiff beds. This equation predicts remarkably well the wavelengths of the internal-buckling folds of the Carmel.

The simplified theory of buckling of multilayered sequences does not account for all the factors recognized in the field but we can understand most of the factors in terms of the theory.

INTRODUCTION

Crenulations of bedding within redbed facies of the Carmel Formation have been reported in many places in southeastern Utah and they occur over an area of several thousand sq.miles. The crenulations have been explained in terms of vertical differential loading, penecontemporaneous slumping of soft sediments, and collapse due to solution of gypsum beds.

Most investigators have visualized the crenulations in the Carmel Formation as chaotic structures. However, regular patterns of crenulations can be seen by focusing on regularities and by ignoring chaotic aspects of the Carmel in Arches National Monument, near Moab, in southeastern Utah (Fig. 1). Many of the crenulations appear to be folds caused by shortening of relatively well-bedded rocks of the Carmel Formation, which is confined above and below by relatively massive sandstones. The primary purpose of my investigation is to determine if we can understand idealized patterns of folds within the Carmel in terms of theoretical folding or interlayered soft and stiff materials.

The folds at Arches Monument present an excellent opportunity to understand a field occurrence of folding in terms of theory. They are low-amplitude folds so that the usual approximation of plate and beam theory that deflections are small should be applicable here. Also, the boundary conditions are as well-exposed as any ever will be in nature, in the many miles of cliff exposures of folded rocks and of undeformed overlying and underlying rocks.

Theories of folding of layered rock are rapidly developing under the able leadership of Hans Ramberg and Maurice Biot. These investigators analyzed the problem of buckling of competent layers of elastic and viscous substances and showed that the wavelengths of resultant folds are related to

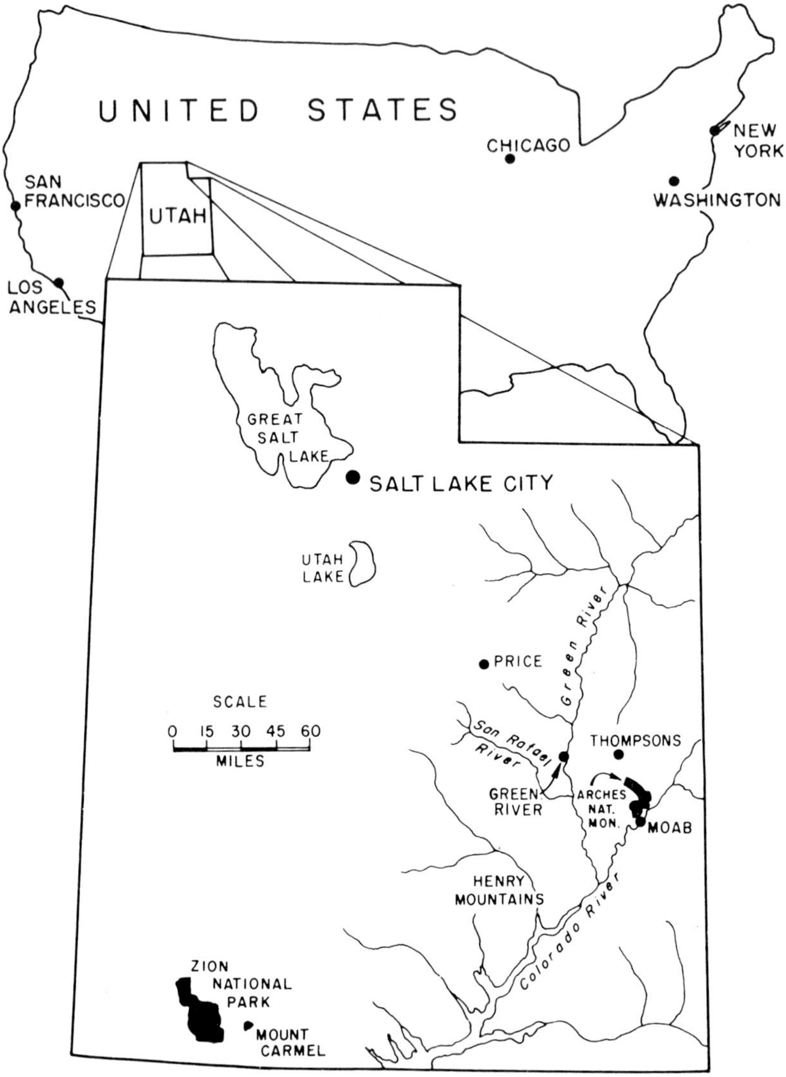

Fig. 1. Area of crenulated beds in southeastern Utah.

thicknesses of the competent layers (Ramberg, 1959, 1960, 1963b; Biot, 1961). Sherwin and Chapple (1968) compared wavelengths and thicknesses of quartz veins folded within phyllite and slate and found that, with slight modifications, the theories of Biot and Ramberg account for the forms actually observed.

Folded competent members that are isolated within incompetent materials are uncommon in rocks, however, except for a few buckled quartz veins and ptygmatic folds. Consequently Biot (1963, 1964, 1965a, 1965c) and Ramberg (1963a, 1964) began to analyze behavior of multilayers, which are groups of many layers, and to predict patterns of folds under various conditions. Potter (1968) reported that crinkled bands of mica within sandstone beds have assumed a pattern that is similar to the pattern that Biot has termed internal buckling. The resemblance between the theoretical and the actual patterns is not particularly apparent in Potter's figures, however.

Biot's theory of the multilayer is a special case of a much more general theory, which he has developed to describe finite deformation of initially stressed, anisotropic materials ranging from elastic through viscoelastic (Biot, 1963, 1965a, 1966, 1967). His development of the theory involves a large number of conceptual and mathematical manipulations so that it is rather difficult to follow. An elastic analog of his theory of viscous multilayers is presented here, employing principles of elastic beam and plate theory, which are relatively easily understood and visualized. One purpose of presenting details of the derivations is to encourage other investigators to understand and to apply the Biot and Ramberg theories and to develop new theories in order to explain fold patterns seen in the field. Another purpose is to make easier the task of tailoring the theories for particular applications, by emphasizing many of the assumptions that have led to the simplified theory of the multilayer.

The theory developed in the following pages is based upon the assumption that the folded rock layers behaved as elastic plates at the time the folds began to grow. However, the conclusions of the theory are not restricted to elastic materials. Biot (1961, 1963, 1964, 1965a, 1967) and Ramberg (1963a, 1964) have shown repeatedly that conclusions derived from the type of analysis presented here are valid insofar as rock behaves viscously, viscoelastically or elastically. Accordingly, I have used the general term "stiff" to imply either layers with high elastic moduli or layers with large viscosity coefficients and the general term "soft" to imply either low elastic moduli or viscosity coefficients.

The folded beds of the Carmel at Arches Monument were somewhat crumpled before the main episode of folding that produced the forms we see today, so that theory is developed to determine the influence of initial deflections on the buckling of stratified media. The influence of initial deflections on folding was clearly recognized by Bailey Willis in 1880's, but it has been largely ignored by geologists since that time.

The theory of the buckling of multilayered sequences does not account for all the factors we can recognize in the field at Arches Monument, but we can understand many of the factors in terms of the theory. In this respect the theory is of great value to an observer of the folds.

CRENULATIONS WITHIN CARMEL AT ARCHES NATIONAL MONUMENT

Some fascinating crenulations are well exposed in the Carmel Formation near Moab, Utah. They are visible in the cliffs around the peripheries of numerous buttes and mesas in Arches National Monument, a few miles north

Fig. 2. Crenulations within Carmel Formation at Windows Section of Arches Monument (location *1*, Fig. 3).

of Moab. For example, Fig. 2 shows a series of crenulations at the base of a cliff in the Windows Section of the Monument (see Fig. 3, location *1*). They are in the Carmel Formation, which comprises the crenulated beds between the massive, light-toned rocks forming the apron at the base of the cliff and the massive rocks forming the upper part of the cliff. The rocks above the Carmel are part of the Entrada Sandstone and those below are part of the Navajo Sandstone.

The light-toned Navajo Sandstone at the base of the cliff shown in Fig. 2 is cross-bedded on a grand scale as is typical of this formation. The top of the Navajo is a nearly perfect plane, separating light-toned Navajo and dark-toned Carmel. The top of the Carmel is presumably the base of the massive sandstone forming the upper part of the cliff.

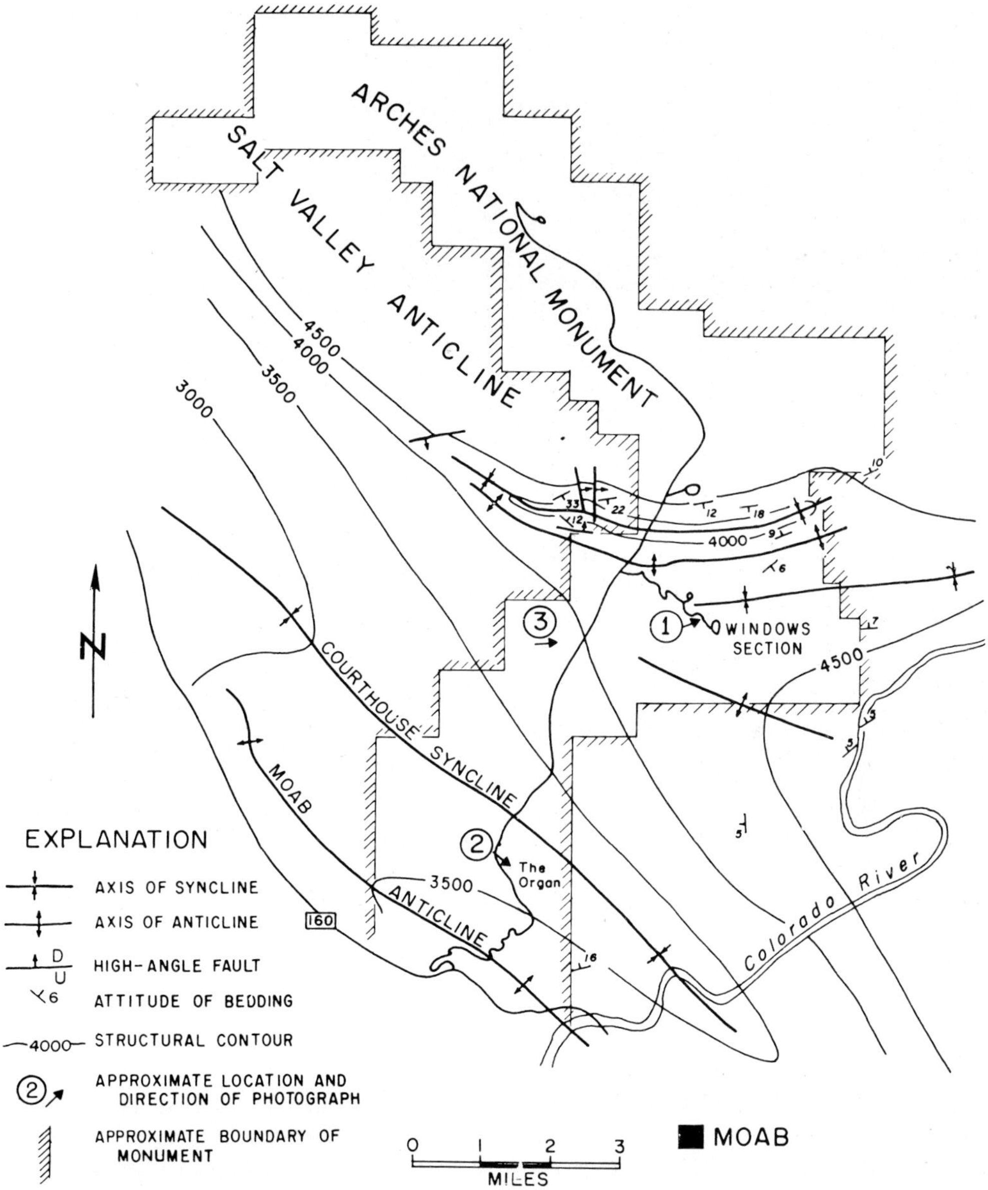

Fig. 3. Large structures in area of Arches National Monument, southeastern Utah. Structural contours are on base of Wingate Sandstone (after Baker et al., 1954). Approximate locations and directions of photographs are indicated by numbered arrows. *1* = Fig. 2; *2* = Fig. 4; *3* = Fig. 5.

The bedding within the lower half of the Carmel, which is estimated to be 60—70 ft. thick here, is distinctly wavy. In the right-hand end of the cliff there is a relatively resistant bed, about 3—4 ft. thick, separating relatively

Fig. 4. Gentle waves in bedding of Carmel Formation at The Organ, Arches Monument (location *2*, Fig. 3).

softer beds about 1 ft. thick. Above the upper and below the lower soft beds are resistant beds that range in thickness from about 6—12 ft. and that change continuously in thickness in a repetitive fashion along the cliff face; thus the thickness of the lower, relatively resistant bed changes from 12—6 to 12—8 and then to 12 ft., from right to left along the cliff face. Immediately above the upper, thin, relatively soft layer, faint bedding planes are seen to bend downward where the soft layer bends downward, but the amount of downward bending decreases higher within the Carmel until at the top of the Carmel the bedding planes are essentially flat.

Slightly left of center in the cliff face shown in Fig. 2, there are two thin, relatively resistant beds, each bounded by thinner, relatively soft beds. Each of these resistant beds and surrounding beds are deflected upwards and downwards into patterns similar to those described above.

Thus, there are two types of distortions of bedding planes within the right-hand half of the cliff shown in Fig. 2: (1) upward or downward deflection without appreciable thickening or thinning of beds; (2) thickening or thinning of beds, resulting in diminishing amplitudes of upward and downward deflections of bedding planes with increasing distances normal to bedding planes.

Both types of distortions of bedding planes are repeated at nearly uniform intervals along the cliff face shown in Fig. 2. In this sense, both types of distortions are periodic.

The gradual diminishing amplitudes of upward and downward deflections of bedding planes within the Carmel is more clearly seen in The Organ, one of several tabular rock bodies that project abruptly through the desert floor like dorsal fins of a school of sharks. The western end of The Organ is shown in Fig. 4, which shows gnarly, light-toned Navajo in the foreground, well-bedded Carmel above, and massive Entrada on top. The sharp downwarp slightly right of center and the more gentle upward to its left gradually decrease in amplitude skyward from a maximum of about 6 ft. at the top of the thick, poorly bedded unit in the center of the Carmel to essentially zero at the base of the very thick, massive bed of Entrada.

ORIGINS OF CRENULATIONS

The crenulations we see in the Carmel Formation and in the lower part of the Entrada Sandstone probably had several origins and they have been explained in terms of several mechanisms. Some investigators have suggested that the crenulations are due to slumping or to differential loading of water-saturated muddy sediments, whereas others maintain that the crenulations are collapse features related to solution of gypsum beds. The origin probably is different for different facies of the Carmel.

The Carmel Formation is a highly varied mixture of clastic and chemical depositional products, consisting of sandstones, siltstones, shales, limestones and gypsum. It was first described and named by Gregory and Moore (1931), who examined exposures near Mount Carmel, a town near Zion National Park, in southwestern Utah (Fig. 1). In the type area, it can be divided into a lower unit, composed of red shale and sandstone, and an upper unit, composed of limestone with some shale and gypsum (Baker et al., 1936, p. 6). In San Rafael Swell, about 50 miles west of Arches Monument, a lower unit consists of a red, limy sandstone, overlain by limestones and limy shales and an upper unit consists of shales and gypsum. The gypsum is commonly highly contorted. Eastward from San Rafael Swell, the Carmel thins and becomes predominantly red shales and sandstones, which are highly contorted (Baker et al., 1936, p. 6).

In the vicinity of Arches National Monument, the Carmel is predominantly red muddy sandstone which is thin-bedded to massive. It is generally softer than the underlying Navajo Sandstone and the overlying Entrada Sandstone so that it commonly forms a bench between the two formations (Dane, 1935). In the Windows section of the monument, the unit I have interpreted to be Carmel consists of about 70 ft. of interbedded muddy sandstone and sandy siltstone. The sandstone beds vary in thickness horizontally and vertically. They range in thickness from a few inches to 10 or 12 ft. The thickness of a given sandstone bed may vary by as much as 50% in the horizontal direction. This horizontal variation in thickness is caused by discontinuous siltstone beds and by waviness of one surface of some sandstone beds (e.g., see Fig. 2). The siltstone beds generally are thin, being a few inches thick, and are discontinuous. The bedding is more apparent from a distance than close to the outcrop. At close quarters a cliff face appears to be quite massive and the continuity of the siltstone beds is not readily apparent. The siltstone and sandstone beds both respond as hard substances to the blow of a hammer, but the siltstone appears to be more brittle and more highly fractured so that it has been weathered preferentially (see Fig. 2).

Stokes and Holmes (1954), who examined crenulations within the Carmel in the vicinity of San Rafael Swell, postulated that the crenulations are related to solution and flowage of gypsum beds: "The gypsum and associated beds are usually disturbed and contorted and in sections where no gypsum is present the equivalent interval shows disturbances suggesting that the gypsum has been removed by solution allowing the overlying beds to settle in an irregular manner" (Stokes and Holmes, 1954). Similarly, Steed (1954), who studied the Circle Cliffs anticline, about 90 miles southwest of the monument, reported that crenulations within the Carmel were caused by flowage of gypsum. The crenulations in the monument probably are unrelated to the solution or the flowage of gypsum beds, however, because the facies of the Carmel that contains significant thicknesses of gypsum terminates several tens of miles west of the monument.

Most investigators, who have examined the eastern facies of the Carmel, apparently visualized the crenulations as chaotic features that formed during or soon after deposition (Gilluly and Reeside, 1928; Dane, 1935; Baker et al., 1936; McKnight, 1940). According to Gilluly and Reeside (1928, p. 74): "The crenulations . . . are surely depositional features, possibly due to subaqueous erosion and flow." Dane (1935, p. 100) imagined that " . . . the initiating cause of the movements lies in the plastic adjustment and gliding of the water-soaked sandy mud of the Carmel under differential loading".

Some of the irregularities of bedding within the Carmel certainly are due to depositional and erosional processes. For example, local and irregular thickening and thinning of individual sandstone beds probably is a result of scour by currents. Other types of irregularities of bedding, however, have definite patterns, which probably reflect a different origin. For example, the periodic or wavelike thickening and thinning of a lower sandstone bed of the Carmel, shown in Fig. 2, is remarkably regular. The overlying, thin sandstone bed, sandwiched between the two thin siltstone beds, follows faithfully the contour of the waves in the underlying sandstone beds, indicating vertical deflection. I have interpreted this feature and many similar ones in Arches National Monument as folds, caused by lateral shortening of interbedded relatively soft and relatively stiff layers of sandstone and siltstone.

Apparent lack of folding in underlying and overlying units is not a valid argument against lateral shortening and resultant buckling in the Carmel, as has been suggested by some investigators. The underlying unit, the Navajo Sandstone, is a relatively massive sandstone unit of about 150 ft. thickness. The overlying unit is also quite massive and is probably more than 200 ft. thick. In contrast, the Carmel consists of interbedded sandstone and siltstone in which the layers range from a few inches to a few feet in thickness. We know from previous theoretical, field, and experimental investigations (e.g., Biot, 1961; Currie et al., 1962; Ramberg, 1963) that the amount of shortening and of loading required to fold thick beds is greater than the amount required to fold thin beds. Therefore, the thin beds of the Carmel would be expected to fold under lower loads and with less shortening than adjacent thick beds and, as a result, the amount of shortening may have been inadequate to fold the Navajo and the Entrada. These units may have simply shortened while the Carmel buckled into folds. Also, the wavelength of folds increases approximately with the thickness of the bed being folded, so that folds involving both the Navajo and the Entrada would be expected to have a wavelength of at least one-half mile. Folding of these units probably would not be visible in the small outcrops shown in Figs. 2 and 4.

The pattern of folds within the Carmel is rather complex because the bedding is complex. If the beds were originally nearly perfectly flat plates of uniform thickness and if all the plates had the same thickness and physical properties, we would expect a simple fold pattern to emerge when the beds were folded. In nature, of course, such ideal conditions are lacking and, indeed, as we can see in Fig. 2, the bedding within the Carmel is not simple; the sandstone beds change in thickness laterally and the siltstone beds are discontinuous. In addition, the beds probably contained initial dips and warps that were formed during sedimentation and compaction of the sediments and that were buckled into folds. Each of these factors contributed to the irregularities in the pattern we see in the Carmel. We will analyze some of them in later pages when we discuss the theory of buckling of interlayered materials.

SOURCES OF SHORTENING OF CARMEL

The sources of the shortening that folded the Carmel are unknown, but they are probably multiple. One possible source has been suggested by several investigators: Local lateral movement of soft sediments shortly after deposition, the movements being caused by erosion and removal of lateral support of the sediments or to slight "differential loading" (Gilluly and Reeside, 1928; Dane, 1935; McKnight, 1940). Another source of the lateral shortening might be lateral sliding from low slopes created by upward movement of salt masses in various salt anticlines in the area of the monument. Finally, the shortening that produced the folds might have occurred at the time the broad flexures formed in the area of the monument (Fig. 3).

The small folds in Arches Monument are in the northwestern part of an area called the Paradox fold and fault belt by Kelley (1958, p. 31). The belt is centered on La Sal Mountains, east of Moab, Utah, and is about 65 miles wide and 150 miles long. It is elongated along a northwest-southeast axis. The fold and fault belt, according to Kelley, is characterized by open folds that trend northwesterly. In the cores of some of the open anticlines are salt piercement structures, one of which underlies Salt Valley, which is crossed by the monument (Fig. 3), and another of which underlies Moab Valley.

The crenulations in the Carmel Formation may be an expression of compression of the rocks in the fold and fault belt. According to Kelly (1958) and to Shoemaker et al. (1958), several periods of arching are recorded in the area. The salt anticlines began to grow between Late Permian and Middle Triassic times and they continued to grow until Late Jurassic time, as is indicated by thinning and pinchout of formations ranging from Late Triassic to Late Jurassic age (Shoemaker et al., 1958, p. 39). After Early Montana time, broad open folds formed in rocks between and around as well as on the older salt anticlines. The anticlines that formed over the salt anticlines are wider and somewhat different in form than the salt anticlines, themselves, so that they probably were caused by lateral compression rather than by rising of the salt (Shoemaker et al., 1958). More precise dating of the regional shortening that produced the folds has not been possible, but, according to Shoemaker et al. (1958, p. 53), the deformation usually is considered to have been between latest Cretaceous and Eocene times. The Carmel is Early to Middle Jurassic in age (Stokes and Holmes, 1954, p. 37) so that if the folds within it are indeed compressional features, they might be minor flexures produced by the regional shortening that formed the broad folds in the Paradox fold and fault belt. Thus the crenulations within the Carmel and the folds with a wavelength of several miles might be different orders of folds produced by the same shortening and, as such, similar to those produced experimentally by Ramberg (1964, Fig. 2).

AREAL EXTENT OF CRENULATIONS IN CARMEL

The types of crenulations we see in the Carmel Formation in Arches National Monument might be restricted to the belt of faults and open folds described by Kelley (1958). The areal extent was not determined in detail, but I did see them about 15 miles southwest of the monument, near Dead Horse Point, and about 30 miles southeast of the monument, a few miles south of LaSal Junction. Dane (1935, p. 96) mentioned contorted bedding in his description of a measured stratigraphic section south of Ryan Creek, about 6 miles east of the monument.

Dane (1935, p. 90) reported that the Carmel grades laterally eastward from limestones, shales and gypsum in San Rafael Swell into red, thinbedded sandstones and sandy shales, the bedding of which is contorted at many places. According to descriptions of stratigraphic sections measured by Gilluly and Reeside (1928), contorted bedding in sandstones of the Carmel and lower Entrada occurs on the eastern edge of San Rafael Swell, about 15 miles west of Green River, and between Green River and Moab (Fig. 1). Throughout most of San Rafael Swell, however, contorted bedding is restricted to gypsum beds (e.g., see the measured sections, 1, 4, 16, 22 23 and 24 of Gilluly and Reeside, 1928). Thus the types of crenulations visible at Arches Monument probably are restricted to an area southeast of San Rafael Swell, which is about 50 miles northwest of Arches Monument.

The bedding of the Carmel is irregular and contorted throughout most of an area bounded on the west by Green River, on the southeast by Colorado River and on the northeast by a line approximately through Moab and Thompsons (McKnight, 1940, p. 87). According to McKnight, who mapped this area, the Carmel is largely reddish-brown muddy sandstone with some reddish sandy mudstone. Thus the crenulated red beds are found at least 30 miles southwest of Moab.

Similar crenulations apparently are absent in the Kaiparowits region, immediately east of Zion National Park and about 150 miles southwest of the monument (Fig. 1). Gregory and Moore (1931), who studied that region, did not mention crenulations of bedding of either the Carmel Formation or the lower part of the Entrada Sandstone. Indeed, their photographs of thinly bedded Carmel and Entrada show remarkably even and planar bedding (Gregory and Moore, 1931, plates 10B, 11A, 12B).

The Carmel Formation has two facies in the vicinity of the Henry Mountains, about 100 miles southwest of Moab, and both facies contain crenulated beds (Hunt, 1953). It is largely gypsum, limestone, and gypsiferous sandstone along the Waterpocket fold, west of the mountains. On the eastern side of the mountains it is represented by interbedded red shaly sandstone and massive sandstone. Hunt reported that beds in the sandy and shaly facies on the eastern side of the Henry Mountains are deformed into a series of low domes and shallow basins that are commonly 50—100 ft. in diameter and that have about 10 ft. of relief. The strongest contortions, according to Hunt, are in the gypsiferous facies, on the western side of the mountains.

No crenulations are mentioned in the description of a measured section of Entrada and Carmel about 30 miles northeast of Moab (Gilluly and Reeside, 1928). This is approximately the southwestern limit of the Paradox fold and fault belt of Kelley (1958).

CONDITIONS OF CARMEL AT TIME OF FOLDING

Several lines of evidence indicate the conditions of the Carmel Formation at the time the crenulations were formed. Some of the crenulations must have formed after the Entrada Sandstone was deposited upon the Carmel, because in many places they extend through the Carmel and into the overlying Entrada. They diminish in amplitude upwards into the Entrada but, according to Dane (1935, p. 99), they extend as far as 50 ft. into the Entrada in some places. For example, Fig. 5 shows gentle crenulations in the Carmel Formation about two or three miles west of the Windows section of Arches Monument (Fig. 3, location *3*). The Carmel is the relatively well-bedded rock in the lower half of the cliff and the Entrada is the relatively massive sandstone, above. A downwarp in the beds of the Carmel near the center of the figure is reflected as a downwarp in the base of the massive Entrada, above. A similar downwarp can be seen near the left-hand edge of the figure. Both downwarps seem to diminish in amplitude within the Entrada, so that a few tens of feet into the massive sandstone of the Entrada, the downwarping is invisible.

Several investigators have mentioned that upturned bedding planes within some of the crenulated rocks were truncated and that horizontally bedded rocks cut across the tops of the crenulations (e.g., see Dane, 1935, p. 99; McKnight, 1940, p. 88). I saw no evidence of truncation of beds within the Carmel or of Carmel beds by overlying Entrada beds within Arches Monument, but I did not search specifically for this phenomenon. The evidence cited by Dane and McKnight seems to indicate that part of the deformation we see in the Carmel could have occurred before the Entrada was deposited, but other evidence indicates that much of the deformation occurred after the Entrada was deposited.

Fig. 5. Folds extend upward into relatively massive Entrada Sandstone about two miles west of Windows section of Arches Monument (location *3*, Fig. 3).

According to Dane (1935, p. 99), the rolling contact between the Entrada and the Carmel is offset by small faults in some places. The faults gradually die out upwards in gentle flexures of the Entrada, which in turn, diminish in amplitude upwards until they disappear. Some of them extend some 200 ft. into the Entrada, so that the cover on the Carmel at the time the crenulations formed was at least 200 ft. thick (Dane, 1935, p. 100).

The marked changes of thickness of sandstone beds associated with the crenulations in the Carmel probably is an indication that the muddy sand of the Carmel was soft at the time the crenulations formed. For example, it is difficult to conceive that the thick sandstone bed near the base of the Carmel, shown in Fig. 2, could have been firm at the time units overlying it were thrown into the series of waves we see today. The thick sandstone ranges in thickness back and forth between about 6 and 12 ft. over relatively short distances. Similar thickening and thinning is visible in a thick sandstone bed near the middle of the Carmel at The Organ (Fig. 4). Dane also believed that the Carmel was soft at the time the crenulations formed because he referred to the "water-soaked sandy mud of the Carmel" in discussing the origin of the crenulations.

The underlying sands of the Navajo must have been relatively firm at the time the crenulations formed. Dane reported that the contact between the Navajo and the Carmel, or where the Carmel is missing, the Navajo and the Entrada, is sharp and was undisturbed by whatever produced the crenulations. A sharp, planar contact between the Navajo and the Carmel is clearly visible as a line separating light- and dark-toned rocks shown in Fig. 2.

ANALYSIS OF FOLD PATTERNS WITHIN CARMEL FORMATION

General conditions of rocks

Some folds in the Windows section of Arches National Monument have an especially simple pattern, so that we will analyze them first. The folds shown at the right-hand side of Fig. 2 are gentle waves with a wavelength of about 40 ft. and a maximum amplitude, from trough to crest, of about 6 ft. The amplitudes die out upwards and downwards so that within a vertical distance of about 40 ft. the folds originate, grow to maximum amplitude, and die out (Fig. 6A). The contact between the Navajo Sandstone and the Carmel Formation is a markedly planar surface, so that we can represent that surface by an undeflecting, rigid boundary (Fig. 6C). The upper boundary of the folds is less well-defined than the lower one, but the bedding within the Carmel is essentially flat about 40 ft. above the base of the Carmel, so that we can imagine that the rocks above also behaved as nearly a rigid unit during folding.

The Carmel Formation is an interlayered sequence of thin sandy siltstone beds and thicker sandstone beds, as is shown in Fig. 6A. The siltstone beds were probably relatively soft and the sandstone beds relatively stiff at the

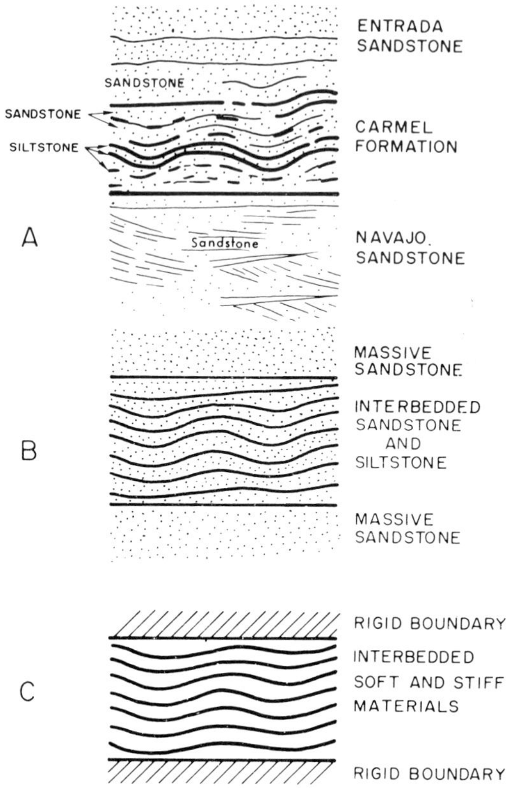

Fig. 6. Stages of idealization of folds in Windows section of Arches Monument. A. Map of sandstone and siltstone beds exposed in cliff face. B. Entrada and Navajo sandstones, idealized by massive sandstones, and Carmel Formation, idealized as uniformly interbedded sandstones and siltstones. C. Behavior of massive sandstones idealized by rigid boundaries and behavior of Carmel idealized by interbedded soft and stiff elastic materials.

time the folds formed. The siltstone beds probably provided places for sliding adjustments between adjacent sandstone beds. This conclusion is supported by Dane's observation that, where the Carmel is absent and where the basal Entrada is massive sandstone, without siltstone, he observed none of the bedding irregularities that are so common elsewhere (Dane, 1935, p. 100). Thus, we might idealize the sequence of beds within the Carmel as thin, interbedded soft and stiff layers, called a *multilayer* by Biot (1963), confined above and below by thick, rigid layers (Fig. 6B,C). This assumption of many soft and stiff layers can be only a first approximation to the actual situation, because the siltstone beds are generally discontinuous layers that thicken and thin laterally. We will reevaluate this assumption in later pages.

We will represent the soft and stiff layers by soft and stiff elastic plates, which are bonded together so that they cannot slip relative to each other without deforming internally. An elastic plate has a strictly defined mathematical meaning, and therefore it corresponds rather poorly to a real rock layer. However, Biot has shown repeatedly that viscous and viscoelastic plates assume the same types of buckling forms as elastic plates, so that our analysis actually is applicable to a wide range of physical behaviors of rock; our conclusions are valid insofar as rocks behave elastically, viscously, or viscoelastically (Biot, 1961, 1963, 1964, 1965a, 1967; Ramberg, 1963a, 1964). Plastic yielding and brittle fracture of rocks are excluded, but these behaviors probably did not play an important role in the folding of the rocks we are describing. Plastic behavior of buckling materials is manifested in sharpening of fold hinges, which is not apparent in the Carmel. The fracture patterns in the Carmel seem to be unrelated to the folds. The layers of the Carmel were probably initially somewhat wavy rather than perfectly flat as is assumed in plate theory, but we will investigate effects of waviness on folding in later pages.

Biot theory of buckling of multilayers

Among the many important contributions M.A. Biot has made to our understanding of folding processes is his analysis of the bending of sequences of interlayered soft and stiff beds. He has shown that the theory of bending of such sequences can be approximated by equations that are special cases of equations he has derived for the deformation of initially stressed, anisotropic, elastic, viscous and viscoelastic materials (Biot, 1964, 1965a, 1965b). His theory is an exceedingly powerful tool for geologists, who usually are trying to understand mechanical systems that are too complex to be treated by the theories developed by engineers and physicists to explain behaviors of simple materials under simple loading conditions. For example, whereas engineers usually deal with problems such as structural design involving atmospheric pressures, geologists typically deal with problems involving materials that are under high pressures or that were under high pressures when they formed. Biot considers effects of such pressures when he evaluates behaviors of various ideal materials under initial stresses.

Biot (1965a) introduces a large number of concepts in rapid succession when he explains his theory of initially stressed materials, so that his developments of the general theory and of the theory of the multilayer are rather difficult to follow. Following is a theory that is essentially equivalent to his theory of the multilayer, derived by methods employed in traditional beam and plate theories, which are relatively easy to understand. I hope that the following derivations will clarify some of the methods Biot applies and will encourage other geologists to assimilate Biot's more general theory, which should find applications in the solution of a wide range of geological problems.

We will idealize the beds of the Carmel at Arches Monument by a multilayer, consisting of alternating layers of soft and stiff materials (Fig. 7A).

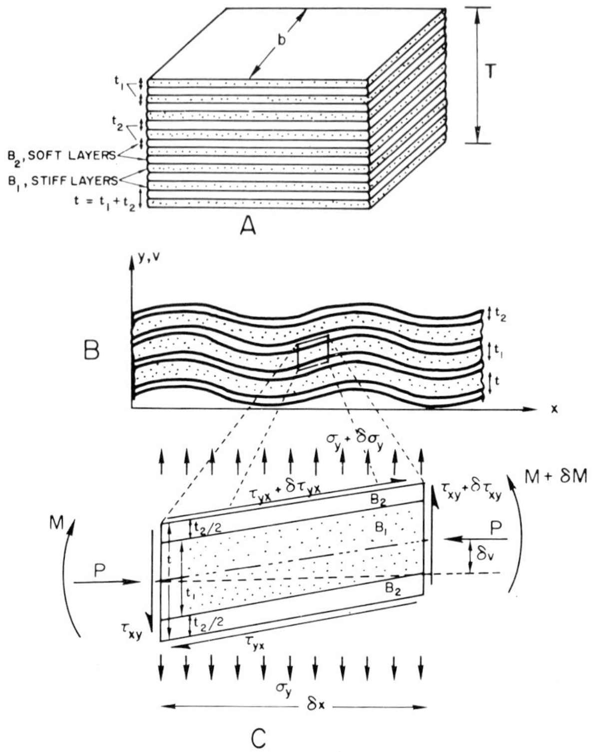

Fig. 7. Forces, stresses and bending moments acting on element of multilayer. A. Multilayer consists of interbedded soft and stiff materials with thicknesses and elastic moduli of t_2, B_2 and t_1, B_1, respectively. Total thickness of multilayer is T. B. Position of element within bent multilayer. C. Element of multilayer of thickness t and length δx.

The soft layers have a thickness of t_2 and an elastic modulus of B_2, where:

$$B_2 = \frac{E_2}{1 - \nu_2^2} \tag{1}$$

Here E is Young's modulus and ν is Poisson's ratio. Similarly, the stiff layers have a thickness of t_1 and a modulus of B_1. The total thickness of the system of beds, that is of the multilayer, is T (Fig. 7A).

Differential equation of deflection of multilayer

Suppose that the multilayer is slightly deformed under the action of axial loads, shear stresses, normal stresses, and bending moments (Fig. 7B). Following Biot (1965c), we define the fundamental element of the multilayer as a unit length, δx, width, b, and thickness, t, equal to the combined thickness of one stiff and one soft layer (Fig. 7C). Thus:

$t = t_1 + t_2$

Acting on the element are shear stresses, τ_{yx} and τ_{xy}, normal stresses, σ_y, axial loads, P, and bending moments, M (Fig. 7C). The stresses and moments change across the width or thickness of the element so that, for example, the normal stress in the vertical, y-direction is σ_y at the base of the element and slightly larger, $\sigma_y + \delta\sigma_y$, at the top of the element.

In order to derive relations among the various stresses, forces, and moments, we sum forces and moments and equate the sums to zero. Thus, summing moments about the center of the right-hand end of the element, $[\Sigma M = 0]$:

$$-M + M + \delta M + P(\delta v) + \tau_{xy}(bt)\delta x + \sigma_y(\delta x)\, b\left(\frac{\delta x}{2}\right) - (\sigma_y + \delta\sigma_y)\delta x\, b\left(\frac{\delta x}{2}\right)$$

$$- \tau_{yx}\,\delta x\, b\left(\frac{t}{2}\right) - (\tau_{yx} + \delta\tau_{yx})\,\delta x\, b\left(\frac{t}{2}\right) = 0$$

which reduces to:

$$\delta M + \tau_{xy}\,\delta x\, bt - \tau_{yx}\,\delta x\, tb - \frac{\delta\tau_{yx}\,\delta x\, tb}{2} + P\delta v - \delta\sigma_y\, b\,\frac{(\delta x)^2}{2} = 0$$

Some of the terms in the latter equation are smaller than others, so that to a first approximation we can ignore tham. Thus:

$$\tau_{yx}\,\delta x >> \frac{\delta\tau_{yx}\,\delta x}{2} \rightarrow 0$$

and:

$$\delta\sigma_y(\delta x)^2 \rightarrow 0$$

These two terms are of "second order".

To a first approximation, therefore, the equation of equilibrium of moments becomes:

$$\delta M + \tau_{xy}\,\delta x\, tb - \tau_{yx}\, t\delta x b + P\delta v = 0$$

Dividing through by δx:

$$\frac{\delta M}{\delta x} + (\tau_{xy} - \tau_{yx})\, tb + P\frac{\delta v}{\delta x} = 0 \tag{2}$$

If we take the limit of $\delta M/\delta x$ and $\delta v/\delta x$ as δx approaches zero, we have the differential equation:

$$\frac{\delta M}{\delta x} + (\tau_{xy} - \tau_{yx})tb + P\frac{\partial v}{\partial x} = 0 \tag{3}$$

Eq. 3 is essentially equivalent to the first of Biot's (1965c) eq. 1 (p. 833). The differential equation derived in standard beam and plate theory for the relation between the bending moment and the deflection of an axially loaded beam or plate is slightly different (Timoshenko and Gere, 1961):

$$\frac{dM}{dx} + Q + P\frac{dv}{dx} = 0$$

Here Q is equivalent to $\tau_{xy}tb$ in eq. 3, so that the only additional term in eq. 3 is the shear force, $\tau_{yx}tb$. The remaining difference between the two equations is that here we use partial derivatives instead of total derivatives. The reason for this change is that the vertical displacement of a multilayer is a function of both the x- and y-coordinates, whereas the vertical displacement is usually considered to be a function of only the distance along the members, the x-coordinate, in standard beam theory.

We can (see Fig. 7C) relate the shear stress, τ_{xy}, to the normal stress, σ_y, by summing forces in the y-direction acting on the small element of the multilayer, $(\Sigma F_y = 0)$:

$$-\sigma_y \delta x b + (\sigma_y + \delta\sigma_y)\delta x b - \tau_{xy} tb + (\tau_{xy} + \delta\tau_{xy}) tb = 0$$

or:

$$\delta\sigma_y \delta x + \delta\tau_{xy} t = 0 \tag{4}$$

We are assuming that the bending of the multilayer is so slight that the shear stress, τ_{yx}, is parallel to the x-direction. If we assume that the vertical normal stress, σ_y, varies continuously in the vertical direction, then the relation:

$$\delta\sigma_y = \frac{t\delta\sigma_y}{\delta y} \tag{5}$$

exists, so that eq. 4 becomes:

$$\frac{\delta\sigma_y}{\delta y} = -\frac{\delta\tau_{xy}}{\delta x}$$

or in the limit:

$$\frac{\partial\sigma_y}{\partial y} = -\frac{\partial\tau_{xy}}{\partial x} \tag{6}$$

Eq. 6 and eq. 3 can be combined by differentiating eq. 3 with respect to x and by substituting eq. 6 into eq. 3:

$$\frac{\partial^2 M}{\partial x^2} - \left(\frac{\partial\sigma_y}{\partial y} + \frac{\partial\tau_{yx}}{\partial x}\right) tb + P\frac{\partial^2 v}{\partial x^2} = 0 \tag{7}$$

Elastic constants of multilayer

We need to substitute functions of displacement, v, for the variables M, σ_y, and τ_{yx} into eq. 7 in order to put the differential equation into a form that we can solve. The functions we will derive will relate stresses and displacements to the elastic constants, B_1 and B_2, of the stiff and soft layers, respectively. Each layer of the multilayer has homogeneous and isotropic elastic properties. That is, the elastic properties are the same everywhere and in every direction within each type of layer in the multilayer. But when several layers with different elastic properties are sandwiched together the composite behaves as an anisotropic material; the elastic properties parallel to bedding are different from the elastic properties normal to bedding.

We will assume that each element of a multilayer has homogeneous elastic properties, which are sort of "average" properties. Each element, however, has anisotropic properties in the sense that the elastic moduli, B_p, parallel to bedding, and B_n, normal to bedding, may be different. The values of B_p and B_n depend upon the degree of bonding of the layers of the multilayer and upon the differences of the elastic properties of the individual layers (e.g., see Biot, 1966).

According to standard beam and plate theory (e.g., see Timoshenko and Gere, 1961), the bending moment produced by fiber stresses within a bent, homogeneous, isotropic, elastic beam or plate is:

$$M = BI \frac{\partial^2 v}{\partial x^2} \tag{8}$$

where M is bending moment, B is elastic modulus, defined in eq. 1, I is moment of inertia, which depends solely upon the cross-sectional shape of the member, and $\partial^2 v/\partial x^2$ is the curvature of the member. Actually, the exact expression for curvature contains a slope term, $\partial v/\partial x$, but if deflection is small, and slopes are less than about 20°, the second derivative is an adequate approximation to the curvature, which is:

$$\text{curvature} = \frac{1}{\rho} = \frac{\left|\dfrac{\partial^2 v}{\partial x^2}\right|}{\left[1 + \left(\dfrac{\partial v}{\partial x}\right)^2\right]^{3/2}} \tag{9}$$

Here ρ is the radius of curvature.

For a homogeneous, isotropic plate:

$$B = \frac{E}{1 - \nu^2}$$

and:

$$I = \frac{bh^3}{12}$$

where h is the thickness of the plate. The problem is to derive expressions for B and I for a multilayer.

If we assume that the resistance to bending of a multilayer arises from

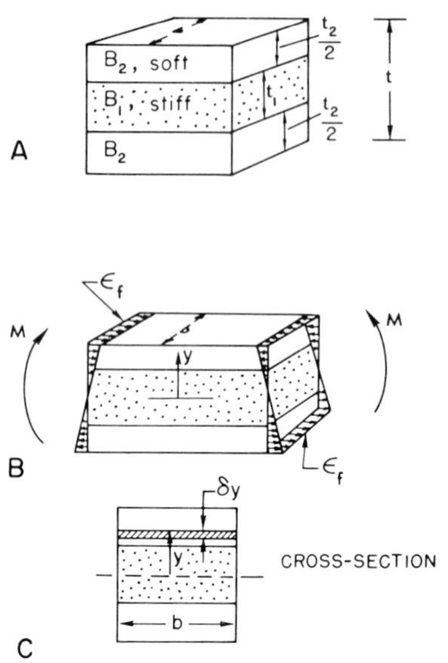

Fig. 8. Distortion of element of multilayer. A. Element of multilayer. B. Linear distribution of fiber strains is assumed. C. Cross-section of element.

both the soft and stiff layers, we can determine the modulus and the moment of inertia of the multilayer as follows. Fig. 8A shows a small element of the multilayer, consisting of one soft and one stiff layer, half of the soft layer being on each side of the stiff layer. If the element is bent, the soft and stiff layers are strained in the horizontal direction, that is, there are *fiber strains.* We will assume that the fiber strain is linearly distributed relative to the center of the stiff layer so that it increases linearly from zero at the center to a maximum at the top and bottom edges of the element. The fiber strain is related to the radius of curvature, ρ, of the element and to the distance from the neutral plane of the element, by the relation (Fig. 8B):

$$\epsilon_f = \frac{y}{\rho} \tag{10}$$

Also, the bending moment, M, is related to the fiber stress, σ_f, by the equation (Fig. 8c):

$$M = b \int \sigma_f y \, dy \tag{11}$$

However, the fiber stress is related to the fiber strain by the equations:

$\sigma_f = \epsilon_f B_1$, within stiff layers

and: (12)

$\sigma_f = \epsilon_f B_2$, within soft layers.

Therefore, combining eq. 10, 11 and 12, the bending moment is:

$$M = b\int \epsilon_f B_1 y\,dy + b\int \epsilon_f B_2 y\,dy = \frac{B_1 b}{\rho}\int y^2\,dy + \frac{B_2 b}{\rho}\int y^2\,dy \tag{13}$$

in which the first integral applies to the stiff layer and the second to the soft layer. Integrating through the stiff and soft layers, we find that:

$$M = \frac{2B_1 b}{\rho}\int_0^{\frac{t_1}{2}} y^2 dy + \frac{2B_2 b}{\rho}\int_{\frac{t_1}{2}}^{t/2} y^2 dy$$

where t is the combined thickness of one stiff and one soft layer: Thus:

$$M = \frac{B_1 b}{12\rho}\,t_1^3 + \frac{B_2 b}{12\rho}\,[t^3 - t_1^3]$$

or:

$$M = \frac{t_1^3 b}{12\rho}\left\{B_1 + B_2\left[\left(\frac{t}{t_1}\right)^3 - 1\right]\right\} \tag{14}$$

Therefore, if we define two new constants:

$$I_p = \frac{t_1^3 b}{12}\text{, moment of inertia of stiff layer} \tag{15}$$

and:

$$B_p = B_1 + B_2\left[\left(\frac{t}{t_1}\right)^3 - 1\right]\text{ elastic modulus parallel to bedding *} \tag{16}$$

we can write the equation relating the bending moment to the radius of curvature in the standard form:

$$M = \frac{B_p I_p}{\rho} = B_p I_p\,\frac{\partial^2 v}{\partial x^2} \tag{17}$$

For a multilayer, then, the differential equation of equilibrium, eq. 7, becomes:

$$B_p I_p\,\frac{\partial^4 v}{\partial x^4} - \left(\frac{\partial \sigma_y}{\partial y} + \frac{\partial \tau_{yx}}{\partial x}\right) tb + P\frac{\partial^2 v}{\partial x^2} = 0 \tag{18}$$

Now we will relate the shear stress, τ_{yx}, to the shear modulus and to the deflection of the multilayer. We will assume that the soft and stiff layers are subjected to the same shear stress when the multilayer is bent and when the

* In other chapters this has been simplified to $B_p = B_1$.

layers attempt to slip relative to each other. The resulting shear strain in the stiff layer is $(\gamma_{yx})_1$ so that the stress—strain relation for shear in the stiff layer is:

$$(\gamma_{yx})_1 = \frac{1}{G_1}\tau_{yx} \tag{19}$$

where G_1 is the shear modulus of elasticity of the stiff layer. Similarly, the shear strain in the soft layer is:

$$(\gamma_{yx})_2 = \frac{1}{G_2}\tau_{yx} \tag{20}$$

where G_2 is the shear modulus of the soft layer.

The total horizontal displacement due to shear of the surfaces of the unit element of the multilayer, shown in Fig. 8B, is the "average" shear strain, γ_{ave}, times the thickness, t, of the unit element. It is equal to the sum of the displacements in the soft and stiff layers, so that:

$$t\gamma_{\text{ave}} = t_1(\gamma_{yx})_1 + t_2(\gamma_{yx})_2$$

or:

$$\gamma_{\text{ave}} = \frac{t_1}{t}(\gamma_{yx})_1 + \frac{t_2}{t}(\gamma_{yx})_2 \tag{21}$$

Substituting eq. 19 and 20 into eq. 21:

$$\gamma_{\text{ave}} = \frac{\tau_{yx}}{t}\left[\frac{t_1}{G_1} + \frac{t_2}{G_2}\right]$$

so that we can define an "average" shear strain of the multilayer as:

$$\gamma_{\text{ave}} = \frac{\tau_{yx}}{G_a} \tag{22}$$

where the "average" shear modulus is:

$$G_a = \frac{1}{\left[\dfrac{t_1}{tG_1} + \dfrac{t_2}{tG_2}\right]} = \frac{tG_1G_2}{G_2t_1 + G_1t_2} \tag{23}$$

Biot (1965a) calls G_a the "sliding modulus".

The shear strain, γ_{yx}, can be defined in terms of displacements, u and v, in the x- and y-directions, respectively:

$$\gamma_{yx} = \left(\frac{\partial u}{\partial y} + \frac{\partial v}{\partial x}\right) \tag{24}$$

Let us define the average shear strain in the same way so that $\gamma_{\text{ave}} = \gamma_{yx}$ in eq. 24. If the multilayer is deflected mostly in the y-direction, so that the layers move upwards but not horizontally, the displacement in the x-direction, u, is zero, and [1]:

$$\frac{\partial u}{\partial y} = 0$$

The average shear strain reduces to:

$$\gamma_{\text{ave}} = \frac{\partial v}{\partial x} \tag{25}$$

the change in vertical displacement with respect to the horizontal direction. Substituting eq. 25 into eq. 22, the shear stress becomes:

$$\tau_{yx} = G_a \frac{\partial v}{\partial x}$$

Thus the differential equation, eq. 18, becomes:

$$B_p I_p \frac{\partial^4 v}{\partial x^4} - \left(\frac{\partial \sigma_y}{\partial y} + G_a \frac{\partial^2 v}{\partial x^2}\right) tb + P \frac{\partial^2 v}{\partial x^2} = 0 \tag{26}$$

Now we have a differential equation in which all but one of the terms are expressed as functions of the displacement, v.

The differential equation can be completely expressed in terms of displacement in the y-direction if we can relate the vertical normal stress, σ_y to the deflection, v. In order to derive an expression for σ_y in terms of v, we will have to assume a general boundary condition for our problem of folding. If folding can occur only by compressing the multilayers, the vertical normal stress imposed by neighboring layers on any layer is:

$$\sigma_y = B_n \epsilon_y$$

where B_n is a modulus, which we will derive, and where the vertical normal strain is:

$$\epsilon_y = \frac{\partial v}{\partial y}$$

---

[1] In some applications the change of displacement, u, with respect to the y-direction might be appreciable, but constant. If $\partial u/\partial y$ is constant, the differential equation, eq. 26, is unchanged so that the solution we will derive in following pages will be applicable. Such a situation might arise where a multilayer is confined between two thick, competent layers that are being folded on a larger scale than are the multilayers. As the two thick layers bend they slip relative to each other, transferring a shear strain through the multilayer. The result would be a differential horizontal displacement of the crests of the minor folds in the multilayer, presumably producing drag flexures.

Thus:

$$\frac{\partial \sigma_y}{\partial y} = B_n \frac{\partial^2 v}{\partial y^2} \tag{27}$$

If we assume that vertical strains * are the same throughout an element of the multilayer:

$$\epsilon_y = \frac{(\sigma_y)_1}{B_1} = \frac{(\sigma_y)_2}{B_2}$$

Then the weighted average of the vertical normal stress, σ_y, for the soft and stiff layers is:

$$\sigma_y = \frac{t_1}{t}(\sigma_y)_1 + \frac{t_2}{t}(\sigma_y)_2$$

Thus:

$$\sigma_y = \epsilon_y \left(\frac{t_1}{t} B_1 + \frac{t_2}{t} B_2\right)$$

and we can define B_n, the elastic modulus normal to bedding, as (Biot, 1965a):

$$B_n = \left[\frac{t_1}{t} B_1 + \frac{t_2}{t} B_2\right] \tag{28}$$

When we substitute the expression for the constraint of the adjacent layers, eq. 27, into the differential equation, eq. 26, we have an equation that is entirely in terms of the vertical displacement, v:

$$B_p I_p \frac{\partial^4 v}{\partial x^4} + (P - tb\, G_a) \frac{\partial^2 v}{\partial x^2} = tb\, B_n \frac{\partial^2 v}{\partial y^2} \tag{29}$$

The coefficients are all constants, independent of the three variables, x, y and v.

Review of assumptions

Before we proceed to solve differential eq. 29 and to derive the buckling form of multilayers confined by rigid boundaries, we will state explicitly the assumptions we used in order to derive the differential equation. Some of the assumptions probably will have to be changed to solve problems other than the one of interest here so that they should be clearly understood.

(1) The soft and stiff layers of the multilayer are ideal elastic substances.

* See Appendix at the end of this chapter.

Biot has shown that the differential equations and their solutions are essentially the same for elastic and viscous substances.

(2) Strains are infinitesimal.

(3) Deflections are small enough so that $(\partial v/\partial x)^2$ is much smaller than unity (see eq. 9).

(4) The thicknesses of the individual beds, t_1, t_2, and of the unit element of the multilayer, t, are small relative to the total thickness of the multilayer.

(5) Buckling is caused solely by the axial load. There are no external shear forces applied to the boundaries of the multilayer.

(6) The fiber strains in an element consisting of one stiff and one soft layer are linearly related to the distance from the center of the stiff layer. This assumption led to the following definitions of the modulus and the moment of inertia of the multilayer.

$$I_p = \frac{t_1^3 b}{12}$$

$$B_p = B_1 + B_2 \left[\left(\frac{t}{t_1} \right)^3 - 1 \right]$$

If we had assumed, on the contrary, that the stresses are linearly related to the distance from the center of the stiff layers, the horizontal strains would have to be discontinuous. This is necessary because the constants of proportionality between stress and strain are different in the two materials. Thus, if the strains are continuous, the stresses are discontinuous; if the stresses are continuous, the strains are discontinuous (see Biot, 1966).

(7) There is high friction between layers so that the shear stress is transferred from one layer to the other. Further, the shear stress is the same in the soft and stiff layers. With these assumptions we derived the shear modulus, which is Biot's "slide modulus", given by eq. 23. If we had assumed, instead, that the shear strain was the same in adjacent soft and stiff layers, we would replace eq. 21 with:

$$(\tau_{xy})_{\text{ave}} = \frac{t_1}{t} (\tau_{yx})_1 + \frac{t_2}{t} (\tau_{yx})_2$$

where $(\tau_{yx})_1 = \gamma_{yx} G_1$ and $(\tau_{yx})_2 = \gamma_{yx} G_2$, so that the elastic modulus would be:

$$G_a = \frac{t_1}{t} G_1 + \frac{t_2}{t} G_2 \qquad (30)$$

instead of the value given by eq. 23.

(8) We assumed that the vertical normal strains are the same in adjacent soft and stiff layers of an element. If we had assumed, instead, that the vertical normal stresses are the same in the soft and stiff layers of an element:

$$\sigma_y = (\epsilon_y)_1 B_1 = (\epsilon_y)_2 B_2$$

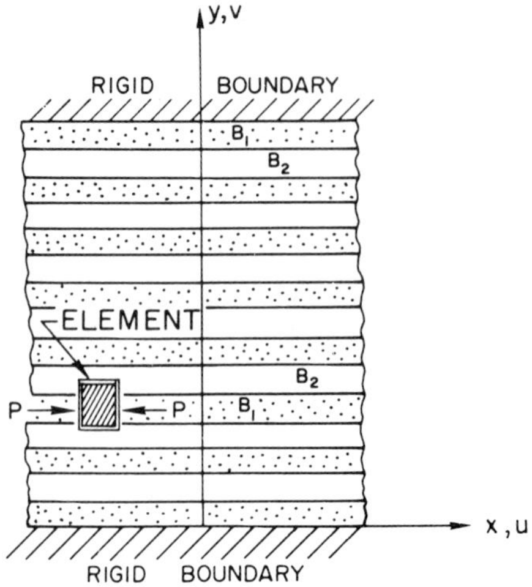

Fig. 9. Boundary conditions of multilayer. Multilayer is confined above and below by rigid boundaries so that vertical deflections are zero at boundaries. P is axial load acting on each element.

and:

$$(\epsilon_y)_{\text{ave}} = \sigma_y \left[\frac{t_1}{tB_1} + \frac{t_2}{tB_2}\right]$$

so that B_n would have been: *

$$B_n = \frac{tB_1 B_2}{t_1 B_2 + t_2 B_1} \qquad (31)$$

rather than the value given by eq. 28.

Buckling of multilayer between rigid boundaries

We now have a differential equation that we can solve in order to describe folds of the type shown in Fig. 2, which shows crenulated beds of the Carmel Formation, near Moab, Utah. Field relations of the folds seem to indicate that parts of the Carmel were crenulated while they were trapped between rocks forming nearly rigid boundaries (Fig. 6A). A uniform horizontal load of magnitude P is assumed to have acted on each member in the multilayer. The boundary conditions, which we have deduced from field

* See Appendix at the end of this chapter.

observations, are illustrated in Fig. 9. The boundary conditions for the displacements are:

$$\begin{aligned} &\text{at } x = 0,\ v = 0 \text{ for all } y \\ &\text{at } y = T, \text{ and } y = 0,\ v = 0 \text{ for all } x \end{aligned} \tag{32}$$

We will use the method of *separation of variables* (Sokolnikoff and Redheffer, 1958) to solve the differential equation, eq. 29:

$$B_p I_p \frac{\partial^4 v}{\partial x^4} + (P - tG_a b) \frac{\partial^2 v}{\partial x^2} = tbB_n \frac{\partial^2 v}{\partial y^2}$$

By this method, we select an expression for the displacement:

$$v = f(x)g(y) \tag{33}$$

which is a function of x, $f(x)$, and a function of y, $g(y)$. The product of the two functions is the displacement. The problem is to find the functions.

Substituting eq. 33 into eq. 29:

$$B_p I_p g(y) \frac{\partial^4 f(x)}{\partial x^4} + (P - tbG_a b)g(y) \frac{\partial^2 f(x)}{\partial x^2} = bB_n tf(x) \frac{\partial^2 g(y)}{\partial y^2}$$

Dividing each side of the equation by the product $f(x)g(y)$:

$$B_p I_p \frac{1}{f(x)} \frac{\partial^4 f(x)}{\partial x^4} + (P - tbG_a b) \frac{1}{f(x)} \frac{\partial^2 f(x)}{\partial x^2} = btB_n \frac{1}{g(y)} \frac{\partial^2 g(y)}{\partial y^2} \tag{34}$$

Thus we have "separated the variables". The terms on the left-hand side of the equation depend only upon x and the one on the right-hand side depends only upon y. The right-hand side must equal the left-hand side of eq. 34 for any value of x and y, so that each side of eq. 34 must be constant. Let the constant be $-C_0$. Then we can rewrite eq. 34 as two equations:

$$B_p I_p \frac{1}{f(x)} \frac{d^4 f(x)}{dx^4} + (P - tbG_a) \frac{1}{f(x)} \frac{d^2 f(x)}{dx^2} = -C_0 \tag{35}$$

and:

$$tbB_n \frac{1}{g(y)} \frac{d^2 g(y)}{dy^2} = -C_0 \tag{36}$$

We can write the differential as total derivatives becuase $f(x)$ is independent of y and $g(y)$ is independent of x, according to our definitions of the functions.

We will solve eq. 36 first. Try:

$$g(y) = e^{my}$$

as a solution. Then eq. 36 becomes:

$$btB_n m^2 e^{my} + C_0 e^{my} = 0$$

so that if m and y are finite:

$$m = \pm \sqrt{\frac{-C_0}{btB_n}}$$

It follows that:

$$g(y) = C_3 \exp \left[\sqrt{\frac{-C_0}{btB_n}}\, y\right] + C_4 \exp\left[-\sqrt{\frac{-C_0}{btB_n}}\, y\right]$$

where, for example, $\exp[y]$ means e^y. In terms of trigonometric functions (Churchill, 1960):

$$g(y) = C_1 \sin \sqrt{\frac{C_0}{btB_n}}\, y + C_2 \cos \left[\sqrt{\frac{C_0}{btB_n}}\, y\right] \tag{37}$$

is the general solution of eq. 36. If the term under the brackets is positive, therefore, the displacement:

$$v = f(x)\, g(y)$$

is a sinusoidal function of y.

We can evaluate the constant, C_0, with the boundary condition that at $y = T$, $v = 0$. Thus:

$$v = 0 = C_{15} f(x) \sin \left(\sqrt{\frac{C_0}{tbB_n}}\, T\right)$$

Except for the trivial solution that $f(x) = 0$ the term:

$$\sqrt{\frac{C_0}{btB_n}}\, T$$

must be some multiple of π. Thus:

$$n\pi = \sqrt{\frac{C_0}{tbB_n}}\, T$$

or:

$$C_0 = \left(\frac{n\pi}{T}\right)^2 B_n bt \tag{38}$$

where n is 1, 2, 3 ... We see that C_0 is always a positive quantity, because b, t and B_n are always positive.

Now we will solve eq. 35. Try:

$$f(x) = e^{mx}$$

for a solution. Then eq. 35 becomes:

$$B_p I_p m^4 e^{mx} + (P - tb\,G_a) m^2 e^{mx} + C_0 e^{mx} = 0$$

or:

$$m^4 + \left(\frac{P - tb\,G_a}{B_p I_p}\right) m^2 + \frac{C_0}{B_p I_p} = 0$$

Solving for m^2:

$$m^2 = -r \pm \sqrt{r^2 - s}$$

where:

$$r = \frac{P - b\,G_a t}{2B_p I_p}$$

and: (39)

$$s = \frac{C_0}{B_p I_p}$$

Solving for m:

$$m_1 = -m_2 = \sqrt{-r + i\sqrt{s - r^2}}$$

$$m_3 = -m_4 \sqrt{-r - i\sqrt{s - r^2}} \qquad (40)$$

We must know certain things about the quantities under the radical signs in eq. 40 in order to further evaluate the solution. The type of solution will depend upon whether the expression under the smaller radical sign is positive or negative. First we will assume that $s - r^2 > 0$.
Then eq. 40 can be rewritten in the form (Hetenyi, 1946):

$$m = \pm(\alpha \pm i\beta) \qquad (41)$$

where:

$$\alpha = \sqrt{\sqrt{\frac{s}{4}} - \frac{r}{2}}$$

and: (42)

$$\beta = \sqrt{\sqrt{\frac{s}{4}} + \frac{r}{2}}$$

We could verify that eq. 41 is equivalent to eq. 40 by squaring and then taking the square root of eq. 41. The general solution to the differential equation, eq. 35, is, therefore:

$$f(x) = C_5 e^{m_1 x} + C_6 e^{-m_1 x} + C_7 e^{m_3 x} + C_8 e^{-m_3 x}$$

or:

$$f(x) = (C_9 e^{\alpha x} + C_{10} e^{-\alpha x}) \cos\beta x + (C_{11} e^{\alpha x} + C_{12} e^{-\alpha x}) \sin\beta x \tag{43}$$

as we have shown in earlier pages.

The arbitrary constants in eq. 43, $C_9 \ldots C_{12}$, can be evaluated by means of the boundary conditions, eq. 32:

at $x = 0$, $v = 0 = f(x)g(y) = 0$, for all y

at $y = T$ and $y = 0$, $v = 0$, for all x

If the folds maintain constant amplitudes throughout the length of the buckled layers, we can further state that as $x \to \pm\infty$, v is finite, that is the deflection remains finite even though x becomes very large. This condition requires that α in eq. 42 and 43 equals zero. The alternative, that all four arbitrary constants are zero, is uninteresting.

Thus $\alpha = 0$, and according to the first of eq. 42 and the definitions of r and s in eq. 39:

$$\frac{P - tbG_a}{4B_p I_p} = \sqrt{\frac{C_0}{4B_p I_p}} \tag{44}$$

or:

$$P = 2\sqrt{C_0 B_p I_p} + G_a bt \tag{45}$$

It presumably expresses the critical axial load for buckling of the multilayer.

Because alpha is zero, the solution, eq. 43, reduces to:

$$f(x) = C_{13} \cos\beta x + C_{14} \sin\beta x \tag{46}$$

According to the first of boundary conditions, eq. 32, the deflection is zero at the origin of coordinates. Thus:

at $x = 0$, $v = 0 = f(x)g(y)$

so that according to eq. 46:

$$f(x) = C_{13} + 0 = 0$$

or $C_{13} = 0$. Thus the solution for v, according to eqs. 37, 46 and 33 becomes:

$$v = f(x)g(y)$$

or

$$v = C_4 \sin\beta x \left[C_1 \sin \left(\sqrt{\frac{C_0}{btB_n}}\, y \right) + C_2 \cos \left(\sqrt{\frac{C_0}{btB_n}}\, y \right) \right] \tag{47}$$

However, $v = 0$ at $y = 0$, because the rigid boundary does not deflect, so that C_2 in eq. 47 is zero and eq. 47 reduces to:

$$v = C_{15} \sin \beta x \sin \left(\sqrt{\frac{C_0}{btB_n}}\, y \right) \tag{48}$$

Substituting eq. 38 into eq. 48:

$$v = C_{15} \sin \beta x \sin \left(\frac{n\pi y}{T} \right) \tag{49}$$

we see that our solution predicts that the multilayer will buckle into a series of waves that change in amplitude, sinusoidally, with distance from the rigid boundaries. The maximum deflection is along the center line of the multilayer, equidistant from the rigid boundaries. The form of the theoretical folds is shown in Fig. 10. The theoretical form is remarkably similar to the forms of the actual folds in the Carmel Formation, shown in Fig. 2.

We can calculate the most likely wavelength of the folds by examining β in eq. 49. In order for v to be zero at the origin and at any number of nodal points:

$$\beta x = \frac{2\pi x}{L}$$

because $\sin \pi = 0$. Here L is the wavelength of the folds. Thus, using eq. 42 and 39:

$$\beta = \frac{2\pi}{L} = \sqrt{\sqrt{\frac{s}{4}} + \frac{r}{2}}$$

or:

$$\left(\frac{2\pi}{L} \right)^2 = \sqrt{\frac{C_0}{4B_pI_p}} + \frac{P - btG_a}{4B_pI_p} \tag{50}$$

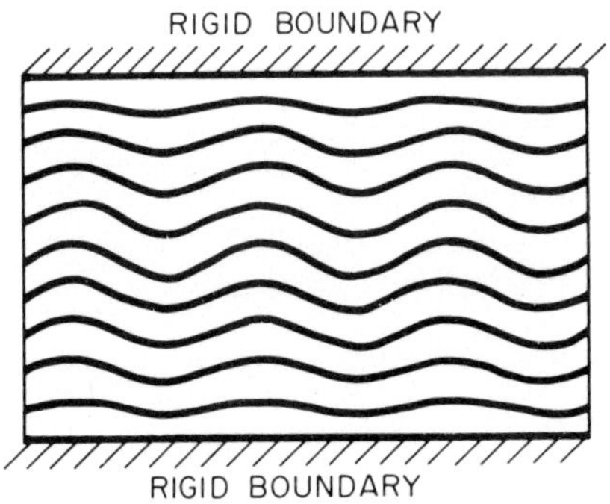

Fig. 10. Theoretical form of multilayer buckled internally. Vertical deflections vary sinusoidally in vertical and horizontal directions.

According to eq. 44, the two terms on the right-hand side of eq. 50 are equal, so that:

$$\left(\frac{2\pi}{L}\right)^2 = 2\sqrt{\frac{C_0}{4B_pI_p}} \tag{51}$$

Substituting the value of C_0 from eq. 38 into eq. 51:

$$\left(\frac{2\pi}{L}\right)^2 = 2\left(\frac{n\pi}{T}\right)\sqrt{\frac{btB_n}{4B_pI_p}}$$

Inverting, taking the square root of each side, and solving for L gives: *

$$L_B = 2\pi\sqrt{\frac{T}{n\pi}}\sqrt[4]{\frac{B_pI_p}{btB_n}} \tag{52}$$

the Biot wavelength, which, presumably, is the wavelength of folds in the multilayer. Waves of this length presumably form when the axial load applied to the ends of the multlayer is given by eq. 45:

$$P_{\text{mult}} = 2\left(\frac{n\pi}{T}\right)\sqrt{btB_nB_pI_p} + G_atb \tag{53}$$

which is equivalent to:

$$P_{\text{mult}} = \left(\frac{2\pi}{L_B}\right)^2 B_pI_p + \left(\frac{nL_B}{2T}\right)^2 btB_n + btG_a$$

because:

$$\left(\frac{n\pi}{T}\right)^2 btB_n = \left(\frac{2\pi}{L_B}\right)^4 B_pI_p$$

We see that the wavelength, L_B, is independent of the shear modulus, G_a, which is the constant of proportionality between the shear stress and the shear strain of soft and stiff layers.

If the elastic modulus or the thickness of the soft layers is much less than the modulus of the stiff layers, the elastic moduli, B_p and B_n reduce to (see eq. 16 and 28):

$$B_p = B_1 + B_2\left[\left(\frac{t}{t_1}\right)^3 - 1\right] \approx B_1, \text{ if } B_1 >> B_2$$

Also:

$$B_n = \left[\frac{t_1}{t}B_1 + \frac{t_2}{t}B_2\right] \approx \frac{t_1}{t}B_1, \text{ if } B_1 >> B_2$$

In this case the Biot wavelength, L_B, reduces to:

$$L_B \approx 2\pi\sqrt{\frac{T}{n\pi}}\sqrt[4]{\frac{B_1t_1^3b}{12\,btB_1(t_1/t)}} = 2\sqrt{\frac{T\pi t_1}{n\sqrt{12}}}$$

* See Appendix at the end of this chapter.

or:

$$L_B \approx 1.90 \sqrt{\frac{Tt_1}{n}} \tag{54}$$

which is identical to the value derived by Biot by different methods (Biot, 1965c).

According to eq. 54, the expected wavelength is solely a function of the thickness, t_1, of the stiff layers, the total thickness, T, of the sequence of layers, and the number, n, of cycles of folds stacked one on top of the other. There is one cycle of folds shown in the field example, Fig. 2, and the theoretical folds, Fig. 10, so that for them, $n = 1$, and eq. 54 becomes:

$$L_B \approx 1.90\sqrt{Tt_1} \tag{55}$$

Comparison of theoretical folds with folds in the Carmel Formation

Our solution, eq. 55, expresses the wavelength that should be most commonly found in natural multilayers, because this wavelength requires the least axial load. We assumed that the layers were initially perfectly straight, but, as will be shown in following pages, if the layers have slight initial deflections, the deflection with the wavelength indicated by eq. 55 would grow in amplitude most rapidly under an increasing axial load.

We can determine if our solution is approximately applicable to the natural situation at Moab, Utah, by measuring the folds in the Carmel, shown in Fig. 2. First we will rewrite eq. 55 in another form:

$$\frac{L_B}{T} \approx 1.90 \sqrt{\frac{t_1}{T}}$$

which is an expression for the ratio of wavelength to total thickness of the folded sequence, T. The ratios of wavelength of folds to total thickness of the folded sequence shown in Fig. 2 range from about 0.9 to 1.3. The sandy siltstone interbeds are quite thin, so that to a first approximation the total thickness, T, divided by the thickness, t_1, of the stiff layers is equal to the number of stiff layers. Therefore:

$$\frac{L_B}{T} \approx 1.90 \sqrt{\frac{1}{\text{no. of stiff beds}}} \approx 0.9 - 1.4$$

In order for L_B/T to range from 0.9 to 1.4, the number of stiff layers should range between about 2 and 3 if the stiff layers are of equal thickness.

The sandstone beds shown in Fig. 2 are of varying thickness at different places in the cliff face, because the sandy siltstone beds are discontinuous, but there appears to be a thick bed near the top and bottom and a thinner bed near the middle of the Carmel Formation. The folded beds comprise two thick beds and one thin bed so that the estimate of 2 to 3 stiff beds predicted by theory seems to describe remarkably well the natural occurrence. The correlation between the theoretical and actual folds is particularly amazing, because the theory we developed assumed that there were many stiff and soft layers being folded, whereas in the part of the Carmel Formation shown in Fig. 2 there are only a few beds.

Biot (1965c) has shown that eq. 55 is valid for the buckling of viscous multilayers as well as elastic multilayers. He also has determined the effect of flow of material within the soft layers. If the ratio of the viscosities of the stiff and the soft layers is less than about 3.5 times the number of layers, the correction for interstitial flow is unnecessary (Biot, 1965c).

In another paper, M.A. Biot also shows that multiples of the pattern of folds shown in Fig. 10 can develop in a thick sequence of beds, where the rigid boundaries are widely separated. In this case, which Biot calls "self confinement", the amplitudes of the folds vary sinusoidally in the vertical direction as well as in the horizontal direction and there are several cycles of folds stacked one on top of the other, with anticlines being developed directly on top of synclines. The equation for the most likely wavelength, L_B, given by eq. 54, should predict the wavelengths of such fold sequences. The denominator, n, in eq. 54 is the number of stacks of folds.

Effects of irregularities of bedding on folding

The Biot theory of buckling of multilayers which we have derived in preceding pages should describe certain buckling behaviors of stacks of layers that are initially flat and planar. In addition, our solutions for the shape of the buckled multilayers predict that the train of fold waves that develops within a multilayer will extend indefinitely in a lateral direction. This latter prediction is not realized in the folds we see in the Windows section of Arches National Monument, where the folds die out altogether in the left-hand side of the cliff face shown in Fig. 2.

There are at least three explanations for dying-out of folds in a lateral direction: (1) high-angle faulting of the beds to the right of the cliff face shown in Fig. 2, providing vertical drag and a series of damped waves (e.g., see Hetenyi, 1946; Biot, 1968); (2) a source of thrust, such as the shouldering aside of sediments by a rising salt mass or an intrusive body, to the right of the view in Fig. 2. Severing and pushing of beds by an intrusive could produce a series of damped waves such as we see in the cliff face shown in Fig. 2 (e.g., see Willis, 1894; Hetenyi, 1946); (3) the beds of the Carmel might not have been uniformly initially deflected at the time the folds formed.

This latter is the most likely cause of the lack of folds in the left-hand side of the cliff shown in Fig. 2, because the other two do not correspond with field observations that can be made by examining rocks to the right of the cliff face. We will discuss the effects of initial deflections on folding processes in some detail because initial deflections can have a marked effect on deter-

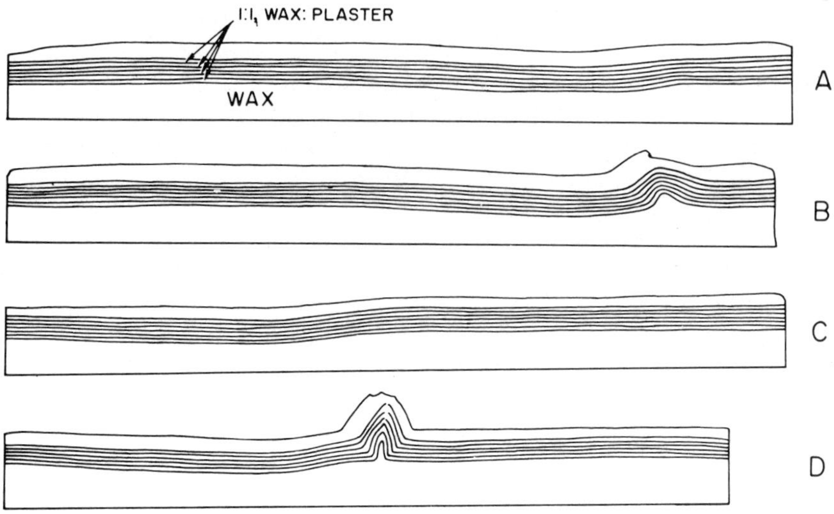

Fig. 11. Localization of buckle by position of initial warp in experimental layers. Original length of layers was 1 m, original thickness was 8.8 cm. Thin, hard layers consisted of one part of wax and one part of plaster. Thick, soft layer was mixture of two parts wax and one part turpentine. Top of layers was loaded by lead shot. A. Initial warp near driving piston, at right end of layers. B. Buckle formed at position of warp. C. Initial warp about 0.5 m from driving piston. D. Buckle formed at position of initial warp. (Drawn from photographs by Willis, 1893, plate LXXIX)

mining sizes and shapes of folds. Probably the best evidence that significant irregularities of bedding were present in the Carmel at the time the Carmel was folded is the occurrence of discordances of bedding within the Carmel and between the Carmel and the Entrada (e.g., see Gilluly and Reeside, 1928; Dane, 1935; McKnight, 1940).

Bailey Willis, in 1894, made an important discovery. He found that initial dips of bedding planes of a sequence being folded can control the position and the relative growth of folds. His main objective was to explain why Nittany Mountain, which is quite far from the assumed push in the southeast that formed the Appalachian folds, should be much larger than its neighbors and even larger than many of the folds many tens of miles toward the southeast. In the process of trying to explain the large size of Nittany Mountain, he discovered a general principle: the amplitudes of folds are largely determined by the amplitudes of the initial deflections that precede the folds. Willis made several experiments with layers of wax in a large oaken squeeze box in order to check his ideas qualitatively. Figs. 11A,C show the initial warps Willis built into this experimental strata, and Figs. 11B,D show folds that grew as the piston of this squeeze box was advanced from right to left. The ends of the experimental strata near the piston were most highly loaded, because of friction between the strata and the box, yet in both experiments the folds formed at the sites of the initial waves (see Willis' paper for additional illustrations).

Probably the most direct way to gain some understanding of the effect of initial warps on folding of strata is to consider the buckling of an isolated elastic strip that is subjected to axial loads. Suppose that the neutral plane of an isolated strip is initially deflected into a sine wave:

$$v_i = \delta_i \sin\left(\frac{2\pi x}{L}\right) = \delta_i \sin\left(\frac{\pi x}{l}\right) \tag{56}$$

where δ_i is the amplitude at the midlength of the strip, L is the wavelength of the initial warp, and ℓ is the length of the strip. This initial deflection, by definition, remains in the strip even when there is no axial load applied to it (Fig. 12A).

When the strip is subjected to an axial load, P, it further deflects an additional amount, v_a, which is a function of x. The problem is to find this function.

In order for the strip to maintain equilibrium, the driving and resisting moments must be equal. The driving moment, M_D, is a function of the total deflection of the member and of the applied axial load, so that:

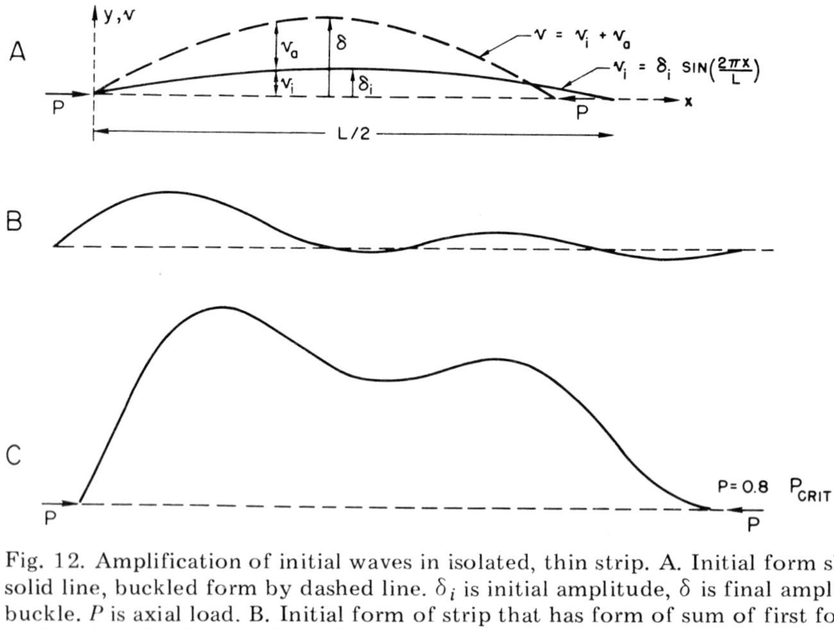

Fig. 12. Amplification of initial waves in isolated, thin strip. A. Initial form shown by solid line, buckled form by dashed line. δ_i is initial amplitude, δ is final amplitude of buckle. P is axial load. B. Initial form of strip that has form of sum of first four sinusoidal modes. Amplitude of each mode is the same. C. Buckled form of strip with initial form shown in B. First mode, with wavelength equal to twice length of strip, dominates even though it is not apparent in initial form.

$$M_D = -P(v_i + v_a) \tag{57}$$

The resisting moment is only a function of the addition deflection, v_a, because the strip maintains its initial curvature when it is supporting no axial loads. Thus:

$$M_R = B_I \frac{d^2 v_a}{dx^2} \tag{58}$$

Summing moments and rearranging the differential equation we find that:

$$\frac{d^2 v_a}{dx^2} + \frac{P}{BI} v_a = \frac{-P\delta_i}{BI} \sin\left(\frac{2\pi x}{L}\right) \tag{59}$$

The solution of the homogeneous part of this equation is:

$$v_a = C_1 \cos\left(x \sqrt{\frac{P}{BI}} + C_2 \sin\left(x \sqrt{\frac{P}{BI}}\right)\right) \tag{60}$$

where C_1 and C_2 are arbitrary constants. A particular solution can be derived by trying:

$$v_a = A \sin\left(\frac{2\pi x}{L}\right) \tag{61}$$

where A is a constant, which can be determined by substituting eq. 61 into eq. 59. Thus:

$$A = \frac{\delta_i}{\left[\left(\frac{2\pi}{L}\right)^2 \left(\frac{BI}{P}\right) - 1\right]} \tag{62}$$

The general solution of the differential equation, eq. 59, is the sum of eqs. 60 and 61:

$$v_a = C_1 \cos\left(x \sqrt{\frac{P}{BI}}\right) + C_2 \sin\left(x \sqrt{\frac{P}{BI}}\right) + \frac{\delta_i}{\left[\left(\frac{2\pi}{L}\right)^2 \left(\frac{BI}{P}\right) - 1\right]} \sin\left(\frac{2\pi x}{L}\right) \tag{63}$$

The ends of the member are fixed in place so that at $x = 0$ and $x = L/2$, the additional deflection, v_a, must equal zero. Accordingly, C_1 and C_2 in eq. 63 must be zero because the deflection must be zero regardless of the magnitude of the axial load, P.

The expression for the additional deflection becomes, therefore:

$$v_a = \frac{\delta_i}{\left[\left(\frac{2\pi}{L}\right)^2 \left(\frac{BI}{P}\right) - 1\right]} \sin\left(\frac{2\pi x}{L}\right) \tag{64}$$

and the total deflection, which is the sum of the additional and the initial deflections, is:

$$v = v_i + v_a = \delta_i \sin\left(\frac{2\pi x}{L}\right) + \frac{\delta_i}{\left[\left(\frac{2\pi}{L}\right)^2 \left(\frac{BI}{P}\right) - 1\right]} \sin\left(\frac{2\pi x}{L}\right)$$

or:

$$v = \frac{\delta_i}{\left[1 - \frac{P}{BI}\left(\frac{L}{2\pi}\right)^2\right]} \sin\left(\frac{2\pi x}{L}\right) \tag{65}$$

Eq. 65 is interesting in several respects. We see that as the axial load, P, approaches the quantity, $BI(2\pi/L)^2$, that the denominator approaches zero, which causes the theoretical deflection to approach infinity. The quantity, $BI(2\pi/L)^2$, is called the critical Euler load and it is the critical load required to cause an initially perfectly straight member to buckle (e.g., see Currie et al., 1962). Therefore, as the axial load acting on an initially deflected strip approaches the critical buckling load for a perfectly straight strip, the deflection becomes boundlessly large. Let us define the critical load as:

$$P_{\text{crit}} = BI\left(\frac{2\pi}{L}\right)^2 \tag{66}$$

because we will want to refer to it later.

Our solution for the deflection of an initially deflected member indicates that as the magnitude of the initial deflection becomes smaller and smaller the magnitude of axial load required for significant deflection becomes larger and larger, until, in the limit, where the initial deflection has zero magnitude, the axial load equals the critical load (Fig. 13). Further, our solution helps us understand Willis' experimental discovery: according to our solution, for a given magnitude of axial load, P, the amount of deflection of the member is proportional to the magnitude of the initial deflection. The larger the initial deflection, the larger the amplitude of the fold that grows at the site of the initial deflection. Our solution also provides a reasonable explanation for the absence of folds on the left-hand side of the cliff face shown in Fig. 2. The parts of the layers shown on the right-hand side of the figure were probably more highly deflected initially than those on the left-hand side.

Neither the initial warps in Willis' experiments nor the initial warps in the beds in the Carmel, however, were simple sinusoidal forms at the time the folds grew at their sites. Regardless of the forms of the initial deflections, unless they were recumbent folds, we can represent their forms by an infinite series of sine or cosine terms (Churchill, 1960; and Kreyszig, 1962). Such a series is:

$$v_i = \delta_i \sin\left(\frac{\pi x}{l}\right) + \delta_2 \sin\left(\frac{2\pi x}{l}\right) + \delta_3 \sin\left(\frac{3\pi x}{l}\right) + \ldots \delta_n \sin\left(\frac{n\pi x}{l}\right) + \ldots \tag{67}$$

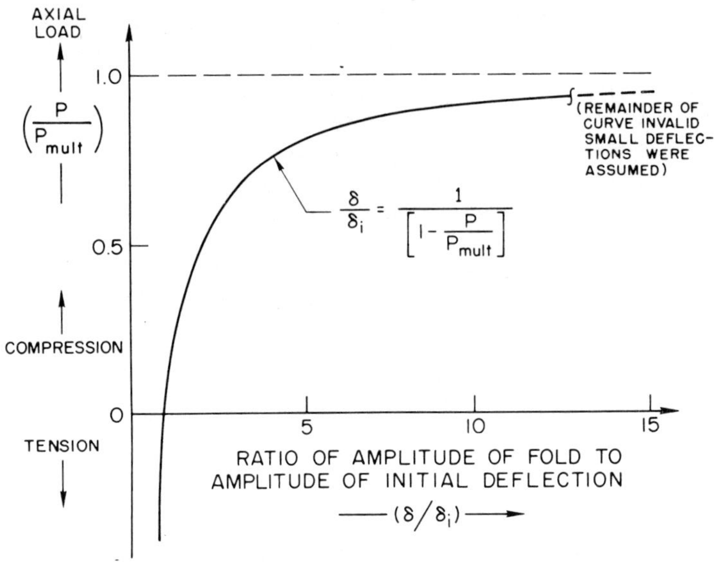

Fig. 13. Relation between amplitude of fold in multilayer and axial loads applied to multilayer. As the axial load approaches the critical load for buckling of initially flat multilayer, P_{mult}, the amplitudes of folds become very large.

on to infinity, where ℓ is the length of the strip that is initially deflected and $\delta_1, \delta_2, \ldots \delta_n, \ldots$ are the amplitudes of the initial deflections. The initial shape of an extremely crooked strip could be described by such a series so that it should adequately describe the initial deflections of a single bed within the Carmel.

If we substitute the series, eq. 67, into the differential equation for deflection of the member, eq. 59, and proceeded much as before toward a solution, we would find that the total deflection would be (e.g., see Timoshenko and Gere, 1961):

$$v = \frac{\delta_i}{\left[1 - \frac{P}{P_{crit}}\right]} \sin\left(\frac{\pi x}{l}\right) + \frac{2^2 \delta_2}{\left[2^2 - \frac{P}{P_{crit}}\right]} \sin\left(\frac{2\pi x}{l}\right) + \ldots \frac{n^2 \delta_2}{\left[n^2 - \frac{P}{P_{crit}}\right]} \sin\left(\frac{n\pi x}{l}\right) \ldots \quad (68)$$

As the axial load approaches the critical load, P_{crit}, in eq. 68, the denominator of the first term will approach zero much more rapidly than will the denominators of succeeding terms. Therefore, the first term usually will predominate and the strip will buckle largely as a unit, that is, in the first mode. The second mode and higher modes might predominate under small axial loads if the amplitudes, $\delta_2 \ldots \delta_n \ldots$, of their initial deflections are much greater than the amplitude, δ_1, of the first mode. As a result, an unconfined, crooked member usually will buckle largely into one-half of a wave, equal to the length of the member. For example, Fig. 12B shows a strip initially deflected in the first through the fourth modes. The first mode is not apparent in the initial form, but when the strip is buckled (Fig. 12C) it dominates the form of the strip.

Confined members, or multilayers, however, usually do not buckle into the first mode. Whereas it is easiest for an unconfined member to buckle into a wavelength equal to twice the length of the member, it is easiest for confined members or multilayers to buckle into other modes (e.g., see Biot, 1961; Currie et al., 1962; Ramberg, 1963).

For example, suppose that a multilayer is initially deflected into the form:

$$v_i = \delta_i \sin\left(\frac{n\pi y}{T}\right) \sin\left(\frac{2\pi x}{L}\right)$$

where δ_i is the amplitude and L is the wavelength, in the horizontal direction, of the initially deflected beds. The differential equation, eq. 29, becomes, for an initially deflected multilayer:

$$B_p I_p \frac{\partial^4 v_a}{\partial x^4} + (P - tbG_a)\frac{\partial^2 v_a}{\partial x^2} - tbB_n \frac{\partial^2 v_a}{\partial y^2} = -P\frac{\partial^2 v_i}{\partial x^2} \tag{69}$$

where v_i is the initial deflection and v_a is the additional deflection, due to the axial loading. The pertinent solution to eq. 69 is:

$$v_a = \frac{\delta_i}{\left[\dfrac{P_{\text{mult}}}{P} - 1\right]} \sin\left(\frac{n\pi y}{T}\right) \sin\left(\frac{2\pi x}{L}\right) \tag{70}$$

$$P_{\text{mult}} = B_p I_p \left(\frac{2\pi}{L}\right)^2 + tbG_a + tbB_n \left(\frac{n\pi}{T}\right)^2 \left(\frac{L}{2\pi}\right)^2 \tag{71}$$

so that the expression for the total deflection, v, is:

$$v = v_i + v_a = \frac{\delta_i}{\left[1 - \dfrac{P}{P_{\text{mult}}}\right]} \sin\left(\frac{n\pi y}{T}\right) \sin\left(\frac{2\pi x}{L}\right) \tag{72}$$

Thus, by analogy with our study of the unconfined member (Fig. 13), the deflection of the multilayer becomes large as the axial load, P, approaches the critical value, P_{mult}, given by eq. 72. This is the load we calculated for buckling of a multilayer consisting of initially flat beds. By taking the derivatives of the critical load with respect to the wavelength, $d\,P_{\text{mult}}/dL$, setting it equal to zero, and solving for the wavelength, we compute the wavelength for which the axial load is a minimum; that is the Biot wavelength:

$$L_B = 2\pi \sqrt{\frac{T}{2\pi}} \sqrt[4]{\frac{B_p I_p}{tbB_n}} \tag{73}$$

which is identical to eq. 52.

Now, suppose that the beds of the multilayer are highly irregular rather than being planar sheets. The initial form of the beds can be described by an infinite series:

$$v_i = \delta_1 \sin\left(\frac{n\pi y}{T}\right) \sin\left(\frac{2\pi x}{L}\right) + \delta_2 \sin\left(\frac{2\pi y}{T}\right) \sin\left(\frac{4\pi x}{L}\right) + \ldots$$

$$\delta_m \sin\left(\frac{n\pi y}{T}\right) \sin\left(\frac{2m\pi x}{L}\right) + \ldots \tag{74}$$

and so on to infinity. Here m is an integer, $m = 3, 4, 5, \ldots$

Proceeding as before, we find that the total deflection, v, is:

$$v = \frac{\delta_1 \sin\left(\frac{n\pi y}{T}\right) \sin\left(\frac{2\pi x}{L}\right)}{1 - \dfrac{P}{\left[B_p I_p \left(\frac{2\pi}{L}\right)^2 + tbG_a + tbB_n \left(\frac{n\pi}{T}\right)^2 \left(\frac{L}{2\pi}\right)^2\right]}} + \ldots$$

$$\frac{\delta_m \sin\left(\frac{n\pi y}{T}\right) \sin\left(\frac{2m\pi x}{L}\right)}{1 - \dfrac{P}{\left[B_p I_p \left(\frac{2m\pi}{L}\right)^2 + tbG_a + tbB_n \left(\frac{n\pi}{T}\right)^2 \left(\frac{L}{2m\pi}\right)^2\right]}} + \ldots \tag{75}$$

where m is 2, 3, ...

The first term in this expression will usually dominate the fold pattern because its denominator approaches zero most rapidly as the axial load increases. This can be shown by differentiating with respect to m the divisor of P in the denominator of the second term of the equation. Let the divisor be P_0:

$$P_0 = B_p I_p \left(\frac{2m\pi}{L}\right)^2 + tbG_a + tbB_n \left(\frac{n\pi}{T}\right)^2 \left(\frac{L}{2m\pi}\right)^2$$

Then:

$$\frac{dP_0}{dm} = 0 = 2B_p I_p \left(\frac{2\pi}{L}\right)^2 m - tbB_n \left(\frac{n\pi}{T}\right)^2 \left(\frac{L}{2\pi}\right)^2 \frac{1}{m^3}$$

so that:

$$m = \frac{L}{2\pi} \sqrt{\frac{n\pi}{T}} \sqrt[4]{\frac{B_n tb}{B_p I_p}} \tag{76}$$

The value of m for which the axial load is a minimum depends upon the value of L. If L is the Biot wavelength, given by eq. 73, then $m = 1$, and the first term in the solution for the total deflection, eq. 75, dominates the total deflection. If L is some multiple of the Biot wavelength, say 3, then m is 3 and the initial deflections with a wavelength equal to the Biot wavelength still dominate. Thus, regardless of the value of L in the infinite series used to describe the initial deflections, the Biot wavelength normally dominates and grows the most rapidly. The exception will be where the Biot wavelength is relatively small in the initial deflection.

Many of the irregularities we see in the fold patterns at Arches Monument and at many other places probably are due to irregularities in patterns of initial deflections in rocks. All rocks contain initial deflections; folding processes sort through and selectively amplify certain ones more than others so that commonly a recognizable and regular pattern of waves emerges even where the pattern of initial deflections was highly complex.

Reason maximum amplitudes of folds are near base of Carmel

The forms of the waves within the theoretical multilayer fail to precisely correspond with the form of the actual folds of the Carmel in another respect. The layer with maximum amplitude is at the center of the theoretical multilayer whereas the beds with maximum amplitudes in the Carmel are near the base of the Carmel. Thus, in the theoretical pattern the amplitudes diminish at the same rate upwards and downwards from the center of the multilayer, but in the Carmel, the amplitudes diminish much more rapidly towards the base of the Carmel than towards the top.

For example, at the Windows section of the monument the lowest thick sandstone bed absorbs all the vertical compression so that the deflections are damped out over a vertical distance of about 10—12 ft. (Fig. 2). Above the zone of maximum amplitude, however, the vertical compression is damped out over a vertical distance of about 30 ft. (Fig. 2).

The folds in the western end of The Organ also diminish in amplitude more rapidly downwards than upwards (Fig. 4). The massive sandstone near the base of the cliff shown in Fig. 4 absorbs a large amount of vertical deflection over a short distance, whereas the bedded rocks above absorb the deflection over a distance of more than twice the thickness of the massive sandstone below.

Even though our theory does not account for this asymmetric absorption of vertical deflection, we can understand it in terms of our theory. The stress normal to the bedding of the multilayers is σ_y, which, according to eq. 27 is equal to the normal modulus, B_n, times the vertical normal strain, ϵ_y, so that:

$$\sigma_y = B_n \epsilon_y = B_n \frac{\partial v}{\partial y}$$

According to our theory, the vertical deflection is (eq. 72):

$$v = \frac{\delta_i}{\left[1 - \frac{P}{P_{\text{mult}}}\right]} \sin\left(\frac{n\pi y}{T}\right) \sin\left(\frac{2\pi x}{L}\right)$$

The vertical normal stress, therefore, is:

$$\sigma_y = \frac{B_n \delta_i \left(\frac{n\pi}{T}\right)}{\left[1 - \frac{P}{P_{\text{mult.}}}\right]} \cos\left(\frac{n\pi y}{T}\right) \sin\left(\frac{2\pi x'}{L}\right) \tag{77}$$

At the midheight of a multilayer, where the amplitude of the folds is maximum, the normal stress is zero, according to this latter equation, because $y = T/2$ at midheight and $\cos(n\pi/2)$ is zero if n is an odd integer. The vertical normal stress is tensile below an anticline in a bed at midheight in a multilayer and it increases to a maximum at the base of the multilayer, according to eq. 77. Between the top of the anticline at midheight and the top of the multilayer, the normal stress is compressive. At midheight the normal stress changes from tensile to compressive so that it is zero.

Two conclusions follow from this analysis: if a multilayer is bounded only on one side so that it has a free surface, we would expect the amplitudes of folded beds within the multilayer to be a maximum at the free surface and to diminish sinusoidally to zero at the bounding, rigid surface below. Similarly, if the beds within the multilayer are not of uniform stiffness with respect to vertical compressibility, that is if the beds in one half of a multilayer are softer than beds in the other half, we would expect the bed with maximum fold amplitude to be not at the center of the multilayer but somewhere within the softer beds. An observation that I made repeatedly in Arches Monument is that the maximum amplitude of the folds is near the base of the Carmel. According to our analysis, this implies that the sandstone beds near the base of the Carmel were softer than the beds near the top of the Carmel. No differences in lithology were obvious in the field, but apparently there were significant differences in physical properties at the time of folding.

Extension of folding into lower part of Entrada

The folds within the Carmel Formation commonly extend upward into the relatively massive Entrada Sandstone. For example, the left side of Fig. 5 shows low-amplitude folds in the well-bedded Carmel and the massive lower Entrada, about two or three miles west of the Windows section of Arches Monument. Dane (1935, p. 99) observed that where the Carmel is well-bedded, the crenulations are usually confined to a group of beds within the Carmel. However, he noted, where the Carmel is an earthy or muddy sandstone without pronounced bedding planes, the crenulations are less common. Where they do occur, the contact between the Carmel and the Entrada has an irregular, rolling appearance and the disturbances commonly extend several tens of feet into the lower part of the Entrada (Dane, 1935, p. 99).

There seem to be two factors that control whether folding will be of the internal variety in which the folds are restricted to members confined between boundaries that are effectively rigid or whether folding will be of the variety in which the confining medium as well as the members will be deflected and in which the deflections die out within the confining medium. One factor is the elasticity or the viscosity contrast between the members and the medium. The other factor is the number of beds within the stack of members.

We can derive a relation between these factors by analyzing expressions for buckling loads required for internal folding and for folding in which the multilayer bends as a unit. The guiding principle is that the type of folds that will form is the type that requires the minimum buckling load. Where beds are initially deflected into both modes, of course, we would expect both modes of folding to occur, but one mode will dominate. The one that will dominate will be the one that requires the minimum buckling load.

Currie et al. (1962) have shown that the critical buckling load for a stack of stiff layers, confined above and below by soft media (Fig. 14A) is:

$$P_{\text{crit}} = \frac{NBt^3 b}{12}\left(\frac{2\pi}{L}\right)^2 + \frac{B_0 bL}{2\pi} \tag{78}$$

where N is the number of layers, L is the wavelength of folding, B is the elastic modulus, t is the thickness and b is the width of each of the layers. B_0 is the elastic modulus of the confining medium. An implicit assumption is

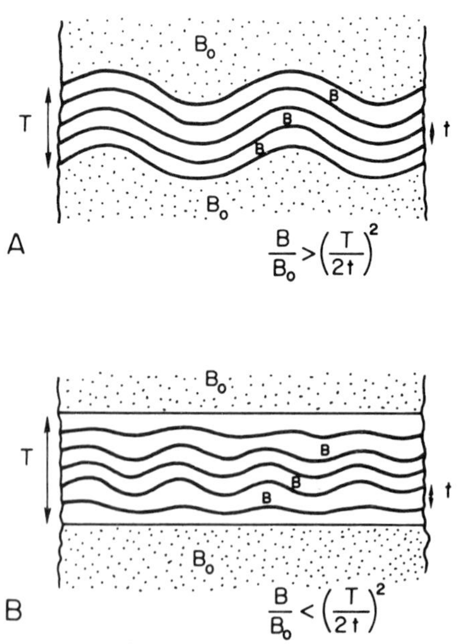

Fig. 14. Two modes of buckling of multilayer confined above and below by elastic media. A. General folding of multilayer. B. Internal folding of multilayer.

that the friction between beds is negiligible, so that when we apply this equation to the folds of the Carmel, we assume that the thin siltstone beds serve as lubricating layers of negligible thickness.

The critical load for internal buckling of a multilayer, in which the thickness of the soft layers is much smaller than the thickness of the stiff layers and in which the friction between beds is negligible, is given by eq. 71 in the following form:

$$P'_{\text{mult}} = \frac{NBbt^3}{12}\left(\frac{2\pi^2}{L_B}\right) + \left(\frac{L_B}{2}\right)^2 \frac{bB_n}{T} \tag{79}$$

Here L_B is the Biot wavelength, eq. 73, T is the thickness of the stack of layers (Fig. 14) and B_n is the modulus normal to bedding. Note that P'_{mult} is the sum of the loads required to buckle the multilayer, whereas P_{mult} is the load on each stiff layer required to buckle the multilayer (Fig. 9).

The wavelength for which the axial load is a minimum for internal buckling is:

$$L_B = 2\sqrt{\frac{T\pi t}{\sqrt{12N}}}$$

for the assumptions we have made here, so that the critical load, eq. 79, becomes:

$$P'_{\text{mult}} = \frac{2bB\pi t}{N\sqrt{12}} \tag{80}$$

The wavelength for which the axial load is a minimum for general buckling, Fig. 14a, is:

$$L = 2\pi T\sqrt[3]{\frac{B}{6B_0N^2}}$$

so that the critical load is:

$$P_{\text{crit}} = \tfrac{3}{2}bt\sqrt[3]{\frac{NBB_0^2}{6}} \tag{81}$$

Now, if the load required for internal folding, P'_{mult}, is greater than the load required for general folding, P_{crit}, the beds should buckle into the general fold pattern. That is, if $P'_{\text{mult}} > P_{\text{crit}}$, which is approximately equivalent to:

$$\frac{B}{B_0} > \left(\frac{T}{2t}\right)^2;\left[\frac{B}{B_0} > 0.307\left(\frac{T}{t}\right)^2\right];\ N = \frac{T}{t} \tag{82}$$

we would expect a fold pattern such as is shown in Fig. 14a. If, however, $P'_{\text{mult}} < P_{\text{crit}}$, or, approximately:

$$\frac{B}{B_0} < \left(\frac{T}{2t}\right)^2 \tag{83}$$

internal folding, shown in Fig. 14B, would be expected.

The quantity T/t in eq. 82 and 83 is the number of beds within the multilayer. If the number of beds is large, folding will be of the internal variety unless the modulus, B_0, (or viscosity) of the confining medium is very low. For example, if there are ten beds, which is about the number within the Carmel shown in Fig. 5, the ratio of T/t in eq. 82 and 83 is 10 so that the ratio of the modulus, B/B_0, would have to be greater than 25 for general folding of the multilayers and less than 25 for internal folding. Thus, if T/t is 10: (1) $B > 25\,B_0$ . . . general folding (Fig. 14A); and (2) $B < 25\,B_0$ . . . internal folding (Fig. 14B).

Each of these factors seems to have controlled patterns of folding in different places within the Carmel Formation and the Entrada Sandstone. For example, the small folds within the Carmel in the Windows section, Fig. 2, are of the internal variety and they seem to have been confined above by a sequence of relatively thick sandstone beds within which the siltstone interbeds are discontinuous laterally. The folds Dane described elsewhere, in the more massive facies of the Carmel, seem to have been controlled by bedding thickness also; where the bedding is thick, the folds extend into the relatively massive Entrada so that the folds are not of the internal variety.

Some fold patterns seem to indicate that the overlying Entrada had a relatively low viscosity or elastic modulus as compared to that of the Carmel. For example, even though the carmel is well-bedded and the Entrada is massive in the left end of the cliff shown in Fig. 5, the deflections extend into the Entrada.

REFERENCES

Baker, A.A., Dane, C.H. and Reeside, Jr., J.B., 1936. Correlation of the Jurassic formations of parts of Utah, Arizona, New Mexico and Colorado. U.S. Geol. Surv., Profess. Paper, 183: 66 pp.

Baker, A.A., Dane, C.H. and McKnight, E.T., 1954. Preliminary map showing geologic structure of parts of Grand and San Juan counties, Utah. U.S. Geol. Surv., Oil Gas Invest., Map OM 169.

Biot, M.A., 1961. Theory of folding of stratified viscoelastic media and its implications in tectonics and orogenesis. Geol. Soc. Am. Bull., 72: 1595—1620.

Biot, M.A., 1963. Theory of stability of multilayered continua in finite anisotropic elasticity. J. Franklin Inst., 276: 128—153.

Biot, M.A., 1964. Theory of internal buckling of a confined multilayered structure. Geol. Soc. Am. Bull., 75: 563—568.

Biot, M.A., 1965a. Mechanics of Incremental Deformations. Wiley, New York, N.Y., 504 pp.

Biot, M.A., 1965b. Similar folding of the first and second kinds. Geol. Soc. Am. Bull., 76: 251—258.

Biot, M.A., 1965c. Further development of the theory of internal buckling of multilayers. Geol. Soc. Am. Bull. 76: 833—840.

Biot, M.A., 1966. Fundamental skin effect in anisotropic solid mechanics. Intern. J. Solids Struct., 2: 645—663.

Biot, M.A., 1967. Rheological stability with couple stresses and its application to geological folding. Proc. Roy. Soc.London, Ser. A., 298: 402—423.

Biot, M.A., 1968. Edge buckling of a laminated medium. Intern. J. Solids Struct., 4: 125—137.

Churchill, R.V., 1960. Complex Variables and Applications. McGraw-Hill, New York, N.Y., 297 pp.

Currie, J.B., Patnode, H.W. and Trump, R.P., 1962. Development of folds in sedimentary strata. Geol. Soc. Am. Bull., 73: 655—674.

Dane, C.H., 1935. Geology of the Salt Valley anticline and adjacent areas, Grand County, Utah. U.S. Geol. Surv. Bull., 863: 184 pp.

Gilluly, J. and Reeside, Jr., J.B., 1928. Sedimentary rocks of the San Rafael Swell and some adjacent areas in eastern Utah. U.S. Geol. Surv. Profess. Paper, 150-D: 61—110.

Gregory, H.A. and Moore, R.C., 1931. The Kaiparowits region. U.S. Geol. Surv., Profess. Paper, 164: 161 pp.

Hetenyi, M., 1946. Beams on Elastic Foundation. Univ. Michigan Press, Ann Arbor, Mich., 255 pp.

Hunt, C.B., 1953. Geology and geography of the Henry Mountains region, Utah. U.S. Geol. Surv., Profess. Paper, 228: 234 pp.

Kelley, V.C., 1958. Tectonics of the region of the Paradox Basin. In: A.F. Sanborn (Editor), Intermountain Assoc. Petrol. Geologists, 9th Ann. Field Conf., pp. 31—38.

Kreyszig, E., 1962. Advanced Engineering Mathematics. Wiley, New York, N.Y., 856 pp.

McKnight, E.T., 1940. Geology of area between Green and Colorado rivers, Grand and San Juan counties, Utah. U.S. Geol. Surv., Bull., 908: 147 pp.

Potter, J.F., 1968. Deformed micaceous deposits in the Downtonian of the Llandeilo region, South Wales. Proc. Geologists Assoc. Engl., 78: 277—287.

Ramberg, H., 1959. Evolution of ptygmatic folding. Norsk Geol. Tidsskr., 39: 99—151.

Ramberg, H., 1960. Relationships between length of arc and thickness of ptygmatically folded veins. Am. J. Sci., 258: 36—46.

Ramberg, H., 1961. Relationships between concentric longitudinal strain and concentric shearing strain during folding of homogeneous sheets of rock. Am. J. Sci., 259: 382—390.

Ramberg, H., 1963a. Fluid dynamics of viscous buckling applicable to folding of layered rocks. Bull. Am. Assoc. Petrol. Geologists, 47: 484—505.

Ramberg, H., 1963b. Evolution of drag folds. Geol. Mag., 100: 97—110.

Ramberg, H., 1964. Selective buckling of composite layers with contrasted rheological properties, a theory for simultaneous formation of several orders of folds. Tectonophysics, 1(4): 307—341.

Sherwin, J.A. and Chapple, W.M., 1968. Wavelengths of single layer folds: a comparison between theory and observation. Am. J. Sci., 266: 167—179.

Shoemaker, E.M., Case, J.E. and Elston, D.P., 1958. Salt anticlines of the Paradox Basin. In: A.F. Sanborn (Editor), Intermountain Assoc. Petrol. Geologists, 9th Ann. Field Conf., pp. 39—59.

Sokolnikoff, L.S. and Redheffer, R.M., 1958. Mathematics of Physics and Modern Engineering. McGraw-Hill, New York, N.Y., 812 pp.

Steed, R.H., 1954. Geology of Circle Cliffs anticline. In: A.W. Grier (Editor), Intermountain Assoc. of Petrol. Geologists, 5th Ann. Field Conf., pp. 99—102.

Stokes, Wm.L. and Holmes, C.N., 1954. Jurassic rocks of south-central Utah. In: B. Grier (Editor), Intermountain Assoc. Petrol. Geologists, 5th Ann. Field Conf., pp. 34—41.

Timoshenko, S.P. and Gere, J.M., 1961. Theory of Elastic Stability. McGraw-Hill, New York, N.Y., 541 pp.

Willis, B., 1894. Mechanics of Appalachian structure. U.S. Geol. Surv., 13th Ann. Rept., 1891—1892, pp. 213—281.

APPENDIX TO CHAPTER 2

The considerations which lead to eq. 28 are invalid. The normal stresses acting normally to layers must be continuous. Therefore, the assumption should be that the normal stress, σ_y, is the same within each stiff and soft layer in the multilayer element, as in the derivation of eq. 31. Accordingly, eq. 31, *not* eq. 28, should be used as a definition of the normal modulus throughout the remainder of the paper. The normal strains will be the same in the stiff and soft layers *only* if the materials are incompressible.

Consequently, the series of approximations leading from eq. 52 to eq. 54 should read:

Suppose that the soft interbeds vanish, $t_2 \to 0$. The the normal modulus, B_n, defined in eq. 31, becomes:

$B_n \to B_1$ and $t \to t_1$

In this case the Biot wavelength, L_B, reduces to:

$$L_B \simeq 2\pi \sqrt{T/n}\, \sqrt[4]{(B_1 t_1^3 b)/(12 b t_1 B_1)} = 2\sqrt{(T\pi t_1)/(\sqrt{12} n)}$$

or:

$$L_B \simeq 1.90 \sqrt{T t_1 / n} \tag{54}$$

However, there can be no folding if the thickness of the soft beds vanishes and the layers are bonded together, as assumed in the analysis, because then the average shear modulus becomes equal to the shear modulus of the stiff layer:

$$G_a = (t_1 + t_2) G_1 G_2 / (t_1 G_2 + t_2 G_1) \to G_1; \text{ as } t_2 \to 0.$$

That is, the multilayer behaves as a single, stiff layer confined by rigid boundaries, a condition which makes folding impossible. In order to remove the analysis from this dead end, we assume that the contacts between stiff layers are frictionless rather than bonded for this part of the analysis. In this case the average shear modulus of the multilayer element is zero and the normal moduli become the normal modulus of the stiff layers. Thus, eq. 54 is not necessarily a special case of eq. 52. Rather, eq. 52 is derived assuming there is perfect bonding between layers and eq. 54 is derived assuming there is perfect slippage between layers. In fact, however, the wavelength, eq. 52, is the same whether the layers are bonded or there is perfect slip, because the average shear modulus does not enter the expression for the wavelength. The condition of the contacts affects the critical axial load, e.g., eq. 53, one term of which contains the average shear modulus, G_a.

The corrections reported in this appendix do not affect the general conclusions drawn in the paper. They do, however, help to clarify certain considerations in the paper.

Chapter 3

IDEALIZED FOLD FORMS AND LINES OF DISCONTINUITY *

SUMMARY

Most folds in stratified rock are similar in form to ideal kink, concentric or chevron folds, in which there are discontinuities in slope or curvature of bedding planes. In this respect most folds appear to be closely related to faults, traces of which can be considered to be lines across which there are discontinuities of displacement of layers. Further, the close association of reverse faults and folds or monoclinal flexures seems to indicate that theories of faulting and folding should be closely related.

The theory of characteristics is a mathematical tool with which we can obtain insights into processes involving discontinuities. Theoretical characteristic lines are directions across which certain variables might be discontinuous and they are directions along which discontinuities propagate. The theory has been widely applied in plasticity theory and in fluid mechanics and theoretical studies of faulting have suggested that faults are analogous to the lines of discontinuity predicted by plasticity theory. Elasticity and viscosity theories, on which theories of folding have been founded, exclude the existence of characteristic lines in the materials unless the equilibrium equations, rheological properties or strains are nonlinear. However, all folding theories are nonlinear to some extent and the theories can be modified so that they predict lines of discontinuity for some conditions of loading and deformation.

Theories of folding will be developed in subsequent papers of this series in order to predict conditions under which characteristic lines can exist in multilayered materials and in order to determine the conditions that must be satisfied across and along the characteristic lines. The theory should help us to recognize lines of apparent discontinuity in natural and experimental folds and study of these lines should provide further understanding of mechanisms of folding.

Experimental studies of folding of a wide variety of materials, including alternating layers of rubber and gelatin, modeling clay and grease or graphite, and potter's clay and rubber or cardboard, suggest that the patterns of folding in these materials begin with sinusoidal forms, transform into concentric or kink forms and then into chevron forms as the multilayers are shortened axially. A suitable theory of folding of multilayers should account for these observations.

---

* Originally published as: Johnson, A.M. and Ellen, S.D., 1974. Part I, Introduction. Tectonophysics, 21: 301—339.

INTRODUCTION TO THE SERIES OF PAPERS

Folds in layered rock commonly are nearly perfect geometric figures, approaching segments of circles in parallel or *concentric* folds, sinusoidal waves in some *similar* folds and sharp angles in some *kink* folds. Partly because of their geometric beauty, folds have attracted the attention of geologists almost from the time of the birth of geological science. The basic idea that folds form as a result of compression parallel to layers is old also. Sir James Hall, in 1788, examined large folds exposed in sea cliffs along the coast of Berwickshire, Scotland, and proposed that the folds resulted from lateral compression of originally flat-lying layers (Hall, 1815). Since that time thousands of folds have been examined and mapped and numerous classification schemes of folds have been proposed. Classifications of fold forms, for example, have been suggested by Van Hise (1896), Willis and Willis (1934), Billings (1954), DeSitter (1956), Whitten (1966), Donath and Parker (1964) and Ramsay (1967), to name a few. Thus, a great deal of attention has been devoted to naming and differentiating forms of folds. Dozens of formation mechanisms for the different fold varieties have been advanced and small folds have been produced experimentally in a wide variety of experimental materials. Bailey Willis (1894) examined large folds in the Appalachian Mountains and performed a series of ingenious experiments with layers of beeswax in order to elucidate certain folding processes which he deduced from field work (e.g., see Johnson, 1970, chapter 2). Some of the most interesting experimental folds have been produced recently by Ramberg (1959), Biot et al. (1961), Currie et al. (1962), Turner (1962), Ramberg (1963a and b), Paterson and Weiss (1966), Ramberg (1967), Donath (1968), Ghosh (1968), Weiss (1968), Cobbold et al. (1971), Ellen (1971), Ramberg and Strömgård (1971), and Roberts and Strömgård (1972).

Each of the experimental and field studies has contributed some understanding of natural folds. Other significant contributions have been made by theoretical analysis. Mechanical theories of folding are surprisingly old (e.g., Smoluchowski, 1909; Tromp, 1937) but theoretical analysis of folding has gained widespread interest in structural geology only recently, beginning with the publications by Ramberg (1960) and Currie et al. (1962). Most of the early theories of folding concentrated on the analysis of buckling of single, stiff layers in soft confining media (Tromp, 1937; Ramberg, 1960 and 1963a; Biot, 1961; Biot and Odé, 1962; Currie et al., 1962; Chapple, 1968). The theories predicted definite wavelength/thickness ratios for folds in isolated layers and, in general, they too have helped us to understand some folds in nature (Ramberg, 1959, 1963b; Ramsay, 1967; Sherwin and Chapple, 1968; Dietrich and Carter, 1969; Johnson, 1970; and Hudleston and Stephansson, 1973).

It is apparent, however, that the theory of folding of single layers tells us almost nothing about most folds, which are in rocks consisting of many, interlayered, stiff and soft beds, that is, consisting of *multilayers*. Thus, theories have been advanced, primarily by Biot and Ramberg, for the folding of multilayers (Biot, 1957, 1960, 1963; Ramberg, 1963a; Biot, 1964; Ramberg, 1964; Biot, 1965a, b and c; Odé, 1966; Ramberg, 1967; Johnson, 1969 and Ramberg, 1970).

The theories advanced by Biot and Ramberg are basically different, reflecting different purposes. Ramberg has been primarily interested in details of deformational shapes of contacts between a few stiff and soft layers in multilayered sequences whereas Biot has been interested in gross deformational shapes of groups of layers in thick multilayered sequences. Ramberg's analyses begin with the assumption that deflections within each layer are sinusoidal. Then interactions at contacts between layers and interactions with external boundaries are considered in order to derive the type and wavelength of sinusoidal folds most likely to develop in the multilayered sequence. The analysis allows one to calculate stresses and displacements at every point in the multilayered sequence. Effects of bending, shear between layers, differences in thicknesses, densities and rheological properties of layers and other factors have been incorporated in Ramberg's theory. The result is that his theory can describe accurately many features of sinusoidal folding as well as growth of diapiric structures in multilayers.

Biot's analysis, on the other hand, begins with assumptions about the gross deformational behavior of a representative element of an entire multilayered sequence. The representative element includes a stiff and a soft layer so that the fundamental element is finite in size. Ramberg's fundamental element is infinitesimal. Biot derives equilibrium equations and integrates the equations to obtain gross displacements throughout the multilayered body. The displacements are not assumed to be sinusoidal although most of the solutions to the equations predict sinusoidal wave forms. Biot's analysis separates stresses and strains into *initial* values, which may be large, and *incremental* values, which must be infinitesimal. Ramberg implicitly uses the same approach. Finally, Biot separates incremental stresses into stresses caused solely by rotation of layers and stresses caused solely by strain of layers. Biot's simplified theory (1965b and c) considers effects of bending moments, as does Ramberg's general theory (1970).

The theory of the behavior of multilayers presented in this series of papers is somewhat different from the theories of Ramberg and Biot. However, we have benefited greatly by studying both theories. Indeed, Biot's (1965a) discussion of a possible condition under which sharp, kink-like folds might be predicted theoretically stimulated our interest in multilayer theory as we studied kink-like folds in cherts of the Franciscan Assemblage (Ellen, 1971).

The basic requirements of a suitable theory of buckling are that the equilibrium equations and the stress—strain equations for a fundamental element are appropriate to the specific problem of interest. The study of the equilibrium equations for problems of finite deformation is quite old. Truesdell (1952) and Truesdell and Noll (1965) indicate that the study of finite elasticity theory is as old as infinitesimal elasticity theory and that accurate equations for stability in finite elasticity were developed by Hadmard in 1903. Truesdell (1952) indicates, further, that there have been many attempts to derive special simplifications of the theory of finite elastic stability. One type of simplification leads to the Biot theory of incremental deformations, which was derived by one method by Biezeno and Hencky (1928) and by a different method by Biot (1939, 1965a) (see Truesdell, 1952, p. 207). We have found the Biot theory to be unsuitable and the theory of equilibrium that we shall follow is clearly explained by Novozhilov (1953), who carefully specifies the assumptions at each step of his derivation.

The second basic requirement for a theory of buckling of multilayers is a suitable stress—strain relation of elements of multilayers. Fortunately, this subject is of considerable interest to engineers because of the wide use of composites and laminates of structural materials in modern technology. One approach to the problem of specifying gross elastic constants of a multilayer is to derive effective moduli for a fundamental element of the multilayer. A fundamental element contains all types of materials in the multilayer so that the multilayer is homogeneous with respect to the element. Thus, the multilayer can be considered to be a homogeneous but anisotropic body. Two methods of this type are the Voigt hypothesis (1928, ref. in Sun, 1971), which assumes that strain components are uniform throughout the inhomogeneous element, and the Reuss hypothesis (1929, ref. in Sun, 1971), which assumes that the stress components are uniform throughout the element. Neither of these hypotheses is acceptable, because one violates equilibrium of stresses and the other violates compatibility of strains so that many modern theories are based on a combination or compromise of the two (Postma, 1955; White and Angona, 1955 and Salamon, 1968). According to Chou et al. (1972, p. 81), the three references cited above lead to equivalent expressions for the effective elastic moduli, although the expressions appear to be different. The combined Voigt—Reuss hypothesis, or "effective modulus" theory, is also used by Biot (1965a), Kafka (1965) and Johnson (1969) and it will be used for some of the derivations in this series of papers.

Sun, Achenbach and Herrmann (e.g. Herrmann and Achenbach, 1967; Sun et al., 1968; Sun, 1971) used another approach, based on an accurate theory for the vibration of plates (Mindlin and Medick, 1959), to derive elasticity constants for elements of multilayers. This method, termed the "effective stiffness" theory, can be far more accurate than the effective-modulus theory, especially for sinusoidal types of deformation. The effective-stiffness theory emphasizes the different properties of the layers whereas the effective-modulus theory tends to mask the differences by averaging. Unfortunately, however, the effective-stiffness theory allows only a limited range of types of deformation and its use appears to be restricted to studies of similar wave forms. We plan to explain and use the effective-stiffness theory in this series of papers.

Other interesting methods of deriving elasticity constants for multilayers have been developed by Mindlin (1964), Bedford and Stern (1971), Kohn et al. (1972) and Bahar (1972).

The purpose of this first paper is to introduce a series of papers concerning the theoretical conditions under which ideal concentric, chevron, sinusoidal and kink folds as well as monoclinal flexures might form. The emphasis here will be on the concept of mathematical characteristic directions and lines of apparent discontinuity, which appear to be physical analogs of characteristic directions, in natural and experimental folds. A second paper will be concerned with nonlinear equations of equilibrium, effective moduli of elastic multilayers, sinusoidal folding and an application of characteristic theory to folds of the concentric family. A third paper will continue with the study of concentric folds but will emphasize kink folding and discuss theoretical conditions under which different types of folds might form. Subsequent papers are planned to describe and interpret some fold patterns in the field and in experiments and to develop theories for homogeneous, anisotropic media and for viscous multilayers.

Our research into processes of folding began in the mid 1960's when S. Ellen spent a summer examining folds in parts of the Appalachian Mountains in Pennsylvania, West Virginia and New Jersey and in parts of Nevada and California. The purposes of the field examinations were to gain first-hand impressions of forms of natural folds and to find places for detailed field work. The detailed field work was done near San Francisco, California, at the north end of the Golden Gate Bridge, where there are excellent exposures of kink-like folds in cherts of the Franciscan Assemblage in road cuts, sea cliffs and quarries.

In order to understand the folds in the Franciscan cherts, a few experiments were made with various types of layered materials and several types of theoretical analyses were attempted. Two theories of kink folding appeared to be promising at the time Ellen wrote his dissertation (Ellen, 1971). One theory considered the buckling of interbedded elastic and plastic layers. The other was a conceptual modification of Biot's theory of the multilayer (Biot, 1965a) which, in its original form, could not account for kink-like folds under conditions of layer-parallel loading.

The presentation here is largely based on theoretical analyses by Johnson while he was on sabbatical leave, visiting the University of Uppsala, Sweden. However, the basic ideas, including the recognition of the importance of discontinuities in slope and curvature and the recognition of the general requirements of a suitable theory, are the result of searching collaboration during the time Ellen did research at Stanford University.

FORMS OF SOME NATURAL FOLDS

Folds in rock layers have almost infinite variety of forms and sizes and they occur in markedly different rock materials. Certain fold shapes recur, however, irrespective of their sizes or the materials involved. Indeed, the

Fig. 1. Natural folds approximately concentric in form. A. Synclinal warp in Tuscarora Quartzite exposed in Delaware Water Gap, New Jersey. The fold form is crudely concentric, with curvature of layers increasing toward the core of the syncline. The exposure is about 10 m wide. B. Antiform fold in amphibolite in Seiad Valley, northern California. Gross form of the fold is concentric. C. Synform and antiform in interbedded sandstone and shale near Alexandria, Pennsylvania. Each thick layer forms an open warp in the synform and a tight bend in the adjacent antiform, as is typical in folds of the concentric variety.

recognition of characteristic shapes of folds has been an important basis for most classifications and names of fold types.

The layers in some folds are gently curved in open, smooth warps. Fig. 1A shows an example of a smooth synclinal warp in quartzite exposed in the Delaware Water Gap of New Jersey. The amount of bending is greatest near

the core of the syncline and bending decreases irregularly with increasing distance from the core. Fig. 1B shows a fold in foliated amphibolite in Seiad Valley, California, and Fig. 1C shows folds in interbedded sandstones and shales near Alexandria, Pennsylvania. A common feature among the folds

C

shown in Fig. 1 is that the average radius of curvature is much greater in layers near the outer arcs of the folds than it is in layers near the cores. Also, a layer bent into an open arc in a synclinal fold is bent more sharply in an adjacent, anticlinal fold, or vice versa, as shown in Fig. 1C. The natural folds shown in Fig. 1 might be classified as parallel or *concentric folds*.

Another type of fold containing smoothly warped layers is shown in Fig. 2, which is a view of folds near Moab, Utah (Johnson, 1969). The folds are confined to the Carmel Formation, a sequence of interlayered thick sandstone beds and thin shale beds, which is overlain and underlain by relatively massive sandstone bodies. The hinges of the folds are rather uniformly spaced along the layers. The radii of curvature are minimal where the fold amplitudes are maximal, slightly below midheight in the sequence of layers, and the radii increase toward the cores of synclines and anticlines. That is, bending decreases from a maximum near midheight in the sequence to zero at the upper and lower contacts of the sequence. The folds shown in Fig. 2 might be classified as *sinusoidal*, implying that there is a definite wavelength and that the layers are bent into crude, sinusoidal wave-shapes. Fig. 3A shows an idealization of the folds in the Carmel Formation. Ideal sinusoidal folds can also be of the *similar* variety, in which the form does not change with distance along axial planes of the folds. Fig. 3B shows an ideal *similar*, sinusoidal fold. The thicknesses of layers are greater in hinges than along limbs of the ideal similar fold.

Fig. 2. Folds within Carmel Formation, in Arches National Monument, Utah. The Carmel Formation is overlain by the relatively massive, Entrada Sandstone and underlain by the broadly cross-bedded, Navajo Sandstone. The Carmel Formation is about 10 m thick. The folds are crudely sinusoidal in shape, with maximum amplitudes slightly below midheight in the sequence. The amplitudes of folds diminish toward the upper and lower contacts of the Carmel.

In some folds the curvature is concentrated in hinges, and the limbs are nearly straight, as shown in Fig. 4A, which shows a fold in interbedded chert and shale of the Franciscan Assemblage, central California. The shale layers are markedly thicker in fold hinges than they are along limbs whereas the chert layers maintain rather uniform thicknesses in limbs and hinges. The gross form is *similar* but it is quite different from the ideal, sinusoidal, similar form shown in Fig. 3B. Fig. 4B shows another example of folds with

A

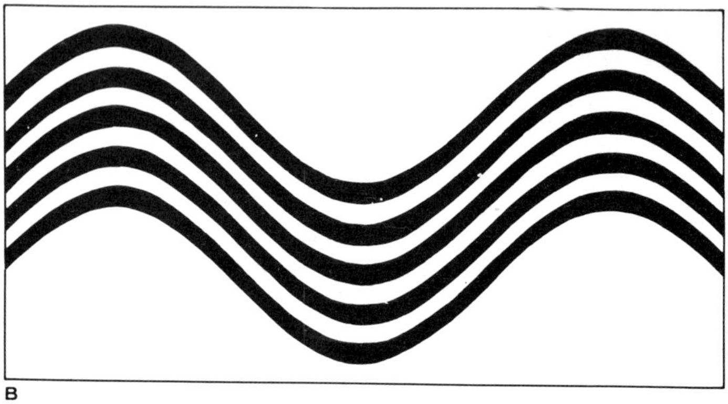

B

Fig. 3. Examples of ideal sinusoidal folds. A. Sinusoidal folds of the *internal* variety, where amplitudes are maximal at midheight and diminish to zero at rigid boundaries. B. Sinusoidal folds of the *similar* variety, where amplitudes remain constant through the section and layers thin along limbs of folds.

straight limbs and narrow hinges in the Franciscan Assemblage. The layers form a gross zig-zag pattern in the outcrop. Folds in interbedded sandstone, siltstone and shale, in West Virginia (Fig. 5A) have quite narrow hinges and the limbs are nearly straight. Layers in folded amphibolite in northern California (Fig. 5B) show the same pattern although this rock also contains folds of concentric form, such as the one shown in Fig. 1B. The hinges of the folds shown in Fig. 5B are not as sharp as those shown in Fig. 4 and 5A, but the folds all have relatively straight limbs. These might be classified as *kink* or *chevron* folds.

THEORETICAL CHARACTERISTIC DIRECTIONS

The kink-like folds shown in Fig. 4 and 5 are quite different from the smooth folds shown in Fig. 1—3. Ideal sinusoidal folding produces deflection which is variable along a layer but which is *continuous*. In contrast, the warping of a layer involved in ideal kink and chevron folding is *discontinuous*; a segment of a layer may be flat-lying, yet, another adjacent segment may be abruptly tilted upwards or downwards around a narrow bend. The forms of the ideal folds and flexures are markedly different but all, except those of the sinusoidal family, involve discontinuities. Almost all theoretical research into folding processes has been devoted to analyses of conditions of continuous, sinusoidal folding. The theories deal with continuous processes and are therefore amenable to a wide variety of mathematical techniques (e.g., Ramberg, 1963a; Biot, 1965a; Chapple, 1968; Dietrich and Carter, 1969). Unfortunately, but understandably, there are only a few mathematical techniques available for analyzing discontinuous processes. One of these techniques, the method of characteristics, is discussed in following paragraphs. We first discuss the theory and the types of discontinuities that can be represented with the theory. Then we illustrate the positions of lines of discontinuity in several ideal folds and flexures. Finally, we indicate lines in experimental and natural folds which we believe to correlate with the lines of discontinuity in the ideal folds.

A reason for first discussing the theory of characteristics is that the theory provides a new way to examine folds. Ideal flexures and faults are not easily correlated with real ones. For example, is a fault "really" a plane of discontinuity in displacement? Generally a fault is a zone rather than a plane and it is no simple matter to determine whether individual beds are cut off or merely are greatly attenuated across the fault zone. Similarly, are the folds in the Carmel Formation (Fig. 2) "really" sinusoidal in form? Theories of folding, cited above, assume that the folds are continuous, sinusoidal waves. However, suppose that there were some theoretical reason to expect there to be certain discontinuities within many folds. Is it then possible that one could reinterpret the folds in the Carmel and recognize certain discontinuities in them? Indeed it is, as we show in later pages. Study of the theory of characteristics leads to the search for possible discontinuities in folds just as the study of sinusoidal folding theory has led to the recognition of wavelengths in folds.

Let us probe slightly more deeply into the concepts of continuity and discontinuity because these concepts form the foundation of our theory of folding. The following analogy is based on a discussion in a book by Abbott (1966), who presents a rigorous yet understandable discussion of the theory of characteristics.

The movement of traffic along city streets is discontinuous with respect to both time and space. We shall focus on the spatial discontinuities. Fig. 6 shows a pattern of intersecting streets in a certain city. Movement of automobiles is spatially discontinuous, being restricted to certain paths, the streets, and normally excluded from intervening areas, the buildings. This concept of streets is analogous to the concept of mathematical *characteristic directions*. A mathematical characteristic direction is a path along which something propagates. Thus, in the analogy, the streets are physical "characteristics" along which automobiles propagate. Although quantities propagate *along* mathematical characteristics, the same quantities may be discontinuous *across* them. Thus, in the analogy, the automobiles can travel along but not across streets, except at intersections.

The theory of characteristics has been given the most consideration by students of wave phenomena, where the characteristics are defined in terms of temporal as well as spatial coordinates (e.g., Sokolnikoff and Redheffer, 1958; Courant, 1962; Fox, 1962; Green, 1964 and Abbott, 1966). However, the theory also has found important applications in plasticity theory (Hill, 1950; Prager and Hodge, 1951; Johnson and Mellor, 1962 and Abbott, 1966).

First-order equations

In order to develop a mathematical concept of characteristic directions we will consider a simple, first-order partial differential equation. Most aspects

Fig. 4. Folds in interbedded chert and shale of the Franciscan Assemblage, exposed at the north end of the Golden Gate Bridge, San Francisco, California. The chert beds protrude in outcrops and the shale beds recede, apparently an expression of differential weathering. The thicknesses of chert beds average about 5 cm. A. Typical fold in Franciscan chert, where limbs are nearly straight and bending is concentrated in narrow hinges. Thicknesses of chert beds, measured normally to bedding, are essentially constant around hinges whereas shale beds thicken markedly in hinge zones. B. Zig-zag pattern of kink folds. Original, unfolded layering dips toward right in view.

of the theory of characteristics can be illustrated by analysis of the simple equation:

$$\partial z/\partial x = \partial z/\partial y \tag{1}$$

which we will interpret in terms of horizontal map coordinates, x and y, and

of altitude, z. Let us suppose that eq. 1 describes the simple, structural configuration of a layer in an area so that, for some reason based on physical analysis, the surface of the layer must satisfy eq. 1. Then, eq. 1 requires that the apparent slopes (dips), $\partial z/\partial x$ and $\partial z/\partial y$, as observed in vertical cross-sections parallel to the x- or y-axis, respectively, must be equal. A possible configuration for the layer is shown in map view in Fig. 7A and in oblique block diagram in Fig. 7B. As shown in that figure, if the y-coordinate points north, the layer strikes N45W and dips toward the southwest.

Now, any number of planes could be drawn in Fig. 7 with the same attitude as the one shown there. The governing equation, eq. 1, merely requires that the strike of the bedding must be N45W. The dip is not specified by the equation. However, if the dip is specified at any point, that is, if the

Fig. 5. Kink-like folds. A. Folds in interbedded sandstone, siltstone and shale of the Portage Series, near Beverly, West Virginia. Exposure is about 10 m wide. B. Kink bands in foliated amphibolite, Seiad Valley, northern California. Scale indicated by coin.

value of $\partial z/\partial x$ or $\partial z/\partial y$ is specified at one point, it might be specified everywhere. But it need not be, as we will show.

Suppose that the dip of the layer is known to be toward the northeast at a point in the southwestern quadrant and toward the southwest at a point in the northeastern quadrant (Fig. 8A). The attitudes everywhere must satisfy eq. 1 according to our first assumption. Therefore, in part of the area the layer dips southwest and in part it dips northeast. The strike must be N45W so that a possible configuration of the layer is shown in cross section a—a' in Fig. 8B. The layer is sharply bent into a fold and the dip is discontinuous across the hinge of the fold. The hinge line projects to the surface (the x,y-plane) as a *characteristic direction*, indicated by a dotted line in Fig. 8A.

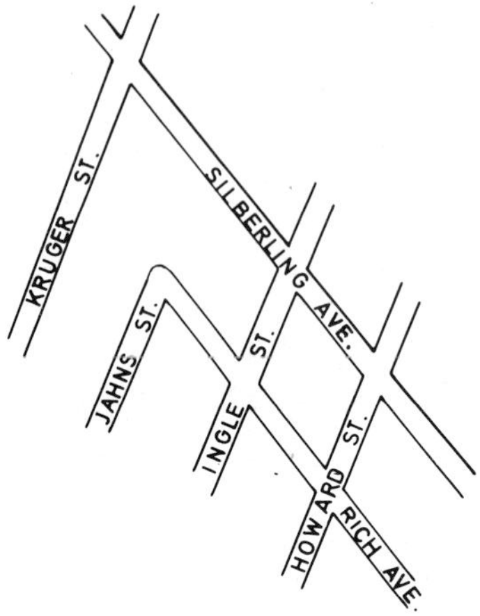

Fig. 6. Pattern of intersecting streets in a certain city. The streets are analogous to theoretical characteristic directions.

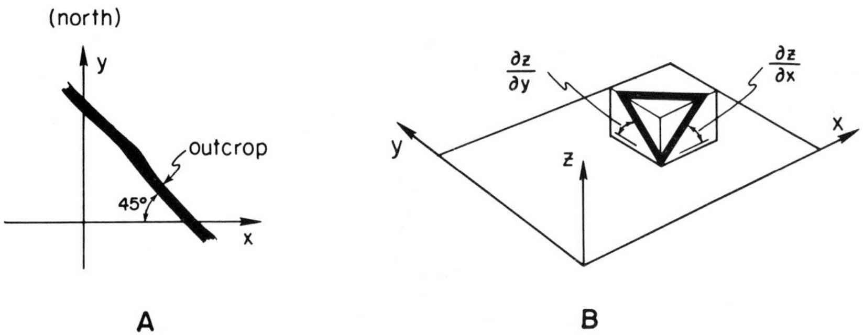

Fig. 7. Configuration of a uniformly dipping stratum that satisfies eq. 1.

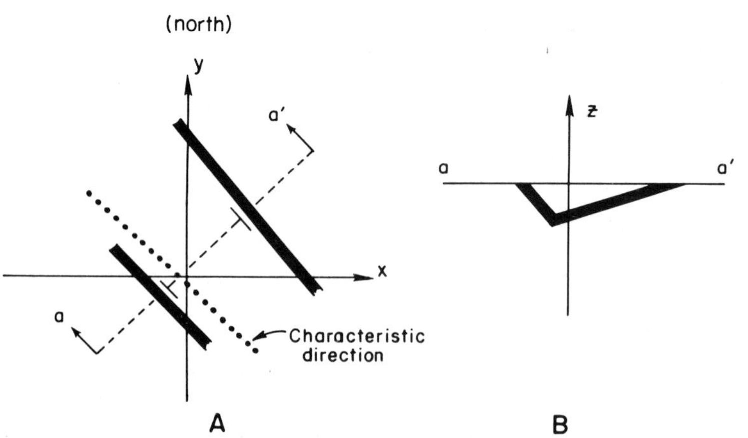

Fig. 8. Fold form and a characteristic direction that satisfy eq. 1.

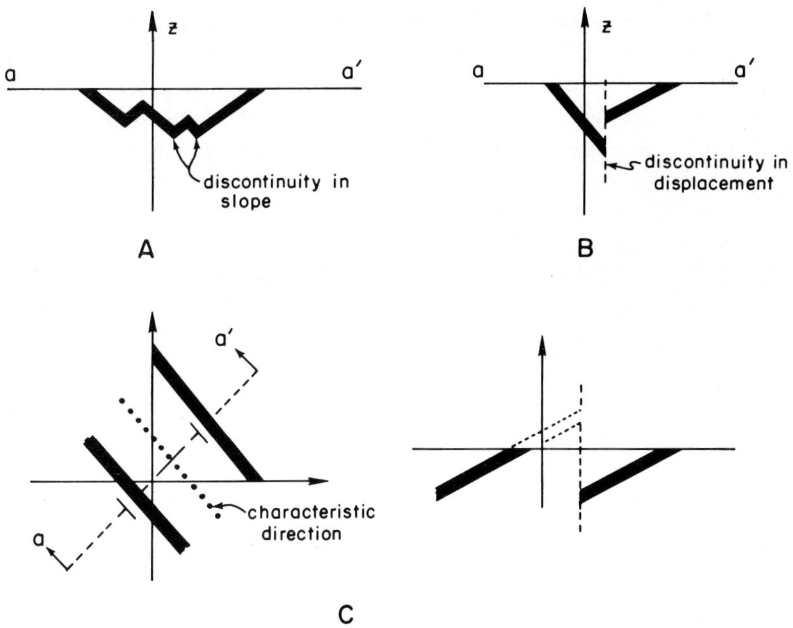

Fig. 9. Three kinds of discontinuity that satisfy eq. 1. A. Multiple discontinuities in slope. B. Discontinuities in slope and displacement. C. Discontinuity in displacement.

Thus, even though eq. 1 is satisfied everywhere, there may be a discontinuity in the variables $\partial z/\partial x$ and $\partial z/\partial y$ and these derivatives may be undefined and indeterminant at *characteristic lines*. The term "may be" is critical here. The variables do not have to be discontinuous anywhere.

Another important feature of the theory of characteristics is illustrated in Fig. 9A, which was constructed using the same initial data as that used in constructing Fig. 8B. Thus, there may be any number of discontinuities in slope (Fig. 9A) but, in order for eq. 1 to be satisfied, the discontinuities must trend N45W, parallel to the strike of the layer, that is, they must project on a line with a slope of $dy/dx = -1$ in an x, y-plane. Further, all derivatives of higher order than those in the governing equation may also be undefined at the characteristic lines and may be discontinuous across the characteristic lines. Thus, the higher derivatives, $\partial^2 z/\partial x^2$, $\partial^3 z/\partial x^3 \ldots \partial^2 z/\partial y^2 \ldots$ are undefined where the first derivatives in eq. 1 are discontinuous.

Equally significant, the same characteristic lines may be sites of discontinuities in derivatives lower than those in the governing equation. In the example we are considering here, the characteristic line might be the site of a discontinuity in altitude, z, expressed as a fault (Fig. 9B and 9C). Again, eq. 1 is satisfied everywhere but in the example shown in Fig. 9B both the slope and the altitude are discontinuous across the characteristic line whereas only the altitude is discontinuous in the example shown in Fig. 9C. Such discontinuities in derivatives of lower order than those of the governing differential equations are called "weak solutions" by Courant (1962, p. 418, 635), who discusses conditions under which weak solutions are possible.

One might well be concerned that the theory of characteristics provides so much flexibility that nearly anything can be discontinuous across characteristic lines. We will show, as we develop our theory of folding, that this is, indeed, a significant problem. One must introduce additional governing equations, based on physical reasoning, in order to limit the possible types of discontinuities. This flexibility of the theory of characteristics is most encouraging to structural geologists, however, who know that discontinuities in flexure and displacement are closely related in many structures, such as overturned folds and thrust faults or monoclinal flexures and normal or reverse faults.

The characteristic directions derived intuitively in preceding paragraphs can be derived quickly by mathematical analysis. Let us now generalize eq. 1 slightly and determine the characteristic directions mathematically. Replace eq. 1 with:

$$\frac{\partial z}{\partial x} + A\,\frac{\partial z}{\partial y} = B \qquad (2)$$

where A and B are constants or are functions of x, y and z. The total derivative of z is defined in the calculus as:

$$\frac{\partial z}{\partial x}\,\mathrm{d}x + \frac{\partial z}{\partial y}\,\mathrm{d}y = \mathrm{d}z \qquad (3)$$

Eq. 2 and 3 can be considered algebraic equations which we can solve simultaneously for the variables $\partial z/\partial x$ and $\partial z/\partial y$. Accordingly, using the method of determinants:

$$\partial z/\partial x = \frac{\begin{vmatrix} B & A \\ \mathrm{d}z & \mathrm{d}y \end{vmatrix}}{\begin{vmatrix} 1 & A \\ \mathrm{d}x & \mathrm{d}y \end{vmatrix}} = Z_x/\mathbf{D}$$

$$\partial z/\partial y = \frac{\begin{vmatrix} 1 & B \\ \mathrm{d}x & \mathrm{d}z \end{vmatrix}}{\begin{vmatrix} 1 & A \\ \mathrm{d}x & \mathrm{d}y \end{vmatrix}} = Z_y/\mathbf{D} \qquad (4)$$

The variables are indeterminant if the determinants, Z_x, Z_y, and $\mathbf{D}$, in each of eq. 4 are zero. Thus, for example, if:

$$Z_x = 0 = \begin{vmatrix} B & A \\ \mathrm{d}z & \mathrm{d}y \end{vmatrix}$$

$$Z_y = 0 = \begin{vmatrix} 1 & B \\ \mathrm{d}x & \mathrm{d}z \end{vmatrix} \qquad (5)$$

$$\mathbf{D} = 0 = \begin{vmatrix} 1 & A \\ dx & dy \end{vmatrix} \tag{5}$$

the variables $\partial z/\partial x$ and $\partial z/\partial y$ are indeterminant. It follows from eq. 5 that:

$$\begin{aligned} dz/dy &= B/A \\ dz/dx &= B \\ dy/dx &= A \end{aligned} \tag{6}$$

The last of eq. 5 and 6 express the slope of the characteristic direction in the x, y-plane. In order for eq. 2 and 1 to be identical, $A = -1$ and $B = 0$. Thus, the slope of the characteristic direction is -1 in the x, y-plane and the intuitive and mathematical derivations have provided the same result (compare the last of eq. 6 and Fig. 8A). Characteristics are lines for first-order equations and the first two of eq. 6 express the change in altitude of those lines as functions of x and y (e.g., Abbott, 1966, p. 62). In the problem being considered here, however, $B = 0$ and the characteristic lines are horizontal.

Higher-order equations

The first-order equations we have been working with thus far are simple to visualize graphically so that the mathematical derivation of the characteristic direction was not necessary. However, the mathematical method can be applied with confidence once the concept of characteristic direction has been grasped and the mathematical approach generally is much simpler than the graphical approach.

As an example of the treatment of higher-order equations we shall discuss the equations that describe the plane strain of incompressible elastic materials, first according to the theory of linear elasticity (e.g., Timoshenko and Goodier, 1970) and then according to the general, nonlinear theory of elasticity (e.g., Truesdell and Noll, 1965). We will show that characteristic directions cannot develop in situations described by the linear theory but that they can develop, under certain conditions which we will discuss in a subsequent paper, where a nonlinear theory is appropriate.

Linear elasticity theory

We assume plane-strain conditions so that the equilibrium equations are in the form (e.g., Johnson, 1970, p. 256):

$$\begin{aligned} \partial \sigma_{xx}/\partial x + \partial \sigma_{yx}/\partial y + \gamma_x &= 0 \\ \partial \sigma_{xy}/\partial x + \partial \sigma_{yy}/\partial y + \gamma_y &= 0 \end{aligned} \tag{7}$$

where γ_x and γ_y are body forces per unit of volume in the x- and y-directions, and σ are stress components. The material is assumed to be incompressible so that the stress—strain relations are (e.g., Timoshenko and Goodier, 1970):

$$\sigma_{xx} = \bar{\sigma} + N(\partial u/\partial x)$$

$$\sigma_{yy} = \bar{\sigma} + N(\partial v/\partial y) = \bar{\sigma} - N(\partial u/\partial x) \qquad (8)$$

$$\sigma_{xy} = \sigma_{yx} = G[(\partial u/\partial y) + (\partial v/\partial x)]$$

where $N = 2G$ and $\bar{\sigma} = (\sigma_{xx} + \sigma_{yy})/2$

G is the shear modulus, and u and v are displacements in the x- and y- directions, respectively. Also:

$$(\partial u/\partial x) = -(\partial v/\partial y) \qquad (9)$$

is a statement that the material is incompressible.

Let us assume that the body forces, γ, are constant in the material. Then differentiate the first of eq. 7 with respect to y, the second with respect to x, and substitute eq. 8 into the results. Now subtract the second resulting equation from the first and use eq. 9 to eliminate v in one term. The result:

$$2(N-G)\frac{\partial^3 u}{\partial x^2 \partial y} + G\frac{\partial^3 u}{\partial y^3} - G\frac{\partial^3 v}{\partial x^3} = 0 \qquad (10)$$

is independent of the mean stress, $\bar{\sigma}$. It is a third-order partial differential equation, but it can be reduced to a first-order equation, the standard form given in eq. 2, by introducing new variables:

$$p = (\partial^2 u/\partial x^2)$$

$$q = (\partial^2 u/\partial x \partial y) \qquad (11)$$

$$r = (\partial^2 u/\partial y^2)$$

$$s = (\partial^2 v/\partial x^2)$$

According to the definitions in eq. 11 and to eq. 9, the following are identities:

$$(\partial s/\partial y) \equiv -(\partial p/\partial x)$$

$$(\partial p/\partial y) \equiv (\partial q/\partial x) \qquad (12)$$

$$(\partial q/\partial y) \equiv (\partial r/\partial x)$$

The next step in the procedure is to introduce eq. 11 into eq. 10 and to write expressions for the total differentials, analogous to eq. 3, for the variables, p, q, r and s. Thus:

$$2(N-G)(\partial q/\partial x) + G(\partial r/\partial y) - G(\partial s/\partial x) = 0$$

$$(\partial q/\partial x)\mathrm{d}x + (\partial r/\partial x)\mathrm{d}y = \mathrm{d}q \qquad (13)$$

$$(\partial r/\partial x)\mathrm{d}x + (\partial r/\partial y)\mathrm{d}y = \mathrm{d}r$$
$$(\partial p/\partial x)\mathrm{d}x + (\partial q/\partial x)\mathrm{d}y = \mathrm{d}p \tag{13}$$
$$(\partial s/\partial x)\mathrm{d}x - (\partial p/\partial x)\mathrm{d}y = \mathrm{d}s$$

The identities in eq. 12 were used to eliminate certain terms in eq. 13.

Now, according to the theory of characteristics, if the coefficients in the basic partial differential equation, eq. 10, have certain ranges of values, the derivatives, $(\partial^3 v/\partial x^3)$, $(\partial^3 u/\partial y^3)$ and $(\partial^3 u/\partial y \partial x^2)$ are indeterminant and might be discontinuous along characteristic lines. If the coefficients have a different range of values, the derivatives are everywhere continuous functions of the coordinate system and there can be no real characteristic lines and no discontinuities in the third derivatives. We can solve eq. 13 simultaneously for the five variables, $(\partial q/\partial x)$ ... $(\partial s/\partial x)$. With this in mind we would make the following interpretation with respect to folding theory. Where the values of the coefficients are such that the derivatives are indeterminant, the derivatives might be discontinuous and the types of folds might not be of the sinusoidal family. However, if the coefficients are such that one can determine values of the derivatives, the derivatives must be continuous everywhere and the types of folds must belong to the continuous or sinusoidal family.

In order to determine these conditions, and, therefore, to determine the types of solutions possible for eq. 10, we solve for the *characteristic directions* of eq. 10. We do this by the method of determinants, as we did with eq. 2 and 3. The denominator, **D**, of the determinants is of primary interest here because it provides essential information about the slope of the characteristic directions and about the existence of real characteristic lines. Thus, the denominator of the determinants for the partial derivatives in eq. 13 is:

$$\mathbf{D} = \begin{vmatrix} 0 & 2(N-G) & 0 & G & -G \\ 0 & \mathrm{d}x & \mathrm{d}y & 0 & 0 \\ 0 & 0 & \mathrm{d}x & \mathrm{d}y & 0 \\ \mathrm{d}x & \mathrm{d}y & 0 & 0 & 0 \\ -\mathrm{d}y & 0 & 0 & 0 & \mathrm{d}x \end{vmatrix} = 0 \tag{14}$$

which reduces to:

$$\left(\frac{\mathrm{d}y}{\mathrm{d}x}\right)^4 + \frac{2(N-G)}{G}\left(\frac{\mathrm{d}y}{\mathrm{d}x}\right)^2 + 1 = 0 \tag{15}$$

Now, eq. 15 is of fourth degree so that there may be as many as four distinct characteristic directions and the slopes, $\mathrm{d}y/\mathrm{d}x$, of the directions are solutions of eq. 15. There may be no real characteristic directions, however. Eq. 15 was derived by assuming that the deformation of the body can be described in terms of linear elasticity theory. According to eq. 8, the normal modulus, N, is $2G$. The moduli are always positive quantities. Therefore, eq. 15 becomes:

$$\left(\frac{dy}{dx}\right)^4 + 2\left(\frac{dy}{dx}\right)^2 + 1 = 0$$

which has two unreal roots. That is, the values of (dy/dx) are imaginary and there are no real characteristic directions in such a material. The linear elasticity theory, therefore, would predict that all folds must be of the sinusoidal, or continuous, variety.

Nonlinear elasticity theory

All theories of folding developed thus far (e.g., Currie et al., 1962; Biot, 1965a; Ramberg, 1970) assume certain nonlinearities. The nonlinearities have three primary sources (Novozhilov, 1953). The equations of equilibrium, analogous to eq. 7, can be nonlinear because of relatively large deformations or rotations. The equations for strains are nonlinear if strains are finite. Finally, the stress/strain (or strain-rate) relations, analogous to eq. 8, are nonlinear if the material properties are nonlinear functions of stress, strain, stress-rate or strain-rate. Folding theories have been restricted to certain nonlinearities in the equations of equilibrium and they are all based on specialized assumptions, some of dubious validity, concerning the forms of the nonlinearities. We will be discussing much more general nonlinearities in subsequent papers of this series because it is the nonlinearity that allows one to predict real characteristic directions.

Here we will introduce, essentially unexplained, nonlinear equilibrium equations presented by Truesdell and Noll (1965, p. 264) for incremental deformations of an initially strained, nonlinear elastic material. The initial strains, the equations of equilibrium and the elastic properties are all nonlinear but the incremental strains are assumed to be infinitesimal and linear. Thus, the equations represent essentially the ultimate in nonlinearity. A remarkable result is that the equations are as simple in appearance as are the equations of linear elasticity. The nonlinearities are confined to the coefficients:

$$a_{xx}\left(\frac{\partial^2 v}{\partial x^2}\right) + b_{xy}\left(\frac{\partial^2 u}{\partial y^2}\right) + (a_{xy} + b_{yx})\left(\frac{\partial^2 v}{\partial x \partial y}\right) = 0$$

$$a_{yy}\left(\frac{\partial^2 v}{\partial x^2}\right) + b_{yx}\left(\frac{\partial^2 v}{\partial y^2}\right) + (a_{yx} + b_{xy})\left(\frac{\partial^2 u}{\partial x \partial y}\right) = 0 \tag{16}$$

where u and v are incremental displacements in the x- and y-directions, respectively, and the coefficients, $a_{xx} \ldots b_{yx}$, depend on the magnitudes of initial strains applied to the body before the incremental deformations took place as well as on nonlinear elasticity coefficients (Truesdell and Noll, 1965, p. 263, eq. 70.3 and 70.4).

Now, if the nonlinear material is assumed to be incompressible:

$$(\partial u/\partial x) = -(\partial v/\partial y)$$

and eq. 16 can be reduced to a single equation of form identical to that of eq. 10 for a linear elastic material:

$$(a_{xx}-a_{xy}-a_{yx}-b_{xy})\left(\frac{\partial^3 u}{\partial x^2 \partial y}\right)+b_{xy}\left(\frac{\partial^3 u}{\partial y^3}\right)-a_{yy}\left(\frac{\partial^3 v}{\partial x^3}\right)=0 \tag{17}$$

Accordingly, the denominator for the determinants of the third derivatives is the same as the determinant in eq. 14, except that the G in the fourth column is replaced by b_{xy}, the $-G$ in the fifth column is replace by $-a_{yy}$ and the term in the second column of the first row is replaced by the coefficient of $(\partial^3 u/\partial x^2 \partial y)$ in eq. 17. Thus, the equation for the characteristic directions is:

$$\left(\frac{\mathrm{d}y}{\mathrm{d}x}\right)^4-\left(\frac{a_{xy}+a_{yx}+b_{xy}-a_{xx}}{a_{yy}}\right)\left(\frac{\mathrm{d}y}{\mathrm{d}x}\right)^2+\frac{b_{xy}}{a_{yy}}=0 \tag{18}$$

We postpone a detailed analysis of eq. 18 and simply note here that there are certain ranges of the coefficients, that is, certain conditions of initial strains, for which there are real characteristic directions and for which discontinuities in third derivatives can exist in a homogeneous, nonlinear elastic material subjected to large deformations.

The foregoing discussion illustrates the approach we will follow, in subsequent papers, in deriving theoretical conditions for various kinds of ideal folding of multilayered, isotropic or anisotropic materials. We have simply introduced the theory of characteristics here, leaving most details until they are required to develop the theory of folding in subsequent papers.

LINES OF DISCONTINUITY IN IDEAL FOLDS

Thus far we have shown that many natural folds cannot be described even approximately in terms of wavelengths or of continuous bending, as would be expected if theories of sinusoidal folding adequately explained processes of folding, and we have shown that the method of characteristics is a mathematical tool with which we might be able to describe processes of development of discontinuities of various kinds. Now we will identify lines of discontinuity in various kinds of ideal folds. Before proceeding, however, we should clarify certain terms. We restrict the term *characteristic directions* to directions deduced by analysis of partial differential equations that describe a theoretical process. We restrict the term *lines of discontinuity* to the mathematical directions and to directions in diagrams of ideal folds, faults and monoclinal flexures. Finally, we restrict the term *lines of apparent discontinuity* to lines drawn on maps or photographs of natural or experimental folds. We presume, of course, that there are certain correlations between lines of apparent discontinuity and characteristic directions. Nevertheless, one can map lines of apparent discontinuity in the field without reference to a theory for the development of such discontinuities.

Traces of ideal faults provide perhaps the most familiar geological examples of lines of discontinuity and the lines can be correlated with theoreti-

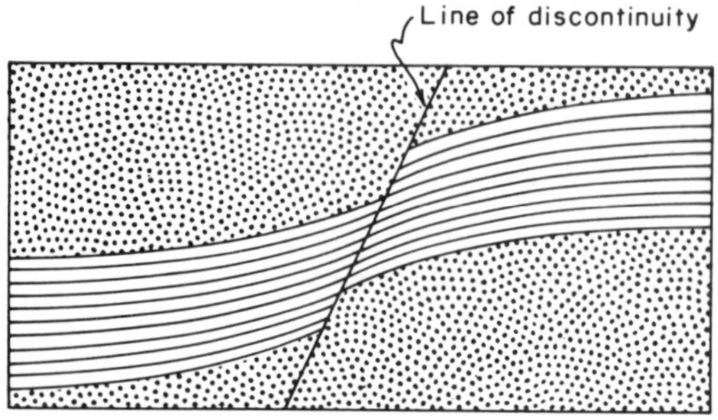

Fig. 10. Ideal fault cutting layers. A fault trace is a line of discontinuity and is analogous to a characteristic direction. Displacements change continuously with distance along layering on each side of the fault plane but they are discontinuous at the fault plane itself.

cal characteristic directions in plastic materials (e.g., Hill, 1950; Odé, 1960). The lines of discontinuity are the fault surfaces, themselves (Fig. 10). Displacement, v, on each side of the fault is continuous up to the fault plane. Across the fault plane, however, displacement is discontinuous and is equal to the offset produced by the fault. Furthermore, additional displacement propagates along the fault plane. Finally, places where active faults cross each other are distorted in a most confusing way and a great deal of energy apparently is dissipated there. Each of these features of ideal and natural faults and fault movements is analogous to the mathematical concept of characteristic directions.

The lines of discontinuity in ideal concentric folds connect the core of a syncline with the cores of adjacent anticlines (Fig. 11). The curvature and radius of curvature of bedding planes are discontinuous across the lines thus defined. If v is displacement of layers normal to bedding planes and x is horizontal distance, parallel to original bedding planes, then the first approximation to the curvature, $\partial^2 v/\partial x^2$, and its inverse, the radius of curvature, are discontinuous across the lines of discontinuity. The radius of curvature is constant for each bedding plane between lines of discontinuity.

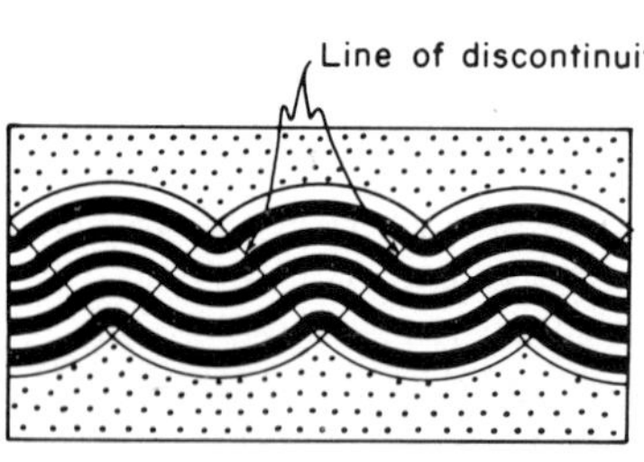

Fig. 11. Lines of discontinuity in ideal concentric folds. Curvatures and radii of curvature of bedding planes are discontinuous across the lines of discontinuity.

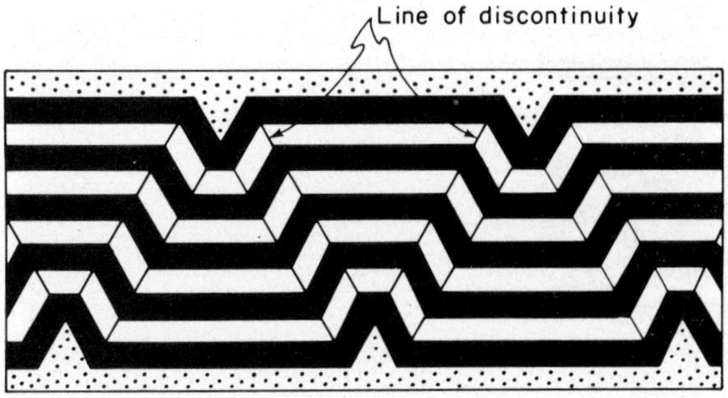

Fig. 12. Lines of discontinuity associated with ideal, conjugate, sharp-kink folds. The slopes or dips of bedding planes are discontinuous.

Thus, for example, the radius of curvature of bedding planes is positive but constant along layers on one side of a line of discontinuity, infinite at the line of discontinuity, and negative and constant along layers on the other side of the line of discontinuity. The layers normally are highly distorted at cores of anticlines and synclines, where the lines of discontinuity intersect and, in the ideal concentric fold (Fig. 11), the slope of one bedding plane is discontinuous there.

The lines of discontinuity in ideal sharp, conjugate, kink folds, such as those shown in Fig. 12, are lines that connect hinges of bends in the layers and bound segments of layers which have been rotated. Thus, the lines of discontinuity are lines across which the slope of layers, $\partial v/\partial x$, is discontinuous. Where the lines of discontinuity in ideal sharp-kink folds intersect, without crossing, the layers are irregularly bent or voids are created.

It is clear that one can recognize at least three types of discontinuity in ideal folds, faults and monoclines. In ideal faults there is a discontinuity in displacement. In sharp-kink folds and monoclines there are discontinuities in slopes of layers. In ideal concentric or parallel folds there are discontinuities in curvature. Finally, in ideal similar folds of sinusoidal form there are no discontinuities at all. Perhaps we could recognize new forms of ideal folds and flexures where change of curvature and change of change of curvature are discontinuous.

The lines of discontinuity in ideal kink and chevron folds are somewhat complicated. The trace of the axial plane of the ideal kink and chevron folds normally is drawn at a line of discontinuity in slope of layering (e.g., see Fig. 12; Paterson and Weiss, 1966; Weiss, 1968; Ellen, 1971 and Cobbold et al., 1971). Sharp hinges require that the layers fail at the hinges; otherwise there would be infinite bending moments at the hinges. Further, close examination of hinges of kink folds in the field and in experiments indicates that the hinges normally are rounded (e.g., see Fig. 4, also, see figures in Weiss, 1968; Ghosh, 1968 and Cobbold et al., 1971). Therefore, as shown in Fig. 13, hinges of ideal conjugate kink folds could be bounded by lines of

Fig. 13. Ideal conjugate, kink folds with rounded hinges which can be associated with lines of gross discontinuity in curvature of neutral planes of the stiff layers.

gross discontinuity in curvature, which relates them to ideal concentric folds. However, the lines of discontinuity in ideal concentric and ideal kink folds are significantly different. The lines of gross discontinuity in the ideal kink folds shown in Fig. 13 parallel the hinge lines but within individual stiff layers they bound arcs of circles in the hinges. It is these latter lines which are the same as the lines of discontinuity in the ideal concentric fold.

Ideal concentric and conjugate kink folds involve conjugate lines of discontinuity but some ideal folds have asymmetric lines of discontinuity in curvature. Fig. 14 shows ideal monoclinal flexures which involve four parallel lines of discontinuity in gross curvature of layers. The figure shows the overburden of an ideal laccolithic intrusion, uplifted along a pair of monoclinal flexures. Monoclinal flexures similar to these are partly preserved in the overburden of the Trachyte Mesa laccolith in the Henry Mountains of Utah (Johnson, 1970, p. 40; Pollard and Johnson, 1973) and of the Shonkin Sag laccolith in the Highwood Mountains of Montana (Hurlbut, 1939; Johnson and Pollard, 1971). The overburdens in the natural occurrences are strongly flexed over the peripheries and are essentially flat-lying over the main bodies of the intrusions. The shapes of the surfaces of the laccoliths have thus far defied theoretical analysis (Johnson and Pollard, 1971).

Possible ideal fold forms that can be correlated with two or three asymmetric directions of discontinuity in curvature are shown in Fig. 15. Fig. 15A shows "asymmetric" concentric folds and Fig. 15B shows possible forms of kink folds where there are three directions of gross discontinuity in curvature, a possible pattern of kink folds where a narrow kink band reflects, perhaps at a stiff boundary, as a broadened kink band. The three directions

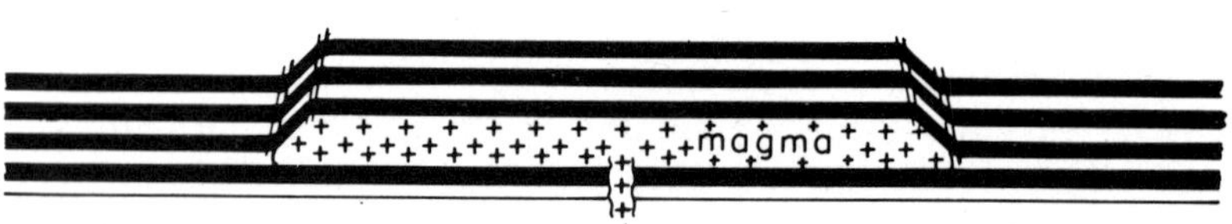

Fig. 14. Monoclinal flexures in the overburden at each end of an ideal laccolithic intrusion.

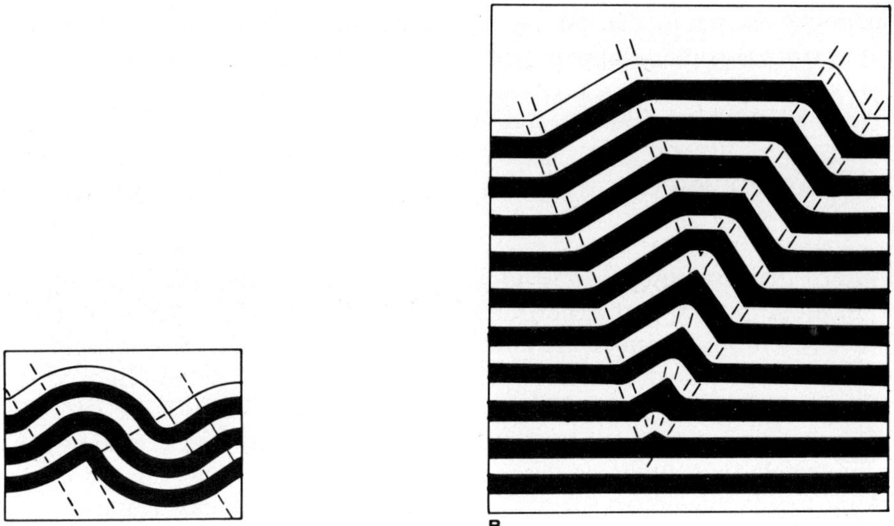

Fig. 15. Ideal fold patterns that might be correlated with asymmetric lines of discontinuity. A. Asymmetric concentric folds. B. Possible fold pattern formed by an area of two and an area of three directions of discontinuity. Pattern depicts the reflection of a narrow kink band as a wider kink band, perhaps at an unusually stiff layer of at a rigid boundary.

Fig. 16. Ideal fold pattern at the terminal region of a kink band propagating diagonally through a multilayered body. There are three directions of discontinuity.

of discontinuity shown in Fig. 16 depict the terminal region of a single kink band propagating through a multilayer. This ideal kink pattern closely resembles sketches of terminal regions of isolated kink bands (e.g., Anderson, 1968, p. 213; Weiss, 1968, p. 309; Ellen, 1971).

LINES OF APPARENT DISCONTINUITY IN NATURAL AND EXPERIMENTAL FOLDS

The lines of discontinuity discussed in the preceding paragraphs can be imagined in most natural and experimental folds and flexures. We have produced forms resembling concentric and kink folds in multilayers consisting of potter's clay and rubber or cardboard (Johnson, 1970, p. 318; Ellen, 1971). Ghosh (1968) and Cobbold et al. (1971) have developed kink and concentric forms in multilayers consisting of modeling clay and grease or graphite, materials that might be described as viscous. The most regular patterns of kink bands have been formed by Weiss (1968) in stacks of paper cards. But experimental folds produced by many other investigators, using widely different materials, resemble ideal kink, concentric or chevron folds (e.g., Willis, 1894; Currie et al., 1962; Turner, 1962; Ramberg, 1963b, 1964; Heard et al., 1965; Donath, 1968 and Roberts and Strömgård, 1972).

Some natural folds

Here we will examine two natural folds and a few experimental folds and illustrate lines of apparent discontinuity in curvature. Fig. 17 shows folds in the Carmel Formation at the Organ, a place in Arches National Monument, Utah (Johnson, 1969). The Carmel Formation is the well-bedded sequence in the center of the view. Below is the relatively massive Navajo Sandstone and above is the relatively massive Entrada Sandstone. The heavy, diagonal lines on the photograph are drawn to represent lines of apparent discontinuity in curvature. The layers of the Carmel shown between the lines are bent into roughly concentric arcs with respect to centers of curvature within the Navajo and the Entrada Sandstones. The folds definitely are not of sinusoidal form because arc lengths of the folds are different at different places within the Carmel Formation. Similar lines of apparent discontinuity in curvature could be readily sketched for the folds shown in Fig. 1.

Fig. 18 shows lines of apparent discontinuity in folds in interbedded shale and chert of the Franciscan Assemblage, near San Francisco, California (Johnson, 1970; Ellen, 1971). The lines of apparent discontinuity are parallel and bound the hinge of the fold. The spacings between the pairs of lines are different at different places and one can imagine that, as the spacing becomes narrower, one could replace the two lines with a single line, representing a discontinuity in the slopes of bedding planes. The hinges of the chert beds are roughly circular arcs and the dashed lines shown in Fig. 18 indicate lines of apparent discontinuity in curvature of individual chert

Fig. 17. Lines of apparent discontinuity in curvature in folds in the Carmel Formation at the Organ, Arches National Monument, Utah. The well-bedded unit at midheight in the view is the Carmel. The dashed lines connect the lines of apparent discontinuity with approximate centers of curvature of layers.

layers. This figure clearly illustrates the difference between lines of gross discontinuity in a sequence of layers and lines of local discontinuity within individual layers.

Some experimental folds

Multilayers with various thickness ratios of soft and stiff, elastic layers were subjected to layer-parallel shortening in order to examine transitions from one fold type to another (Fig. 19). The soft layers were gelatin, with a Poisson's ratio of 0.42 and a Young's modulus ranging from about 10 N/cm^2

Fig. 18. Kink-like folds and lines of apparent discontinuity of curvature of chert beds of the Franciscan Assemblage, near San Francisco, California. Dashed lines in hinge zones show lines of apparent discontinuity within chert beds and solid lines show overall or gross lines of apparent discontinuity in the entire sequence of beds.

to 13 N/cm^2. The stiff layers were rubber strips, with a thickness of 2.2 mm, a Young's modulus of about 500 N/cm^2 and a Poisson's ratio of 0.46. The gelatin was hot when it was poured into the mold containing the rubber strips so that the stiff and soft layers were bonded together, preventing slippage at layer boundaries. The thickness ratio of the multilayers was changed by changing the spacing of the rubber strips. Ten rubber strips were used for each experiment so that the total thickness of the multilayer changed from experiment to experiment. In most experiments the multilayer was bounded

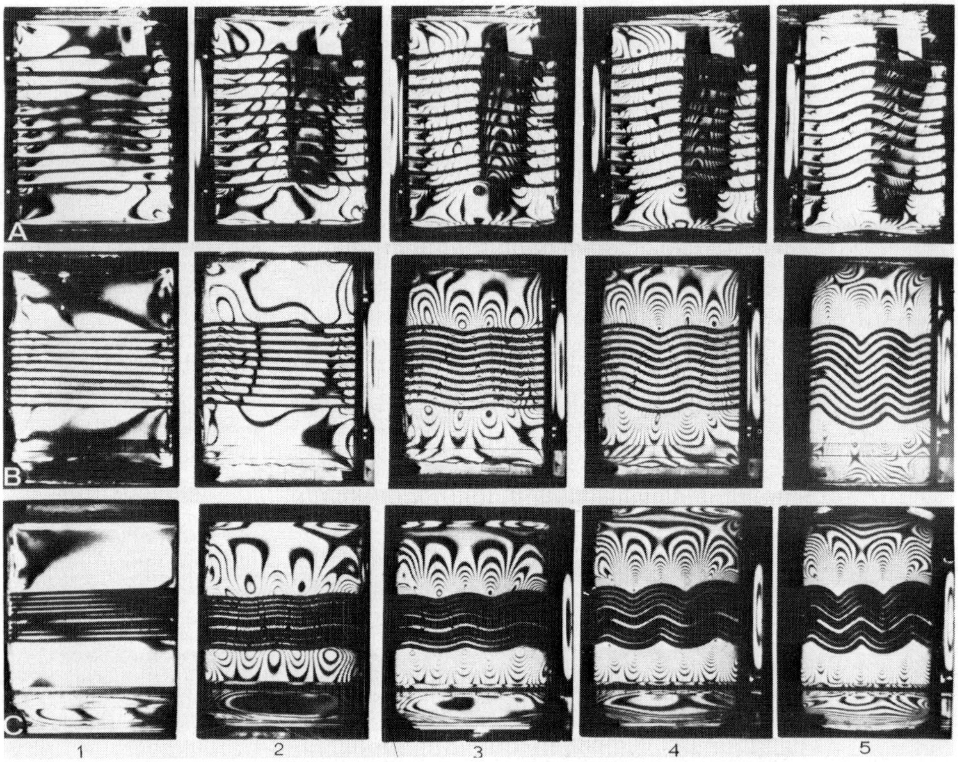

Fig. 19. Experimental fold patterns in multilayers comprised of ten, stiff, black, rubber layers embedded in soft gelatin. Layers loaded axially. Only thickness ratio was varied significantly among the experiments. A. Development of folds as compression increases. Here 1 refers to stiff, rubber layers and 2 refers to soft, gelatin layers. $t_1/t_2 = 0.22$; $E_1/E_2 = 500(\text{N/cm}^2)/10.5(\text{N/cm}^2)$; $\nu_1 = 0.46$ and $\nu_2 = 0.42$. B. Thickness ratio of stiff/soft layers, $t_1/t_2 = 0.44$. Other properties same as in A. C. $t_1/t_2 \simeq 1.0$, $E_1/E_2 = 500(\text{N/cm}^2)/13(\text{N/cm}^2)$.

on each side by thick layers of gelatin so that the multilayer consisted of alternating soft and stiff materials embedded in soft material.

Each multilayer was at room temperature for about 18 hours before it was loaded. Then it was placed in a plexiglass box with five fixed sides and a movable, sixth side, which served as a piston. The thick soft layers were unconfined in the direction normal to layering so that as the multilayer shortened it thickened. The multilayer was shortened by means of two screws, one at each end of the movable side.

Fig. 19 shows fold development in three multilayers subjected to different amounts of axial shortening. The multilayers had ratios of thicknesses of stiff-to-soft layers, t_1/t_2, of 0.22, 0.44 and 1.0 for Fig. 19A, 19B, and 19C, respectively. In each experiment, the folds were first of apparent sinusoidal

Fig. 20. Lines of apparent discontinuity in folds formed in rubber—gelatin multilayer. Same as photographs 4 and 5 of Fig. 19B.

form, as shown in Fig. 19A2, 19B2 and 19C2, but with differing wavelengths, of about 10 cm, 6 cm, and 4 cm, respectively. With further shortening, however, the folds changed to a form closely resembling that of the ideal concentric fold. At large shortening, the layers in the cores of anticlines and synclines assumed chevron forms (Fig. 19B5 and 19C5).

Lines of apparent discontinuity in curvature of layers for two stages of the folding shown in Fig. 19B are shown in Fig. 20A and 20B. The lines are drawn so that they are approximately normal to bedding planes in Fig. 20A and, in early stages of folding, the lines connect inflection points in the stiff, rubber layers. Areas of apparent concentric form diminish markedly as folding progresses and as chevron forms appear and grow in the cores of anticlines and synclines (Fig. 20B). Further, the angle between the conjugate lines of apparent discontinuity increases as folding progresses. The paired, vertical lines of gross discontinuity shown in Fig. 20B appear to be similar to those shown in the natural example in Fig. 18. The vertical lines are oblique to the layers but they bound areas of markedly different curvature and the curvature of stiff layers is essentially constant between pairs of lines and zero elsewhere.

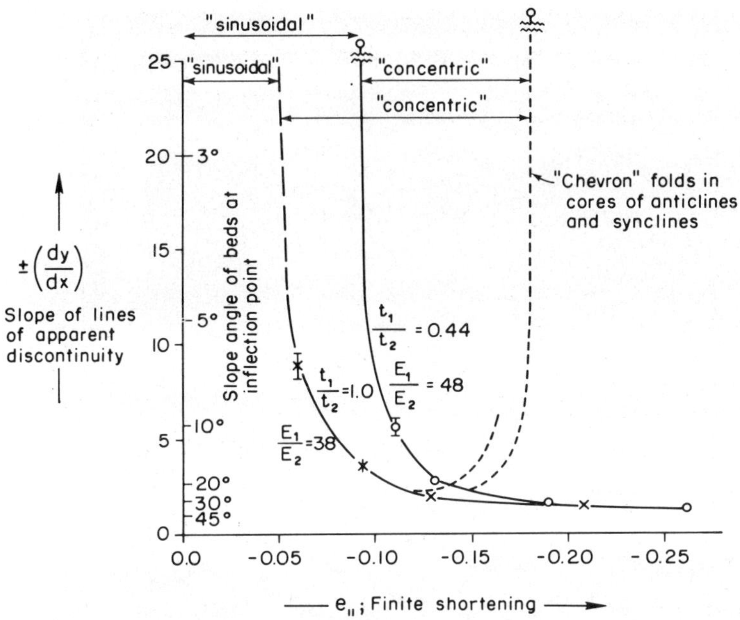

Fig. 21. Relations between slope of lines of apparent discontinuity and overall shortening of multilayered body for experiments shown in Fig. 19B and 19C. Approximate accuracy of measurements of slopes of lines indicated by vertical bars at data points.

The amount of overall shortening parallel to layers was measured for the multilayers shown in Fig. 19B and 19C. Unfortunately, the first photograph for the experiment shown in Fig. 19A was defective so that the initial lengths of the layers is not known accurately enough to measure shortening. The relation between overall shortening and slope of lines of apparent discontinuity, dy/dx, is shown in Fig. 21. The shortening recorded, e_{11}, is the "natural strain" (e.g., Ramsay, 1967):

$$e_{11} = \ln (l'/l)$$

where l is the original length and l' is the deformed length of a specimen. The folding in the multilayer with stiff layers that were 0.44 times the thickness of the soft layers apparently was sinusoidal until the axial shortening reached a magnitude of about −0.09. Then "concentric" folding started, with lines of apparent discontinuity which were essentially vertical. A very small increase in shortening, from −0.09 to −0.11, produced a drastic decrease in the slope of the apparent discontinuities and with further shortening, the slope gradually decreased. The "concentric" pattern in the multilayer with stiff and soft layers of equal thickness, Fig. 19C, began at a smaller value of overall axial shortening, a value of about −0.05.

The short, dashed lines which branch off from each of the solid lines

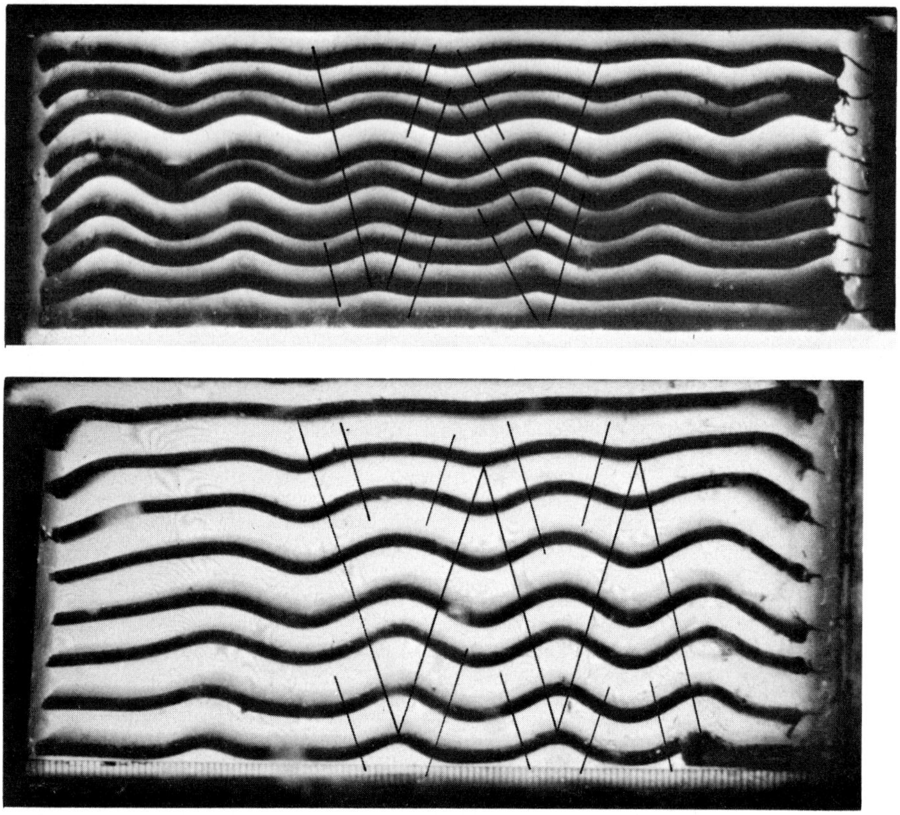

Fig. 22. Experimental fold patterns and lines of apparent discontinuity in multilayers of rubber and gelatin confined above and below by rigid boundaries. Same as multilayers shown in Fig. 19B and 19C, but thick gelatin confining layers have been stripped away. Fold patterns are roughly concentric in middle parts of multilayers. Near edges, however, pattern is more nearly similar to that of kink folds.

shown in Fig. 21 indicate the inception and growth of "chevron" folds in the cores of anticlines and synclines in the two multilayers. In both specimens, "chevron" folds started to develop when the slope of layers at the inflection points was about 30°. The experimental data support Ramberg's theoretical conclusion that chevron folds are a result of relatively high slopes of bedding planes relative to the direction of overall shortening (e.g., Ramberg, 1961, 1963b, and personal communication, 1972).

Another set of experiments was made with the multilayers shown in Fig. 19B and 19C. The thick layers of gelatin on each side of the multilayers were stripped away and the multilayers were placed between aluminum bars which served as a confining medium that could be displaced but not deformed. Then the multilayers were shortened as before.

Fig. 22A and 22B show the two multilayers after they were deformed. As before, the inclined lines represent lines of apparent discontinuity in curvature. The patterns are similar to those in the folds in the Carmel Formation (Fig. 17). The folds near the center of each multilayer might be described as concentric and the arcs there are bounded by symmetric, inclined lines. Near the upper and lower contacts, however, the fold pattern is different. The folds near the boundary most closely resemble incipient kink folds. The layers are straight for relatively large horizontal distances and then the layers are rather sharply bent locally. The short lines on each side of the lines bounding the concentric forms in Fig. 22A and 22B are intended to be lines of discontinuity, bordering incipient kink folds. Perhaps the lines should be paired but this could be determined only if one followed the further development of the folds. The kinds of patterns recognized in Fig. 22A and 22B and the successive development of fold patterns are clearly illustrated in experiments with multilayers of rubber and clay (e.g., Johnson, 1970, p. 315 and 318; Ellen, 1971).

ACKNOWLEDGMENT

The research reported here was financed by the Shell Fund of the Department of Geology, Stanford University, the Penrose Fund of the Geological Society of America, the National Science Foundation, grant no. GA-14899, and the John Simon Guggenheim Memorial Foundation. Some of the research reported here and all the research to be reported in subsequent parts is being financed by the National Science Foundation, grant no. GA-36917. We are grateful to each of these organizations for their generous support. We thank Dr. D. Pollard, University of Rochester, N.Y., and G. McHendrie, Stanford, for criticizing and editing the manuscript. We thank P. Cobbold and M. Biot for pointing out errors in an early form of the manuscript. A. Johnson wishes to thank Professor H. Ramberg and the Institute of Mineralogy, University of Uppsala, Sweden, for providing office and support facilities and an environment conducive to research during the year 1971—1972, when the basic concepts of the theory were formulated. He is grateful to Stanford University for providing him with a sabbatical leave in order to study and do research in Sweden. He thanks Professor H. Ramberg, Dr. P. Chadwick, and Docent O. Stephansson, all at the Institute of Mineralogy, for fruitful discussions of the research. He is especially grateful to his office partner at Uppsala, the late K.-E. Strömgård, for critically reviewing the theory and for help with photoelastic experiments.

The theory to be presented in subsequent parts of this series of papers is the result of a group effort, among E. Honea, G. McHendrie and A. Johnson. Dr. G. Herrmann of the Department of Applied Mechanics at Stanford has been most helpful to us, suggesting fruitful approaches and key references.

REFERENCES

Abbott, M.B., 1966. An Introduction to the Method of Characteristics. American Elsevier, New York, N.Y., 243 pp.
Anderson, T.B., 1968. The geometry of natural orthorhombic systems of kink bands. In: A.J. Baer and D.K. Norris (Editors), Proceedings of the Conference on Research in Tectonics. Geol. Surv. Can. Pap., 68-52: 200—220.
Bahar, L.Y., 1972. Transfer matrix approach to layered systems. Proc. Am. Assoc. Civ. Eng., J. Eng. Mech. Div., EM5: 1159—1172.
Bedford, A. and Stern, M., 1971. Toward a diffusing continuum theory of composite materials. J. Appl. Mech., 38: 8—14.
Biezeno, C.B. and Hencky, H., 1928. On the general theory of elastic stability. Proc. R. Acad., Amsterdam, 31: 569—592.
Billings, M.P., 1954. Structural Geology. Prentice-Hall, Englewood Cliffs, N.J., 514 pp.
Biot, M.A., 1939. Non-linear theory of elasticity and the linearized case for a body under initial stress. Philos. Mag., 28: 468—489.
Biot, M.A., 1957. Folding instability of a layered viscoelastic medium under compression. Proc. R. Soc. London, Ser. A., 242: 444—454.
Biot, M.A., 1960. Instability of a continuously inhomogeneous viscoelastic half-space under initial stress. J. Franklin Inst., 270: 190—201.
Biot, M.A., 1961. Theory of folding of stratified viscoelastic media and its implications in tectonics and orogenesis. Geol. Soc. Am. Bull., 72: 1595—1620.
Biot, M.A., 1963. Internal buckling under initial stress in finite elasticity. Proc. R. Soc. London, Ser. A., 273: 306—328.
Biot, M.A., 1964. Theory of internal buckling of a confined multilayered structure. Geol. Soc. Am. Bull., 75: 563—568.
Biot, M.A., 1965a. Mechanics of Incremental Deformations. Wiley, New York, N.Y., 505 pp.
Biot, M.A., 1965b. Further development of the theory of internal buckling of multilayers. Geol. Soc. Am. Bull., 76: 833—840.
Biot, M.A., 1965c. Theory of viscous buckling and gravity instability of multilayers with large deformation. Geol. Soc. Am. Bull., 76: 371—378.
Biot, M.A. and Odé, H., 1962. On the folding of a viscoelastic medium with adhering layer under compressive initial stress. Appl. Math. Q., 19: 351—355.
Biot, M.A., Odé, H. and Roever, W.L., 1961. Experimental verification of the folding of stratified viscoelastic media. Geol. Soc. Am. Bull., 72: 1621—1630.
Chapple, W.M., 1968. A mathematical theory of finite-amplitude rock folding. Geol. Soc. Am. Bull., 79: 47—68.
Chou, P.C., Carleone, J. and Hsü, C.M., 1972. Elastic constants of layered media. J. Compos. Mater., 6: 80—93.
Cobbold, P.R., Grosgrove, J.W. and Summers, J.M., 1971. Development of internal structures in deformed anisotropic rocks. Tectonophysics, 12: 23—53.
Courant, R., 1962. Methods of Mathematical Physics, Vol. 2. Interscience, New York, N.Y., 830 pp.
Currie, J.B., Patnode, A.W. and Trump, R.P., 1962. Development of folds in sedimentary strata. Geol. Soc. Am. Bull., 73: 461—472.
DeSitter, L.U., 1956. Structural Geology. McGraw-Hill, New York, N.Y., 552 pp.
Dietrich, J.H. and Carter, N.L., 1969. Stress-history of folding. Am. J. Sci., 267: 129—154.
Donath, F.A., 1968. Experimental study of kink-band development in Martinsburg slate. In: A.J. Baer and D.K. Norris (Editors), Proceedings of the Conference on Research in Tectonics. Geol. Surv. Can. Pap., 58-52: 207—249.
Donath, F.A. and Parker, R.D., 1964. Folds and folding. Geol. Soc. Am. Bull., 75: 45—62.

Ellen, S.D., 1971. The Development of Folds in Layered Chert of the Franciscan Assemblage near San Francisco, California. Dissertation, Department of Geology, Stanford University, Stanford, Calif., 330 pp.
Fox, L., 1962. Numerical Solution of Ordinary and Partial Differential Equations. Pergamon, New York, N.Y., 509 pp.
Ghosh, S.K., 1968. Experiments of buckling of multilayers which permit interlayer gliding. Tectonophysics, 6: 207—249.
Green, W.A., 1964. The growth of plane discontinuities propagating into a homogeneously deformed elastic material. Arch. Ration. Mech. Anal., 19: 20—23.
Hall, Sir James, 1815. On the vertical position and convolution of certain strata and their relation with granite. Trans. R. Soc. Edinb., 7: 79—108.
Heard, H.C., Turner, F.J. and Weiss, L.E., 1965. Studies of heterogeneous strain in experimentally deformed calcite, marble, and phyllite. Univ. Calif. Pub. Geol. Sci., 46: 81—152.
Herrmann, G. and Achenbach, J.D., 1967. On dynamic theories of fiber-reinforced composites. Proc. 8th Conf. Structures, Structural Dynamics and Materials, AIAA/ASME, Palm Springs, Calif., 112—118.
Hill, R., 1950. The Mathematical Theory of Plasticity. Clarendon Press, London, 355 pp.
Hudleston, P.J. and Stephansson, O., 1973. Layer shortening and fold-shape development in the buckling of single layers. Tectonophysics, 17: 299—321.
Hurlbut, C.S., 1939. Igneous rocks of the Highwood Mountains, Montana. Bull. Geol. Soc. Am., 50: 1032—1112.
Johnson, A.M., 1969. Development of folds within the Carmel Formation, Arches National Monument, Utah. Tectonophysics, 8: 31—77.
Johnson, A.M., 1970. Physical Processes in Geology. Freeman-Cooper, San Francisco, Calif., 577 pp.
Johnson, A.M. and Pollard, D.D., 1971. Mechanics of Intrusion of Laccoliths. Final report to National Science Foundation, Branner Library, Stanford University, Stanford, Calif., 173 pp.
Johnson,W. and Mellor, P.B., 1962. Plasticity for Mechanical Engineers. Van Nostrand, London, 412 pp.
Kafka, V., 1965. Elastic-plastic deformation of a periodically nonhomogeneous medium. Acta Tech. ČSAV, 4: 404—451.
Kohn, W., Krumhansl, J.A. and Lee, E.H., 1972. Variational methods for dispersion relations and elastic properties of composite materials. J. Appl. Mech., 39: 327—336.
Mindlin, R.D., 1964. Micro-structure in linear elasticity. Arch. Ration. Mech. Anal., 16: 51—78.
Mindlin, R.D. and Medick, M.A., 1959. Extensional vibration of elastic plates. J. Appl. Mech., 26: 561—569.
Novozhilov, V.V., 1953. Foundations of the Nonlinear Theory of Elasticity. Graylock Press, Rochester, N.Y., 233 pp.
Odé, H., 1960. Faulting as a velocity discontinuity in plastic deformation. Geol. Soc. Am., Mem. 79: 293.
Odé, H., 1966. Gravitational Instability of a Multilayered System of High Viscosity. Thesis, University of Utrecht, Utrecht, 96 pp.
Paterson, M.S. and Weiss, L.E., 1966. Experimental deformation and folding in phyllite. Geol. Soc. Am. Bull., 77: 343—374.
Pollard, D.D. and Johnson, A.M., 1973. Mechanics of growth of some laccolithic intrusions in the Henry Mountains, Utah, II. Bending and failure of overburden layers and sill formation. Tectonophysics, 18: 311—354.
Postma, G.W., 1955. Wave propagation in a stratified medium. Geophysics, 20: 780.
Prager, W. and Hodge, P.G., 1951. Perfectly Plastic Solids. Wiley, New York, N.Y., 264 pp.

Ramberg, H., 1959. Evolution of ptygmatic folding. Nor. Geol. Tidsskr., 39: 99—151.
Ramberg, H., 1960. Relationships between arc length and thickness of ptygmatically folded veins. Am. J. Sci., 258: 36—46.
Ramberg, H., 1961. Relationship between concentric longitudinal strain and concentric shearing strain during folding of homogeneous sheets of rock. Am. J. Sci., 259: 382—390.
Ramberg, H., 1963a. Fluid dynamics of viscous buckling applicable to folding of layered rocks. Bull. Am. Ass. Pet. Geol., 47: 484—505.
Ramberg, H., 1963b. Strain distribution and the geometry of folds. Bull. Inst. Geol., Univ. Upps., 42: 1—20.
Ramberg, H., 1964. Selective buckling of composite layers with contrasted rheological properties; a theory for simultaneous formation of several orders of folds. Tectonophysics, 1: 307—341.
Ramberg, H., 1967. Gravity, Deformation and the Earth's Crust. Academic Press, London, 214 pp.
Ramberg, H., 1970. Folding of laterally compressed multilayers in the field of gravity. Phys. Earth Planet. Inter., Part I, 2: 203—232; Part II, 4: 83—120.
Ramberg, H. and Strömgård, K.E., 1971. Experimental tests of modern buckling theory applied on multilayered media. Tectonophysics, 11: 461—472.
Ramsay, J.G., 1967. Folding and Fracturing of Rocks. McGraw-Hill, New York, N.Y., 568 pp.
Roberts, D. and Strömgård, K.E., 1972. A comparison of natural and experimental strain patterns around fold hinge zones. Tectonophysics, 14: 105—120.
Salamon, M.D.G., 1968. Elastic moduli of a stratified rock mass. Int. J. Rock Mech., Min. Sci., 5: 519.
Sherwin, J.A. and Chapple, W.M., 1968. Wavelengths of single-layer folds: A comparison between theory and observation. Am. J. Sci., 266: 167—179.
Smoluchowski, M., 1909. Über ein gewisses Stabilitätsproblem der Elastizitätslehre und der Beziehungen zur Enstehung von Faltenbirgen. Anz. Akad. Wiss. Krakau, Math. Naturw., 2: 3 and 727.
Sokolnikoff, I.S. and Redheffer, R.M., 1959. Mathematics of Physics and Modern Engineering. McGraw-Hill, New York, N.Y., 812 pp.
Sun, C.T., 1971. Microstructure theory of a composite beam. J. Appl. Mech., 38: 947—954.
Sun, C.T., Achenbach, J.D. and Herrmann, G., 1968. Continuum theory for a laminated medium. J. Appl. Mech., 35: 467—475.
Timoshenko, S.P. and Gere, J.M., 1961. Theory of Elastic Stability. McGraw-Hill, New York, N.Y., 541 pp.
Timoshenko, S.P. and Goodier, J.N., 1970. Theory of Elasticity. McGraw-Hill, New York, N.Y., 567 pp.
Tromp, S.W., 1937. On the Mechanism of the Geological Undulation Phenomena in General and of Folding in Particular and their Application to the "Roots of Mountains" Problem. Sijthoff, Leiden, 184 pp.
Truesdell, C., 1952. The mechanical foundations of elasticity and fluid dynamics. J. Ration. Mech. Anal., 1: 125—297.
Truesdell, C. and Noll, W., 1965. The non-linear field theories of mechanics. In: S. Flugge (Editor), Encyclopedia of Physics, Vol. 3, no. 3. Springer, New York, N.Y., 602 pp.
Turner, F.J., 1962. Rotation of crystal lattice in kink bands deformation bands and twin lamellae of strained crystals. Proc. Natl. Acad. Sci., 48: 955—963.
Turner, F.J. and Weiss, L.E., 1965. Deformation kinks in brucite and gypsum. Proc. Natl. Acad. Sci., 54: 359—364.
Van Hise, C.R., 1896. Principles of North American pre-Cambrian geology. 16th Annu. Rep. U.S. Geol. Surv., Part I.

Weiss, L.E., 1968. Flexural slip folding of foliated model materials. In: A.J. Baer and D.K. Norris (Editors), Proc. Conf. Research in Tectonics. Geol. Surv. Can. Pap. 68-52: 294—333.

White, J.E. and Angona, F.A., 1955. Elastic wave velocities in laminated media. J. Acoust. Soc. Am., 27: 311.

Whitten, E.H.T., 1966. Structural Geology of Folded Rocks. Rand-McNally, Chicago, Ill., 663 pp.

Willis, B., 1894. Mechanics of Appalachian structure. 13th Annu. Rep. U.S. Geol. Surv., 1891-2: 213—281.

Willis, B. and Willis, R., 1934. Geological Structures. McGraw-Hill, New York, N.Y., 544 pp.

Chapter 4

INITIAL STRESS AND NONLINEAR EQUATIONS OF EQUILIBRIUM *

SUMMARY

Part of the necessary background for our theory of folding of multilayered elastic materials is a nonlinear theory of equilibrium. The nonlinear theory presented here is for plane-strain conditions and it is a special case of a three-dimensional theory developed by V.V. Novozhilov. The nonlinear theory accounts for finite deformations as well as for finite distortions of bounding surfaces of bodies or elements of bodies. The theory is linearized so that it can be used to describe the initiation of folds but the linearized theory is quite different from the classical linear theory of elasticity. For example, the linearized theory indicates that an elastic material subjected to high uniaxial compression is more easily deformed by simple shear along planes normal than along planes parallel to the direction of compression. Thus, the material behaves as though it were anisotropic with respect to shear, or as though it were a deck of slippery cards layered normal to the direction of axial compression. Further, the compression can become so large that the material has no resistance to shear stress along those planes; the material then behaves much as a plastic except that strains are completely recoverable if the stresses are relaxed. The theory suggests that the problem of estimating rheological properties of rocks at times of folding is even more complicated than we have believed heretofore.

INTRODUCTION

The general plan of this series of papers was outlined in the introduction, Part 1**. Details of the plan have changed somewhat with further research, but we still intend to present a unified theory of sinusoidal, concentric-like and chevron folding and monoclinal flexuring of elastic multilayers in succeeding

---

* Originally published as: Johnson, A.M. and Honea, E., 1975. Part II, Initial stress and nonlinear equations of equilibrium. Tectonophysics, 25: 261—280.

** "Part I" will be used to abbreviate the reference: Johnson and Ellen, 1974. It is Chapter 3 of this book.

papers. This second part will be concerned exclusively with the basic equilibrium equations for elastic bodies. The results are required for all our analyses of folding and should prove to be useful for other problems as well.

The nonlinear theory of equilibrium of elastic bodies deformed three-dimensionally has been developed by several investigators (e.g., Green and Adkins, 1970; Green and Zerna, 1968; Toupin and Bernstein, 1961; Truesdell and Noll, I, 1965*; and Novozhilov, I, 1953) but the theory is quite difficult to follow. The derivations in following pages are for two-dimensional, plane-strain conditions, and we hope that they clarify for the reader as they have for us the fundamental aspects of the nonlinear theory. The power of the nonlinear theory is that it considers large deformations, as well as nonlinear, stress—strain relations, and that it separates these two sources of nonlinearity so that their. distinct effects can be recognized, at least theoretically. Analysis of the theory suggests that some behaviors of materials which may appear to be inelastic can be considered to be manifestations of nonlinear elasticity (Truesdell and Noll, I, 1965). Also, Novozhilov (I, 1953) indicates that elastic—plastic behaviors can be closely approximated by the theory. Thus, the theory should find wide application in geology, where most processes of deformation appear to be nonlinear.

NONLINEAR EQUATIONS OF EQUILIBRIUM

We mentioned in Part I of this series of papers that all theories of folding with which we are familiar incorporate nonlinearities of various types. The nonlinearities have two quite different origins. The stress—strain or stress/strain-rate relations may be nonlinear. Or, the strains and rotations may be nonlinear (finite). Of course, both may be nonlinear. Nonlinear strains and rotations are particularly interesting because they cause the basic equilibrium equations to be nonlinear, even for materials with linear rheological properties.

We begin our analysis with consideration of the effects of finite strains and rotations on the form of the exact equilibrium equations for plane strain of an elastic body. The results of this first part of our analysis for plane strain reduce to a special case of equations derived by Novozhilov (I, 1953), who considered three-dimensional deformation.

Let us consider the equilibrium of stresses acting on an infinitesimal, rectangular element, *efgh*, in its final, *deformed state* (Fig. 1A). Before deformation the sides of the rectangle had the diamond-shaped configuration, $\tilde{e}\tilde{f}\tilde{g}\tilde{h}$, also shown in Fig. 1A. The usual equations of equilibrium apply to the deformed state so that:

---

* References cited in Part I will be noted with an "I", followed by the date, such as I, 1955, and not be repeated in the References at the end of this paper.

NOTATION

Definition of symbols commonly used in text

Symbol	Meaning
A, B	Elasticity coefficients
G	Shear modulus for linear elastic material
λ	Lamé elasticity modulus for linear, elastic material
x, y	Cartesian coordinates in undeformed body
$\tilde{x}$, $\tilde{y}$	Directions of curvilinear, deformed lines that were originally parallel to the x- and y-coordinates
u, v	Components of total displacement in x- and y-directions, respectively
u_0, v_0	Components of incremental displacement
p	Mean incremental stress
$s_{\tilde{x}\tilde{x}}$, $s_{\tilde{x}\tilde{y}}$, $s_{\tilde{y}\tilde{y}}$	Components of incremental stresses referred to curvilinear coordinates
ε_{xx}, ε_{xy}, ε_{yy}	Components of finite strains referred to curvilinear coordinates
$E_{\tilde{x}\tilde{x}}$, $E_{\tilde{y}\tilde{y}}$	Finite extensions referred to curvilinear coordinates
$\sigma_{\tilde{x}\tilde{x}}$, $\sigma_{\tilde{x}\tilde{y}}$, $\sigma_{\tilde{y}\tilde{y}}$	Total stresses referred to curvilinear coordinates
ξ, η	Cartesian coordinates in finally deformed body
$\sigma_{\xi\xi}$, $\sigma_{\xi\eta}$, $\sigma_{\eta\eta}$	Total stresses referred to finally deformed body
X, Y	Cartesian coordinates in body before initial deformations
$\tilde{X}$, $\tilde{Y}$	Curvilinear coordinates in initially deformed body
U, V	Components of initial displacement in X- and Y-directions, respectively
$S_{\tilde{x}\tilde{x}}$, $S_{\tilde{x}\tilde{y}}$, $S_{\tilde{y}\tilde{y}}$	Components of initial stresses referred to curvilinear coordinates after initial and incremental deformations

$$\frac{\partial \sigma_{\xi\xi}}{\partial \xi} + \frac{\partial \sigma_{\eta\xi}}{\partial \eta} + F_{\xi} = 0$$

$$\frac{\partial \sigma_{\xi\eta}}{\partial \xi} + \frac{\partial \sigma_{\eta\eta}}{\partial \eta} + F_{\eta} = 0 \qquad (1)$$

$$\sigma_{\xi\eta} = \sigma_{\eta\xi}$$

where F is the body force in the deformed state. The third of eqs. 1 is required for equilibrium of moments. The ξ- and η-coordinates are parallel to the original x- and y-coordinates. The equilibrium equations, therefore, are identical in appearance to those derived by means of the elementary theory (e.g., Johnson, I, 1970, p. 252). The differences are that eqs. 1 are valid for finite deformations and rotations and that they refer to Cartesian coordinates, ξ, η, in the deformed body, whereas the elementary theory requires that the deformations and rotations be infinitesimal and it refers to Cartesian coordinates, x, y, in the original body, before deformation. Our use of the ξ and η coordinates follows that of Novozhilov (I, 1953) and Biot (I, 1965a). Unfortunately, ξ and η normally represent elliptical coordinates in linear elasticity theory (e.g., Timoshenko and Goodier, I, 1970).

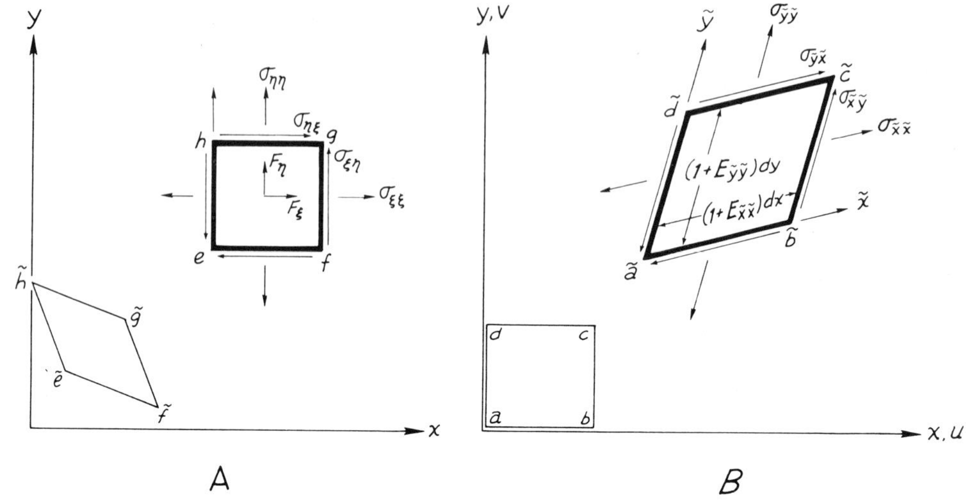

Fig. 1. Components of stress referred to different coordinate systems in a deformed body. A. ξ and η are Cartesian coordinates in a deformed body. Before deformation, the rectangular element, *efgh*, was the diamond-shaped element, $\tilde{e}\tilde{f}\tilde{g}\tilde{h}$. The stress components, $\sigma_{\xi\xi}...\sigma_{\eta\xi}$ act in the deformed body. B. $\tilde{x}$ and $\tilde{y}$ are curvilinear coordinates in the deformed body. $\widetilde{ab}$, for example, represents a line that was parallel to the x-direction before deformation, when it was the line ab. Stress components, $\sigma_{\tilde{x}\tilde{x}}$ and $\sigma_{\tilde{y}\tilde{x}}$, act in the $\tilde{x}$-direction and $\sigma_{\tilde{y}\tilde{y}}$ and $\sigma_{\tilde{x}\tilde{y}}$ act in the $\tilde{y}$-direction. $\sigma_{\tilde{x}\tilde{x}}$, for example, is not a normal stress because it is inclined relative to the face, $\widetilde{bc}$, on which it acts.

Equilibrium equations 1 will be transformed to different coordinate systems than ξ and η for two reasons. First, the stresses and strains need to be related; the components of finite strain, in terms of displacements, u and v, relative to the original coordinates, x and y (e.g., Love, 1944, p. 60), are:

$$\varepsilon_{xx} = (\partial u/\partial x) + (1/2)[(\partial u/\partial x)^2 + (\partial v/\partial x)^2]$$

$$\varepsilon_{yy} = (\partial v/\partial y) + (1/2)[(\partial u/\partial y)^2 + (\partial v/\partial y)^2] \tag{2}$$

$$\varepsilon_{xy} = (1/2)[(\partial v/\partial x) + (\partial u/\partial y) + (\partial u/\partial x)(\partial u/\partial y) + (\partial v/\partial x)(\partial v/\partial y)]$$

Second, according to the nonlinear theory, the stress—strain relations are in terms of stress components that act parallel to curvilinear coordinates, $\tilde{x}$, $\tilde{y}$, which, before deformation, were parallel to the x- and y-coordinates (Figs. 1A and B).

Thus we will first transform the derivatives in equilibrium equations 1 to derivatives relative to the original coordinates x and y. The first of eqs. 1 can be written with the chain rule for partial derivatives as:

$$(\partial\sigma_{\xi\xi}/\partial x)(\partial x/\partial\xi) + (\partial\sigma_{\xi\xi}/\partial y)(\partial y/\partial\xi) + (\partial\sigma_{\eta\xi}/\partial x)(\partial x/\partial\eta) + $$

$$+ (\partial\sigma_{\eta\xi}/\partial y)(\partial y/\partial\eta) + F_\xi = 0 \tag{3}$$

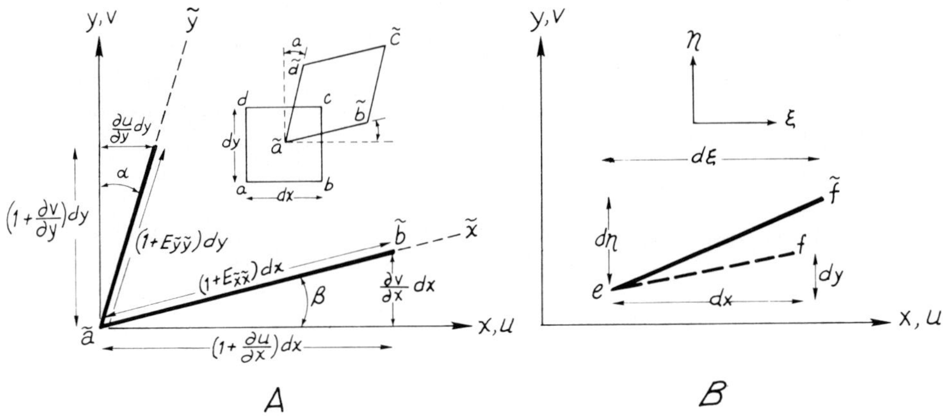

Fig. 2. Deformations of line elements. A. Definitions of finite extensions of line elements which originally were parallel to the x- and y-directions. $E_{\tilde{y}\tilde{y}}$ is the finite extension of side da to side $\tilde{d}\tilde{a}$. $E_{\tilde{x}\tilde{x}}$ is the finite extension of side ab to side $\tilde{a}\tilde{b}$. B. Relations between projections, dx, dy, of original, inclined line element, ef, and projections, $d\xi$, $d\eta$, of final, deformed line element $e\tilde{f}$.

and a similar expression for the second of eqs. 1. In order to derive expressions for the partial derivatives, $\partial x/\partial \xi ... \partial y/\partial \eta$, in eq. 3 we digress briefly to consider finite extensions.

A small element, $abcd$ (Fig. 2A), with sides of length dx and dy is subjected to strain, rotation and translation so that it becomes the parallelogram, $\tilde{a}\tilde{b}\tilde{c}\tilde{d}$. The original side ab of length dx becomes elongated to:

$$\tilde{a}\tilde{b} = (1 + E_{\tilde{x}\tilde{x}})\mathrm{d}x \tag{4a}$$

where $(1 + E_{\tilde{x}\tilde{x}})^2 = [1 + (\partial u/\partial x)]^2 + (\partial v/\partial x)^2$ as shown in Fig. 2A. Similarly, side ad becomes elongated to:

$$\tilde{a}\tilde{d} = (1 + E_{\tilde{y}\tilde{y}})\mathrm{d}y \tag{4b}$$

where $(1 + E_{\tilde{y}\tilde{y}})^2 = [1 + (\partial v/\partial y)]^2 + (\partial u/\partial y)^2$. Here $E_{\tilde{x}\tilde{x}}$ and $E_{\tilde{y}\tilde{y}}$ are the finite extensions of line elements which originally were parallel to the x- and y-directions, respectively. The subscripts of the symbols for finite extensions denote curvilinear axes, $\tilde{x}$ and $\tilde{y}$, instead of Cartesian axes, x and y, to emphasize that lines are reoriented by the deformations. Comparison of eqs. 2 and 4 (a and b) indicates that the finite extensions and finite strains are related by:

$$\begin{aligned}(1 + E_{\tilde{x}\tilde{x}})^2 &= 1 + 2\varepsilon_{xx}\\ (1 + E_{\tilde{y}\tilde{y}})^2 &= 1 + 2\varepsilon_{yy}\end{aligned} \tag{5}$$

Now we can consider the transformation between the Cartesian x- and y-axes before deformation and the Cartesian ξ- and η-axes after deformation. A line element, ef (Fig. 2B), is inclined to the x- and y-axes before deformation and rotation. Its projection onto the original x- and y-axes are dx and dy, respectively. When the body containing the line element is deformed and rotated, line ef becomes line $e\tilde{f}$, the projections of which are, on the ξ- and η-axes:

$$\begin{aligned} d\xi &= \left(1 + \frac{\partial u}{\partial x}\right)dx + (\partial u/\partial y)dy \\ d\eta &= (\partial v/\partial x)dx + \left(1 + \frac{\partial v}{\partial y}\right)dy \end{aligned} \tag{6}$$

Solving eqs. 6 simultaneously for dx and dy:

$$\begin{aligned} dx &= (1/D)\left[\left(1 + \frac{\partial v}{\partial y}\right)d\xi - (\partial u/\partial y)d\eta\right] \\ dy &= (1/D)\left[\left(1 + \frac{\partial u}{\partial x}\right)d\eta - (\partial v/\partial x)d\xi\right] \end{aligned} \tag{7}$$

where:

$$D = \left(1 + \frac{\partial u}{\partial x}\right)\left(1 + \frac{\partial v}{\partial y}\right) - (\partial u/\partial y)(\partial v/\partial x) \tag{8}$$

is one plus the volumetric strain.

Equations 7 are of the form (from the chain rule):

$$\begin{aligned} dx &= (\partial x/\partial \xi)d\xi + (\partial x/\partial \eta)d\eta \\ dy &= (\partial y/\partial \eta)d\eta + (\partial y/\partial \xi)d\xi \end{aligned} \tag{9}$$

and comparison of eqs. 7 and 9 indicates that the partial derivatives are:

$$\begin{aligned} (\partial x/\partial \xi) &= (1/D)\left(1 + \frac{\partial v}{\partial y}\right) \\ (\partial x/\partial \eta) &= -(1/D)(\partial u/\partial y) \\ (\partial y/\partial \xi) &= -(1/D)(\partial v/\partial x) \\ (\partial y/\partial \eta) &= (1/D)\left(1 + \frac{\partial u}{\partial x}\right) \end{aligned} \tag{10}$$

Now we have the expressions required to transform eq. 3, which becomes, if we factor derivatives:

$$(\partial/\partial x)\left[\sigma_{\xi\xi}\left(1+\frac{\partial v}{\partial y}\right)-\sigma_{\eta\xi}(\partial u/\partial y)\right]+(\partial/\partial y)\left[\sigma_{\eta\xi}\left(1+\frac{\partial u}{\partial x}\right)-\sigma_{\xi\xi}(\partial v/\partial x)\right]+$$

$$+\sigma_{\xi\xi}\left\{(\partial/\partial y)(\partial v/\partial x)-(\partial/\partial x)\left(1+\frac{\partial v}{\partial y}\right)\right\}+$$

$$+\sigma_{\eta\xi}\left\{(\partial/\partial x)(\partial u/\partial y)-(\partial/\partial y)\left(1+\frac{\partial u}{\partial x}\right)\right\}+DF_{\xi}=0 \qquad (11)$$

The terms in braces are zero so that eq. 11 reduces to:

$$\frac{\partial}{\partial x}\left[\sigma_{\xi\xi}\left(1+\frac{\partial v}{\partial y}\right)-\sigma_{\eta\xi}(\partial u/\partial y)\right]+\frac{\partial}{\partial y}\left[\sigma_{\eta\xi}\left(1+\frac{\partial u}{\partial x}\right)-\sigma_{\xi\xi}(\partial v/\partial x)\right]+DF_{\xi}=0 \qquad (12a)$$

Similarly, the second of eqs. 1 becomes:

$$\frac{\partial}{\partial x}\left[\sigma_{\xi\eta}\left(1+\frac{\partial v}{\partial y}\right)-\sigma_{\eta\eta}(\partial u/\partial y)\right]+\frac{\partial}{\partial y}\left[\sigma_{\eta\eta}\left(1+\frac{\partial u}{\partial x}\right)-\sigma_{\xi\eta}(\partial v/\partial x)\right]+DF_{\eta}=0 \qquad (12b)$$

Further, moment equilibrium requires that:

$$\sigma_{\xi\eta}=\sigma_{\eta\xi}$$

Equations 12 are accurate equations of equilibrium for stress components referred to Cartesian coordinates in the material in the deformed and rotated state, but the changes are referred to Cartesian coordinates, x and y, in the material in the undeformed state. These equations, clearly, are significantly different from those of the elementary theory of equilibrium.

The next step in the derivation is to replace the stress components, $\sigma_{\xi\xi}...\sigma_{\xi\eta}$, with stress components referred to the curvilinear, $\tilde{x}$- and $\tilde{y}$-axes. The reason for this peculiar transformation is that the finite strains, $\varepsilon_{xx}...\varepsilon_{xy}$, are directly related to the stress components, $\sigma_{\tilde{x}\tilde{x}}...\sigma_{\tilde{x}\tilde{y}}$ (Fig. 1B), by means of a strain-ener-

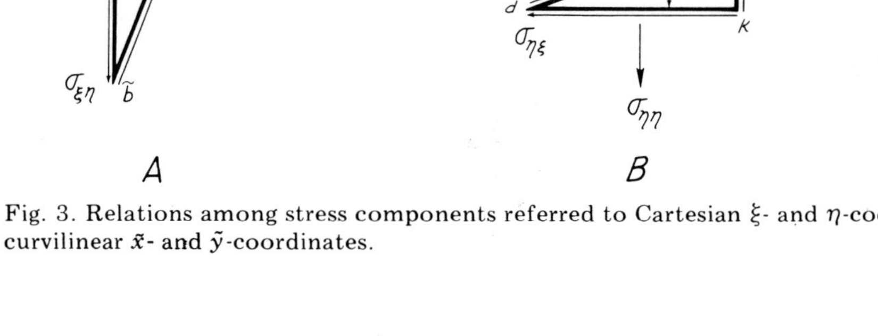

Fig. 3. Relations among stress components referred to Cartesian ξ- and η-coordinates and curvilinear $\tilde{x}$- and $\tilde{y}$-coordinates.

gy function (Novozhilov, I, 1953, p. 110). In effect, the strains, $\varepsilon_{xx}...\varepsilon_{xy}$, are caused by the stresses, $\sigma_{\tilde{x}\tilde{x}}...\sigma_{\tilde{x}\tilde{y}}$. The stress components are not orthogonal in general (Fig. 1B) because the direction of components $\sigma_{\tilde{x}\tilde{x}}$ and $\sigma_{\tilde{y}\tilde{x}}$ is parallel to $\tilde{x}$ and the direction of $\sigma_{\tilde{y}\tilde{y}}$ and $\sigma_{\tilde{x}\tilde{y}}$ is parallel to $\tilde{y}$, and $\tilde{x}$ and $\tilde{y}$ generally are not orthogonal. *

In order to replace the stress components, $\sigma_{\xi\xi}...\sigma_{\eta\xi}$, with the stress components, $\sigma_{\tilde{x}\tilde{x}}...\sigma_{\tilde{y}\tilde{y}}$, we refer to Figs. 3, which are triangular segments of the parallelogram shown in Fig. 1B. Two sides of each triangle are parallel to the ξ- and η-coordinates and the other side is parallel to $\tilde{x}$ or $\tilde{y}$. The angle between the η- and $\tilde{y}$-coordinates is α and the angle between the ξ- and $\tilde{x}$-coordinates is β (Figs. 2A and 3).

Summing forces in the ξ-direction shown in Fig. 3A:

$$\sigma_{\xi\xi}j\tilde{b} - \sigma_{\eta\xi}j\tilde{c} - \sigma_{\tilde{x}\tilde{x}}\tilde{b}\tilde{c}\cos(\beta) - \sigma_{\tilde{x}\tilde{y}}\tilde{b}\tilde{c}\sin(\alpha) = 0 \tag{13}$$

However:

$$j\tilde{b} = \tilde{b}\tilde{c}\cos(\alpha) \quad \text{and} \quad j\tilde{c} = \tilde{b}\tilde{c}\sin(\alpha)$$

so that eq. 13 becomes:

$$\sigma_{\xi\xi}\cos(\alpha) - \sigma_{\eta\xi}\sin(\alpha) = \sigma_{\tilde{x}\tilde{x}}\cos(\beta) + \sigma_{\tilde{x}\tilde{y}}\sin(\alpha) \tag{14}$$

The trigonometric functions of α and β can be derived by inspection of Fig. 2A and they are:

$$\begin{aligned}
\sin(\alpha) &= (\partial u/\partial y)/(1 + E_{\tilde{y}\tilde{y}}) \\
\cos(\alpha) &= \left(1 + \frac{\partial v}{\partial y}\right)/(1 + E_{\tilde{y}\tilde{y}}) \\
\sin(\beta) &= (\partial v/\partial x)/(1 + E_{\tilde{x}\tilde{x}}) \\
\cos(\beta) &= \left(1 + \frac{\partial u}{\partial x}\right)/(1 + E_{\tilde{x}\tilde{x}})
\end{aligned} \tag{15}$$

Thus, eq. 14 becomes:

$$\sigma_{\xi\xi}\left(1 + \frac{\partial v}{\partial y}\right) - \sigma_{\eta\xi}(\partial u/\partial y) = \sigma_{\tilde{x}\tilde{x}}\left[\frac{(1 + E_{\tilde{y}\tilde{y}})}{(1 + E_{\tilde{x}\tilde{x}})}\right]\left(1 + \frac{\partial u}{\partial x}\right) + $$

$$+ \sigma_{\tilde{x}\tilde{y}}\left[\frac{(1 + E_{\tilde{y}\tilde{y}})}{(1 + E_{\tilde{y}\tilde{y}})}\right](\partial u/\partial y) \tag{16}$$

The left-hand side of eq. 16 is the same as the first term in brackets in the first of eqs. 12. The other terms in brackets can be derived similarly by summing forces in the ξ- and η-directions shown in Figs. 3. Accordingly, eqs. 12 be-

* The analysis presented in Chapter 10 is easier to follow.

come replaced by:

$$\frac{\partial}{\partial x}\left[\sigma_{\bar{x}\bar{x}}\left(\frac{1+E_{\bar{y}\bar{y}}}{1+E_{\bar{x}\bar{x}}}\right)\left(1+\frac{\partial u}{\partial x}\right)+\sigma_{\bar{x}\bar{y}}(\partial u/\partial y)\right]+$$

$$\frac{\partial}{\partial y}\left[\sigma_{\bar{y}\bar{y}}\left(\frac{1+E_{\bar{x}\bar{x}}}{1+E_{\bar{y}\bar{y}}}\right)(\partial u/\partial y)+\sigma_{\bar{y}\bar{x}}\left(1+\frac{\partial u}{\partial x}\right)\right]+DF_{\xi}=0$$

$$\frac{\partial}{\partial x}\left[\sigma_{\bar{x}\bar{y}}\left(1+\frac{\partial v}{\partial y}\right)+\sigma_{\bar{x}\bar{x}}\left(\frac{1+E_{\bar{y}\bar{y}}}{1+E_{\bar{x}\bar{x}}}\right)(\partial v/\partial x)\right]+ \tag{17}$$

$$\frac{\partial}{\partial y}\left[\sigma_{\bar{y}\bar{y}}\left(\frac{1+E_{\bar{x}\bar{x}}}{1+E_{\bar{y}\bar{y}}}\right)\left(1+\frac{\partial v}{\partial y}\right)+\sigma_{\bar{y}\bar{x}}(\partial v/\partial x)\right]+DF_{\eta}=0$$

$$\sigma_{\bar{x}\bar{y}}=\sigma_{\bar{y}\bar{x}}$$

These are the basic, nonlinear equilibrium equations. They are complex, obviously, but they incorporate only part of the complexities one can introduce into the general nonlinear theory, which merely assumes that the material is isotropic and that strains are not permanent; the stress—strain relations themselves may be nonlinear also. In fact they normally must be if strains are finite. According to Novozhilov (I, 1953, p. 113), only three parameters are required to completely describe the elastic properties of nonlinear elastic materials; in linear elasticity theory there are two; so, in this respect, the nonlinear theory is not much more complex than the linear theory. Further, eqs. 17 can be simplified somewhat because, as shown by Novozhilov (I, 1953, p. 109), the "stresses", $\sigma^*_{\bar{x}\bar{x}}$ and $\sigma^*_{\bar{y}\bar{y}}$, which are similar to Treftz stresses (Goodier, 1950, p. 384), are required in the stress—strain relations, where:

$$\sigma^*_{\bar{x}\bar{x}}=\sigma_{\bar{x}\bar{x}}(1+E_{\bar{y}\bar{y}})/(1+E_{\bar{x}\bar{x}})$$

$$\sigma^*_{\bar{y}\bar{y}}=\sigma_{\bar{y}\bar{y}}(1+E_{\bar{x}\bar{x}})/(1+E_{\bar{y}\bar{y}}) \tag{18}$$

so that eqs. 17 simplify to:

$$\frac{\partial}{\partial x}[\sigma^*_{\bar{x}\bar{x}}\{1+(\partial u/\partial x)\}+\sigma_{\bar{x}\bar{y}}(\partial u/\partial y)]+\frac{\partial}{\partial y}[\sigma^*_{\bar{y}\bar{y}}(\partial u/\partial y)$$

$$+\sigma_{\bar{y}\bar{x}}\{1+(\partial u/\partial x)\}]+DF_{\xi}=0 \tag{19}$$

$$\frac{\partial}{\partial x}[\sigma_{\bar{x}\bar{y}}\{1+(\partial v/\partial y)\}+\sigma^*_{\bar{x}\bar{x}}(\partial v/\partial x)]+\frac{\partial}{\partial y}[\sigma^*_{\bar{y}\bar{y}}\{1+(\partial v/\partial y)\}+$$

$$+\sigma_{\bar{y}\bar{x}}(\partial v/\partial x)]+DF_{\eta}=0$$

If we were going to solve problems with the complete nonlinear theory we would use eqs. 19. In the following pages, however, we will linearize the equilibrium equations and for this purpose we will use eqs. 17 because they are more explicit.

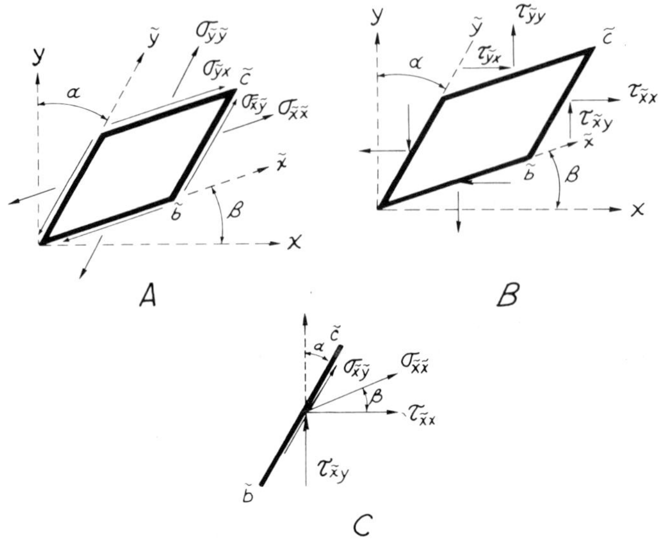

Fig. 4. Components of boundary stresses. A. Components of surface stresses on deformed element, resolved parallel to curvilinear $\tilde{x}$- and $\tilde{y}$-coordinates. B. Components of boundary stresses on same faces of deformed element. The faces originally were parallel to y- or x-coordinate. Boundary stress components are resolved parallel to x- or y-coordinate. C. Relations between surface stresses and boundary stresses acting on one face of element.

STRESS BOUNDARY CONDITIONS

The stresses acting on boundaries, where surface loads but not displacements are prescribed, according to the nonlinear theory, are also different from those derived with the linear theory, because the nonlinear theory accounts for finite distortions of boundary surfaces. In order to derive boundary conditions let us consider face $\tilde{b}\tilde{c}$ of the deformed element shown in Fig. 4A. The face was originally parallel to the y-direction. The *boundary stresses* are designated $\tau_{\tilde{x}y}$ and $\tau_{\tilde{x}x}$ on face $\tilde{b}\tilde{c}$ (Fig. 4B). Novozhilov refers to similar boundary stresses on surfaces of arbitrary orientation as "specific" forces (I, 1953, p. 139); they have dimensions of force per unit of area, the area being that in the *deformed* state. The first subscript, $\tilde{x}$, indicates the face on which the boundary stress acts and the second subscript refers to the direction of the vector, as is typical of stress notation. Summing forces on face $\tilde{b}\tilde{c}$ (Fig. 4C):

$$\tau_{\tilde{x}x} = \sigma_{\tilde{x}\tilde{x}}\cos(\beta) + \sigma_{\tilde{x}\tilde{y}}\sin(\alpha)$$

$$\tau_{\tilde{x}y} = \sigma_{\tilde{x}\tilde{y}}\cos(\alpha) + \sigma_{\tilde{x}\tilde{x}}\sin(\beta)$$

or in terms of eqs. 15 for the circular functions:

$$\tau_{\tilde{x}x} = \sigma_{\tilde{x}\tilde{x}}[\{1 + (\partial u/\partial x)\}/(1 + E_{\tilde{x}\tilde{x}})] + \sigma_{\tilde{x}\tilde{y}}[(\partial u/\partial y)/(1 + E_{\tilde{y}\tilde{y}})]$$

$$\tau_{\tilde{x}y} = \sigma_{\tilde{x}\tilde{x}}[(\partial v/\partial x)/(1 + E_{\tilde{x}\tilde{x}})] + \sigma_{\tilde{x}\tilde{y}}[\{1 + (\partial v/\partial y)\}/(1 + E_{\tilde{y}\tilde{y}})] \qquad (20)$$

similarly:

$$\tau_{\bar{y}x} = \sigma_{\bar{y}\bar{x}}[\{1 + (\partial u/\partial x)\}/(1 + E_{\bar{x}\bar{x}})] + \sigma_{\bar{y}\bar{y}}[(\partial u/\partial y)/(1 + E_{\bar{y}\bar{y}})]$$

$$\tau_{\bar{y}y} = \sigma_{\bar{y}\bar{x}}[(\partial v/\partial x)/(1 + E_{\bar{x}\bar{x}})] + \sigma_{\bar{y}\bar{y}}[\{1 + (\partial v/\partial y)\}/(1 + E_{\bar{y}\bar{y}})] \qquad (20)$$

These are the stress boundary conditions we will use. For a more thorough discussion of boundary conditions see Novozhilov (I, 1953, p. 113 and 138).

INITIAL AND INCREMENTAL STRESSES

The equilibrium equations derived thus far, eqs. 17, are quite accurate and we could use them directly by substituting into them appropriate stress—strain relations for a material of interest and by deriving solutions for the resulting equations analytically or by approximation methods. For two reasons, however, we wish to modify them by separating incremental and initial deformations. In problems involving nonlinear rheological properties it is convenient to solve simplified problems using equations derived by separating small, linear incremental deformation from nonlinear, initial deformation. Many problems involving rock deformation fall into this class. Second, in problems involving materials with linear properties subjected to moderate deformations it is most convenient to separate small, inhomogeneous, incremental deformations from homogeneous, initial deformations. Many stability problems such as those we shall consider in subsequent parts of this series fall into this class.

For example, suppose that a material strains nonlinearly in response to stress so that its stress—strain relation is, appoximately:

$$S_{\bar{x}\bar{y}} = A(\varepsilon_{xy})^{1/n} \qquad (21)$$

where n and A are constants, S is initial stress and ε is initial strain (Fig. 5). Now, suppose that a small or incremental, shear stress, $s_{\bar{x}\bar{y}}$, is added to the initial shear stress, $S_{\bar{x}\bar{y}}$, resulting in an incremental shear strain, ϵ_{xy} (Fig. 5). If the incremental stresses and strains are sufficiently small, the incremental stress—strain relation is essentially linear so that to a close approximation:

$$s_{\bar{x}\bar{y}} = G\epsilon_{xy} \qquad (22)$$

Further, the shear modulus, G, is the local slope * of the finite stress—strain curve and it can be related to the magnitude of the initial shear stress by differentiating eq. 21:

$$G = \partial S_{\bar{x}\bar{y}}/\partial \varepsilon_{xy}$$

The ability to solve problems for materials with nonlinear stress—strain relations such as that shown in Fig. 5 is important in geology because many natural materials display marked rheological nonlinearities.

---

* Incorrect. See Chapter 10, eq. 34g.

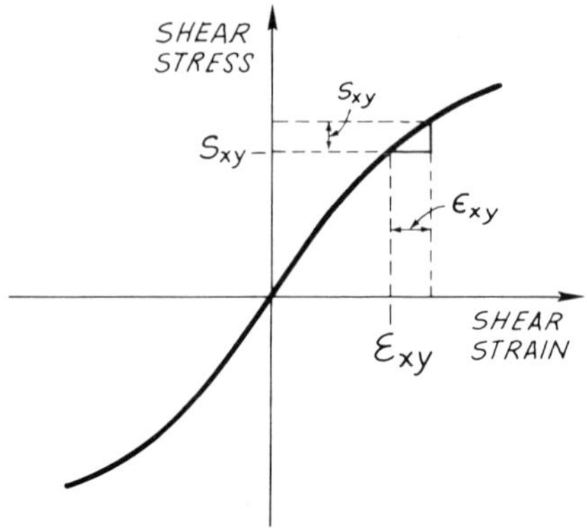

Fig. 5. Idealized, nonlinear relations between shear stress and shear strain. Even though the relation is nonlinear, the behavior of such a material can be described closely in terms of a nonlinear relation between large or initial stresses and strains and a linear relation between small or incremental stresses and strains. The incremental shear modulus, G, is a function of the initial stress.

The second reason for separating initial and incremental deformations is that folding theory is based on an analysis of stability which involves two or more simultaneous equilibrium states. In familiar buckling problems, for example, beams may be in equilibrium both in a straight, shortened form and in a bent or buckled form. One equilibrium state is defined, normally implicity, in terms of initial stresses and strains (e.g., Ramberg, I, 1970; Johnson, I, 1970, chapter 3). The linear theory of elasticity (e.g., Timoshenko and Goodier, I, 1970) would predict that the straight, shortened form is the unique equilibrium form whereas normal buckling theory and the nonlinear theory would predict two or more equilibrium forms if axial loads are sufficiently large. Of course, if there are several possible equilibrium configurations a body can assume, not all of them need be stable so that not all of them are equally likely to occur. Indeed, Novozhilov (I, 1953, p. 153) indicates that, in general, where there are several possible equilibrium configurations, the configurations indicated to be unique by the linear theory of elasticity ordinarily are unstable. In situations where instability is likely, therefore, one should investigate possible nonlinearities in strain and rotation or of rheological properties of the materials and incorporate the most important nonlinearities in the theory used to describe the equilibrium states. This is what we shall begin to do in following pages.

Let us consider states of equilibrium that are slightly different from the state of equilibrium of initial stresses applied to a body. We designate the *initial stresses* referred to curvilinear coordinates by the symbols, $S_{\tilde{X}\tilde{X}}$, $S_{\tilde{X}\tilde{Y}}$, $S_{\tilde{Y}\tilde{X}}$, and $S_{\tilde{Y}\tilde{Y}}$ (Fig. 6B). Now, if the initial stresses are very small, the body is in a state of unique equilibrium, as defined by the linear theory of elasticity, that is, if the rheological properties are linear. If the initial stresses are increased, the solutions remain unique as long as the stresses and strains are smaller

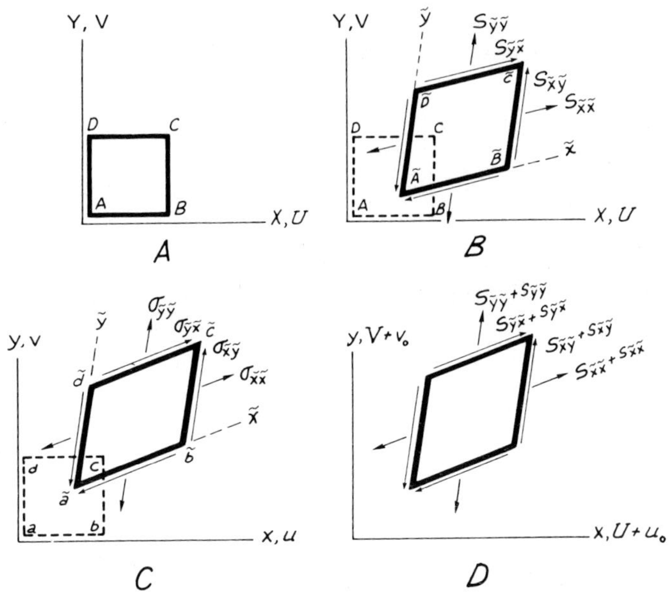

Fig. 6. Sequential deformation of a body. A. Original, undeformed element, *ABCD*, of body. B. Element deformed by initial stresses, *S*, referred to curvilinear coordinates, $\tilde{X}$, $\tilde{Y}$. C. Incrementally deformed element, $\tilde{a}\tilde{b}\tilde{c}\tilde{d}$, with final, total stresses, σ, referred to curvilinear, $\tilde{x}$, $\tilde{y}$, coordinates. D. Components of total stress referred to incrementally deformed element are assumed to be the sum of the components of initial stress, *S*, and incremental stress, *s*, associated with the incremental deformation.

than certain critical values. Where the stresses and strains reach the critical values, however, the linear theory no longer yields all possible solutions; there may be other positions of elastic equilibrium not predicted by the linear theory (Novozhilov, I, 1953).

Suppose that the initial stresses, $S_{\tilde{X}\tilde{X}} \ldots S_{\tilde{Y}\tilde{X}}$ are at certain *critical* values, which will become defined in the course of the analysis. Then, there are two or more possible states of equilibrium, *all compatible with the same initial conditions*. Thus, one configuration is the deformation caused by the initial state itself (Fig. 6B). Other possible configurations are associated with the stress state shown in Fig. 6C, where $\sigma_{\tilde{x}\tilde{x}} \ldots \sigma_{\tilde{y}\tilde{x}}$ represent the final stress components. The problem is to determine relations among the final stress state, Fig. 6C, the intial stress state, Fig. 6B, and the incremental stresses. We know that the final stress state is related to the incremental and initial stresses but the nature of the relationship is not completely clear. Thus, we assume the relationship and it is at this point that the analysis becomes approximate.*

LINEARIZED EQUILIBRIUM EQUATIONS AND BOUNDARY CONDITIONS

The concept of incremental and initial stresses and displacements allows

* This assumption is eliminated in Chapter 10.

one to simplify the nonlinear equilibrium equations (17). Let us assume that:

(1) the components of total stress are the sum of initial and incremental stresses (Fig. 6D):

$$\sigma_{\tilde{x}\tilde{x}} = S_{\tilde{x}\tilde{x}} + s_{\tilde{x}\tilde{x}}$$
$$\sigma_{\tilde{y}\tilde{y}} = S_{\tilde{y}\tilde{y}} + s_{\tilde{y}\tilde{y}} \qquad (23)$$
$$\sigma_{\tilde{x}\tilde{y}} = S_{\tilde{x}\tilde{y}} + s_{\tilde{x}\tilde{y}} = \sigma_{\tilde{y}\tilde{x}}$$

where $s_{\tilde{x}\tilde{x}} ... s_{\tilde{x}\tilde{y}}$ are small, *incremental* stresses and where $S_{\tilde{x}\tilde{x}} ... S_{\tilde{x}\tilde{y}}$ presumably are equivalent to $S_{\tilde{X}\tilde{X}} ... S_{\tilde{X}\tilde{Y}}$. Further, let us define the total displacements as:

$$u = U + u_0$$
$$v = V + v_0 \qquad (24)$$

where U and V are components of *initial* displacement and u_0 and v_0 are components of *incremental* displacement (Fig. 6D).

Let us assume, further, that:

(2) the angles α and β between initial and final positions of an element (Fig. 3A) are small. Thus eqs. 15 become:

$$\cos(\alpha) \simeq 1 \simeq \cos(\beta)$$
$$\sin(\alpha) \simeq \partial u/\partial y \qquad (25)$$
$$\sin(\beta) \simeq \partial v/\partial x$$

Also, let us assume that:

(3) triple products of initial displacement gradients, $\partial U/\partial x ... \partial V/\partial y$, and double products of incremental displacement gradients, $\partial u_0/\partial x ... \partial v_0/\partial y$ are negligible.

Using assumption (2), eqs. 25, and the relations among the circular functions and the displacement gradients, eqs. 15, equilibrium eqs. 17 become, approximately:

$$\frac{\partial}{\partial x}[\sigma_{\tilde{x}\tilde{x}}(1 + E_{\tilde{y}\tilde{y}}) + \sigma_{\tilde{x}\tilde{y}}(\partial u/\partial y)] + \frac{\partial}{\partial y}[\sigma_{\tilde{y}\tilde{y}}(1 + E_{\tilde{x}\tilde{x}})(\partial u/\partial y) +$$
$$+ \sigma_{\tilde{y}\tilde{x}}\{1 + (\partial u/\partial x)\}] + DF_{\xi} = 0$$

$$\frac{\partial}{\partial x}[\sigma_{\tilde{x}\tilde{y}}\{1 + (\partial v/\partial y)\} + \sigma_{\tilde{x}\tilde{x}}(1 + E_{\tilde{y}\tilde{y}})(\partial v/\partial x)] + \frac{\partial}{\partial y}[\sigma_{\tilde{y}\tilde{y}}(1 + E_{\tilde{x}\tilde{x}}) + \qquad (26)$$
$$+ \sigma_{\tilde{y}\tilde{x}}(\partial v/\partial x)] + DF_{\eta} = 0$$

Now, let us consider the products of the stresses multiplied by the finite extensions, $E_{\tilde{x}\tilde{x}}$ and $E_{\tilde{y}\tilde{y}}$. Using assumption (3), and definitions 24, 5 and 2:

$$\sigma_{\bar{x}\bar{x}}(1 + E_{\bar{x}\bar{x}}) = \sigma_{\bar{x}\bar{x}}\sqrt{1 + 2\epsilon_{xx}} =$$

$$= \sigma_{\bar{x}\bar{x}} \sqrt{1 + 2\left(\frac{\partial U}{\partial x} + \frac{\partial u_0}{\partial x}\right) + \left[\left(\frac{\partial U}{\partial x} + \frac{\partial u_0}{\partial x}\right)^2 + \left(\frac{\partial V}{\partial x} + \frac{\partial v_0}{\partial x}\right)^2\right]}$$

Using the first-order approximation of the binomial expansion and rearranging:

$$\sigma_{\bar{x}\bar{x}}(1 + E_{\bar{x}\bar{x}}) \simeq \sigma_{\bar{x}\bar{x}} \left\{1 + (\partial U/\partial x) + (\partial u_0/\partial x) + \frac{\partial U}{\partial x}\frac{\partial u_0}{\partial x} + \frac{\partial V}{\partial x}\frac{\partial v_0}{\partial x} + \right.$$

$$\left. + (1/2)\left[\left(\frac{\partial U}{\partial x}\right)^2 + \left(\frac{\partial V}{\partial x}\right)^2\right]\right\} \qquad (27)$$

The final approximation we will use in the equilibrium equations for our analyses of folding is that the initial and incremental stresses are both linear functions of the respective strains. Thus, for initial stresses:

$$\begin{aligned}
(4)\quad S_{\bar{x}\bar{x}} &= A(\partial U/\partial x) + B(\partial V/\partial y) + \delta P \\
S_{\bar{y}\bar{y}} &= A(\partial V/\partial y) + B(\partial U/\partial x) + \delta P \\
S_{\bar{x}\bar{y}} &= G[(\partial V/\partial x) + (\partial U/\partial y)] \\
P &= (S_{\bar{x}\bar{x}} + S_{\bar{y}\bar{y}})/2
\end{aligned} \qquad (28)$$

and for incremental stresses:

$$\begin{aligned}
s_{\bar{x}\bar{x}} &= A(\partial u_0/\partial x) + B(\partial v_0/\partial y) + \delta p \\
s_{\bar{y}\bar{y}} &= A(\partial v_0/\partial y) + B(\partial u_0/\partial x) + \delta p \\
s_{\bar{x}\bar{y}} &= G[(\partial v_0/\partial x) + (\partial u_0/\partial y)] \\
p &= (s_{\bar{x}\bar{x}} + s_{\bar{y}\bar{y}})/2
\end{aligned} \qquad (29)$$

Here, for *compressible* materials:

$A = 2G + \lambda, \quad B = \lambda, \quad \delta = 0$

and for *incompressible materials*:

$A = 2G, \quad B = 0, \quad \delta = 1$

G is the shear modulus and λ is a Lamé constant.

Substituting the first of eqs. 23 into term 27, and eqs. 28 and 29 into the result, bearing in mind assumption (3), term 27 becomes:

$$(S_{\bar{x}\bar{x}} + s_{\bar{x}\bar{x}})(1 + E_{\bar{x}\bar{x}}) \simeq [A(\partial U/\partial x) + B(\partial V/\partial y) + \delta P][1 + (\partial U/\partial x) + (\partial u_0/\partial x)] +$$

$$+ [A(\partial u_0/\partial x) + B(\partial v_0/\partial y) + \delta p] \simeq S_{\bar{x}\bar{x}}[1 + (\partial U/\partial x) + (\partial u_0/\partial x)] + s_{\bar{x}\bar{x}}$$

The terms in eqs. 26 linearize similarly so that eqs. 26 become:

$$\frac{\partial}{\partial x}\{[S_{\bar{x}\bar{x}}[1 + (\partial V/\partial y) + (\partial v_0/\partial y)] + s_{\bar{x}\bar{x}} + S_{\bar{x}\bar{y}}[(\partial U/\partial y) + (\partial u_0/\partial y)]\} +$$

$$\frac{\partial}{\partial y}\{S_{\bar{y}\bar{y}}[(\partial U/\partial y) + (\partial u_0/\partial y)] + S_{\bar{y}\bar{x}}[1 + (\partial U/\partial x) + (\partial u_0/\partial x)] + s_{\bar{x}\bar{y}}\} + DF_\xi = 0$$

$$\frac{\partial}{\partial x}\{S_{\bar{x}\bar{y}}[1 + (\partial V/\partial y) + (\partial v_0/\partial y)] + s_{\bar{x}\bar{y}} + S_{\bar{x}\bar{x}}[(\partial V/\partial x) + (\partial v_0/\partial x)]\} + \tag{30}$$

$$\frac{\partial}{\partial y}\{S_{\bar{y}\bar{y}}[1 + (\partial U/\partial x) + (\partial u_0/\partial x)] + s_{\bar{y}\bar{y}} + S_{\bar{y}\bar{x}}[(\partial V/\partial x) + (\partial v_0/\partial x)]\} + DF_\eta = 0$$

Most of our analyses of folding will be based on the assumptions that the initial stresses are homogeneous, that is, that they are independent of position and that body forces are negligible so that, for example:

$$\frac{\partial}{\partial x}(\partial U/\partial y) = 0 = \frac{\partial}{\partial x}(S_{\bar{x}\bar{y}})$$

and eqs. 30 become:

$$S_{\bar{x}\bar{x}}(\partial^2 v_0/\partial x\partial y) + \partial s_{\bar{x}\bar{x}}/\partial x + S_{\bar{y}\bar{y}}(\partial^2 u_0/\partial y^2) + 2S_{\bar{x}\bar{y}}(\partial^2 u_0/\partial x\partial y) + \partial s_{\bar{x}\bar{y}}/\partial y = 0$$

$$S_{\bar{x}\bar{x}}(\partial^2 v_0/\partial x^2) + \partial s_{\bar{x}\bar{y}}/\partial x + S_{\bar{y}\bar{y}}(\partial^2 u_0/\partial x\partial y) + 2S_{\bar{x}\bar{y}}(\partial^2 v_0/\partial x\partial y) + \partial s_{\bar{y}\bar{y}}/\partial y = 0 \tag{31}$$

The boundary conditions, which describe the stresses on boundaries originally parallel to the x- or y-direction (Fig. 4B), where surface loading is prescribed but where displacements are not, similarly simplify from eqs. 20, by means of assumption (2), eqs. 25, to:

$$\tau_{\bar{x}x} \simeq S_{\bar{x}\bar{x}} + s_{\bar{x}\bar{x}} + S_{\bar{x}\bar{y}}[(\partial U/\partial y) + (\partial u_0/\partial y)]$$

$$\tau_{\bar{x}y} \simeq S_{\bar{x}\bar{x}}[(\partial V/\partial x) + (\partial v_0/\partial x)] + S_{\bar{x}\bar{y}} + s_{\bar{x}\bar{y}} \tag{32}$$

$$\tau_{\bar{y}\bar{x}} \simeq S_{\bar{x}\bar{y}} + s_{\bar{x}\bar{y}} + S_{\bar{y}\bar{y}}[(\partial U/\partial y) + (\partial u_0/\partial y)]$$

$$\tau_{\bar{y}y} \simeq S_{\bar{x}\bar{y}}[(\partial V/\partial x) + (\partial v_0/\partial x)] + S_{\bar{y}\bar{y}} + s_{\bar{y}\bar{y}}$$

Eqs. 31 and 32 are the *linearized* equations of equilibrium and boundary conditions, respectively. The equilibrium equations are linear in terms of both incremental and initial stresses and strains and the boundary conditions in terms of incremental stresses and strains. However, the equilibrium equations are significantly different from those of the linear theory of elasticity (e.g., Timoshenko and Goodier, I, 1970) and it is the differences that make the theory of interest in studies of folding.

DISCUSSION

The linearized theory provides some fundamental insights into buckling processes in elastic materials that have not been obtained with elementary beam theory. According to beam theory (e.g., Johnson, I, 1970, p. 84 et seq.), the equilibrium of an element of unconfined beam buckled by unaxial loading is governed by the equation,

$$M + P_0 v = 0$$

where M is resistance to bending moment, P_0 is compressive axial loading and v is deflection of the neutral plane (Fig. 7A). According to this relation, the axial load produces a moment which is resisted by bending. In the linearized theory, however, where we consider infinitesimal elements, moment and force equilibrium are satisfied by the boundary stresses. For example, suppose an element is subjected to initial compression and to simple, incremental shear:

$$\partial u_0/\partial y = 0$$

as shown in Fig. 7B. According to Fig. 7B, moment equilibrium requires that:

$$\tau_{\tilde{x}y}(\delta x \delta y) - \tau_{\tilde{x}x}[(\partial v_0/\partial x)\delta x \delta y] - s_{\tilde{x}\tilde{y}}(\delta x \delta y) = 0$$

and eqs. 32 show that this relation is satisfied.

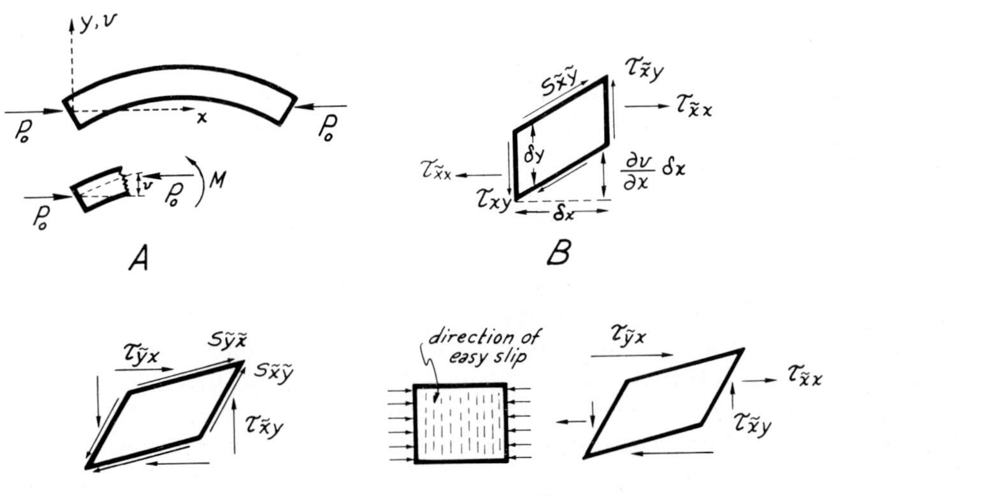

Fig. 7. Some properties of the linearized theory. A. Section of beam buckled by axial forces, P_0. B. Small element of elastic material subjected to incremental shear stress, $s_{\tilde{x}\tilde{y}}$, and to simple shear strain. C. Small element subjected solely to incremental shear stress. D. The same element subjected to initial axial compression and to incremental shear stress.

Another important difference between elementary beam theory and the linearized theory presented here is that the initial stresses, unlike the axial forces in elementary beam theory, behave as though they modify the elasticity moduli of a material. For example, suppose that an element of elastic material (Fig. 7C) is subjected to pure, incremental, shear stresses and to zero initial stresses. Then the boundary shear stresses, $\tau_{\tilde{x}y}$, $\tau_{\tilde{y}x}$, required to deform the material are [eqs. 32]:

$$\tau_{\tilde{x}y} = s_{\tilde{x}\tilde{y}} = G[(\partial u_0/\partial y) + (\partial v_0/\partial x)]$$

$$\tau_{\tilde{y}x} = s_{\tilde{x}\tilde{y}} = G[(\partial u_0/\partial y) + (\partial v_0/\partial x)]$$

and the element deforms in pure shear strain. Now, if there is initial axial stress, $S_{\tilde{x}\tilde{x}}$, the boundary stresses on the element (Fig. 7D) become:

$$\tau_{\tilde{x}y} = (G + S_{\tilde{x}\tilde{x}})(\partial v_0/\partial x) + G(\partial u_0/\partial y)$$

$$\tau_{\tilde{y}x} = G[(\partial v_0/\partial x) + (\partial u_0/\partial y)]$$

That is, the boundary shear stresses are no longer equal even though the element is in moment and force equilibrium. Further, the magnitude of the boundary stress, $\tau_{\tilde{x}y}$, is decreased or increased relative to the stress $\tau_{\tilde{y}x}$ depending upon whether the initial normal stress, $S_{\tilde{x}\tilde{x}}$, is compressive (negative) or tensile (positive), yet the deformation is by pure shear strain (Fig. 7D). Thus, the linearized theory indicates that the element is effectively easier to deform in one sense of simple shear than in the other; it becomes anisotropic with respect to shear (Fig. 7D). * This and related phenomena certainly are the most interesting aspects of the linearized theory. For, if we continue the analysis further, we find that as the magnitude of the initial compression approaches that of the shear modulus, G, in incremental deformation, the material becomes indefinitely easier to shear in simple shear, where $(\partial u_0/\partial y) = 0$ as $S_{\tilde{x}\tilde{x}} + G \rightarrow 0$. We would emphasize that the modulus, G, in this expression is the modulus of incremental deformation so that if a material behaved nonlinearly as shown in Fig. (5) and as most rock does, the incremental modulus might be much smaller than the average modulus for the initial deformation. In any case, if the quantity $S_{\tilde{x}\tilde{x}} + G$ approaches zero, the material behaves much as though it were layered normal to the direction of axial compression or as though it were a deck of slippery cards. In this respect we are reminded of De Sitter's (I, 1956, fig. 67, p. 97) concept of "microlithons".

We shall show evidence in Part III that homogeneous rubber and gelatin, materials probably closely appoximated by the ideal we are studying theoretically, buckle essentially by simple shear if the axial compression is sufficiently large at the time of buckling.

The theoretical analyses rigorously have been restricted to materials that behave elastically, except for a few comments aside that some of the analyses are approximately correct for elastic—plastic materials also. One may well ask, these days when it is popular for we geologists to believe that rocks

* This property is maintained in the exact, first-order theory derived in Chapter 10.

behave as viscous materials, of what value are analyses of folding of elastic materials, especially analyses as complex as the nonlinear theory? Well, we do not know, and we see no evidence that others know, how rocks behave at times of folding. We would suggest that rocks at high levels in the earth's crust would behave roughly as elastic—plastic materials, as many rocks do in triaxial tests, and that our analyses could be useful for understanding folding in such materials. The value of using the relatively complex, nonlinear theory is that, as we show in Part III, the theory provides some first insights into processes of development of fold forms which heretofore have defied theoretical analysis.

We are not prepared to generalize our results to include viscous materials because the analysis is invalid for ideal viscous materials. The reason is simple: A fundamental property of ideal viscous material is that the material does not "remember"; the strain rate is related solely to the stress, not to the history of the strain. The ideal elastic material, in contrast, has a "memory"; it is for this reason alone that the equilibrium equations are nonlinear for elastic materials, as developed here. Thus, the ideal elastic and the ideal viscous materials are fundamentally different and we can expect mechanical similarities which are evident in elementary beam theories of folding to virtually disappear as we consider nonlinear theories (e.g., Truesdell, I, 1952; Truesdell and Noll, I, 1965). An obvious next step is to develop folding theory for viscous materials with the same rigor as folding theory of nonlinear, elastic materials so that the two can be compared and contrasted correctly. Perhaps in this way one can begin to distinguish viscous-like and elastic-like behaviors in the laboratory and in the field.

The linearized theory is the basis for our analyses of folding in Part III, the next in this series of papers, and as we will show, the analyses of folding are incomplete. We can estimate tendencies for deviation of fold forms from sinusoidal shapes but the linearized theory cannot accurately describe the actual changes, even for ideal materials. This reason is twofold. First, the linearized theory does not account for second- or higher-order effects that seem to be critical to a description of the changes. We believe that, in order to incorporate the appropriate nonlinearities, one would have to solve the nonlinear equilibrium equations 19 along with the appropriate stress—strain relations and boundary conditions. Approximate solutions could be obtained by finite-element or finite-difference methods, of course. The second part of the problem is that appropriate stress—strain relations are unknown. Approximate relations could be devised for rubber (e.g., Truesdell and Noll, I, 1965) and perhaps for gelatin, both common experimental materials, through a program of careful experimentation and analysis. The problem of determining appropriate relations for rock is more difficult, in part because standard testing methods, including triaxial tests, are inadequate to determine the constants (Truesdell and Noll, I, 1965). Nevertheless, if the answers are wanted sufficiently intensely, the theoretical bases are available to design experiments and to perform theoretical analyses.

ACKNOWLEDGEMENT

The research reported in this part was financed largely by the National Science Foundation, grant no. GA-36917, but partly by the John Simon Guggenheim Memorial Foundation, N.Y. We thank these organizations for their generous support.

We are grateful to Dr. David Pollard, U.S. Geological Survey, Menlo Park, California, for suggesting many improvements in the organization of the manuscript and for criticizing its content. We thank Professor George Herrmann of the Applied Mechanics Department at Stanford for suggesting that we study Novozhilov's book on nonlinear elasticity theory.

REFERENCES NOT CITED IN CHAPTER 3

Goodier, J.N., 1950. Elastic torsion in the presence of initial stress. J. Appl. Mech. (ASME), 72: 383—387.

Green, A.E. and Adkins, J.E., 1970. Large Elastic Deformations. Clarendon Press, Oxford, 324 pp.

Green, A.E. and Zerna, W., 1968. Theoretical Elasticity. Clarendon Press, Oxford, 457 pp.

Johnson, A.M. and Ellen, S.D., 1974. A theory of concentric, kink, and sinusoidal folding and of monoclinal flexuring of compressible, elastic multilayers, I. Introduction. Tectonophysics, 21: 301—339.

Love, A.E.H., 1944. Mathematical Theory of Elasticity. Dover Publications, New York, 643 pp.

Toupin, R.A. and Bernstein, B., 1961. Sound waves in deformed perfectly elastic materials, acoustical effects. J. Acoust. Soc. Am., 33: 216—225.

Chapter 5

TRANSITION FROM SINUSOIDAL TO CONCENTRIC-LIKE TO CHEVRON FOLDS *

SUMMARY

A basic, sinusoidal solution to the linearized equations of equilibrium for compressible, elastic materials provides solutions to several problems of folding of multilayers. Theoretical wavelengths are comparable to those predicted by Ramberg, using viscosity theory, and to those predicted by elementary folding theory. The linearized analysis of buckling of a single, stiff, elastic layer, either isolated or within a soft medium, suggests that wavelengths computed with elementary beam theory are remarkably similar to those computed with the linearized theory for wavelength-to-thickness ratios greater than about five. This is half the limit of ten normally assumed for use of the elementary theory.

The theory and experiments with deep beams of rubber or gelatin indicate that thick, homogeneous layers folded with short wavelengths assume internal forms strikingly similar to those of the ideal concentric fold. Thus, mechanical layering clearly is not required to produce concentric-like forms.

Further, the theory suggests that "arc and cusp" structure, or "pinches", at edges of deep beams as well as chevron-like forms in single or multiple stiff layers are a result of a peculiar, plastic-like behavior of elastic materials subjected to high normal stresses parallel to layering. In a sense, the elastic material "yields" to form the hinge of the chevron fold, although the strain vanishes if the stresses are released. Accordingly, it may be impossible to distinguish chevron forms produced in elastic—plastic materials, such as cardboard or aluminum and perhaps some rock, from chevron forms produced in purely elastic materials, such as rubber.

Analysis of the theory shows that, just as high axial stresses make straight, shortened multilayers the unstable form and sinusoidal waves the stable form, stresses induced by sinusoidal displacements of the multilayer make the sinusoidal waveform unstable and concentric-like waves the stable form. Thus, concentric-like folds appear to be typical of folded multilayers according to our analysis. Further, where the layers have short wavelengths in the cores of the concentric-like folds, the stiff layers "yield" elastically at hinges and straighten in limbs. Thus the concentric-like pattern is replaced by chevron folds as the multilayer is shortened. In this way we can understand the sequence of events from uniform shortening, to sinusoidal folding, to concentric-like folding, to chevron folding in multilayers composed of elastic materials.

---

* Originally published as: Johnson, A.M. and Honea, E., 1975. Part III, Transition from sinusoidal to concentric-like to chevron folds. Tectonophysics, 27: 1—38.

INTRODUCTION

Part I* of this series of papers indicated the general plan and presented the method of characteristics and lines of discontinuity as a way of describing many ideal folds. Part II* derived nonlinear and linearized equations of equilibrium for elastic materials. Here, in this third part, we will use the theory of equilibrium to derive expressions for wavelengths of sinusoidal folds and to explain the transitions from sinusoidal to concentric-like to chevron folds in multilayers. Our method of solution closely resembles that of Ramberg (I, 1961; I, 1970**) even though Ramberg deals with viscous materials and we deal with elastic materials. We will not use the method of characteristics here, as indicated in Part I.

Much of this third part is devoted to an analysis of sinusoidal folding of single and multiple layers, using the linearized theory of equilibrium developed in Part II. The sinusoidal theory has already received a great deal of attention in geological literature, as documented in Part I, and the types of solutions reported here have already been developed, partly by Biot (e.g., I, 1965a), but largely by Ramberg (I, 1961; I, 1963; I, 1964; I, 1967; I, 1970). Accordingly, the primary contributions of this part to sinusoidal-folding theory are the solution of problems using the linearized elasticity theory and the collection into one place of a group of related solutions which are scattered through the literature. Indeed, we show that all the solutions can be derived from one basic solution to the equilibrium equations.

This third part also is concerned with a way to explain the development of concentric-like and chevron forms of folds, based on higher-order effects of the linearized theory of equilibrium. In the ideal concentric fold, the arcs of the fold are concentric circles, so that a concentric fold is a special case of a parallel fold (e.g., Van Hise, I, 1896; p. 599). By chevron fold, we mean that layers are bent into symmetric, sawtooth-shaped waves, with straight limbs and either rounded or sharp hinges (e.g., Ramberg, I, 1963; Bayly, 1971). DeSitter (I, 1956, p. 183) uses the term "accordian fold" for the same form.

There has been some research into processes of concentric and chevron folding. Ramberg (I, 1963b) discussed the co-existence of concentric-like and chevron-like forms in models of interlayered stiff and soft rubber compressed axially. He anticipated our results, suggesting that, during shortening, the

---

* Part I is the reference, Johnson and Ellen, 1974, and Part II is the reference, Johnson and Honea, 1975.

** References cited in Part I will be noted with an "I", followed by the date, such as I, 1955, and not be repeated in "References cited" here.

Note: Part I = Chapter 3, Part II = Chapter 4.

layers near the free boundaries are subjected to different stress distributions than layers near the center and that the different stress distributions produce the different fold forms. He was unsure, however, what the differences in stresses were.

Biot (1965) presented an approximate theory for the buckling of isolated multilayers and assumed that, where the properties of the stiff and soft materials are nearly the same, the multilayer stack bends concentrically as a unit, as assumed in elementary beam theory. He indicated that, where the properties of the layers are very different, the soft layers behave much as a lubricant and that each of the stiff layers buckles into a sine wave. Biot's theory does not explain concentric-like forms in multilayers with contrasting properties such as those described in Part I, fig. 19.

Bayly (1964, 1970, 1971) has produced chevron-like forms in multilayered viscous materials and has derived expressions for stresses within the limbs and hinges of ideal chevron folds. His theory, which is rather complete, is based in part on the assumption that the folds are of chevron form so that it cannot explain why or how the chevron form itself develops. He noted, however, that the chevron form probably is the type of fold one would expect in an infinitely thick multilayer because chevron-like folds formed near mid-height in experimental multilayers whereas concentric-like, or "arc-and-cusp", forms developed near upper and lower boundaries of experimental multilayers.

Ramsay's (1974) theory of chevron folding also is concerned with growth of chevron forms without consideration of the origin of the forms themselves.

Many investigators (e.g., Van Hise, I, 1896; DeSitter, I, 1956; Donath and Parker, I, 1964) have suggested that mechanisms of folding can be deduced from the fold form, but their explanations of relations between mechanisms and fold forms are difficult to understand. Some (e.g., Van Hise, I, 1896) indicated that parallel (or concentric) folds, characterized by layers of constant thickness in the limbs and hinges of the folds, could have developed only by slippage between layers, or presumably by what is now called flexural slip. Unfortunately, the picture may not be so simple, as we indicated in the Discussion of Part II. We will show experimental folds in the following pages that closely resemble the ideal parallel pattern, produced in deep beams of *homogeneous* rubber or gelatin without layering or surfaces of slip. Grid lines drawn on the deep beams indicate concentric-like fold patterns. As indicated in Part II, layering or slippage between layers is only one of several sources of anisotropy; another is anisotropy to incremental deformation induced by an initial state of stress.

BUCKLING OF ISOLATED, HOMOGENEOUS LAYERS

The analysis of buckling of isolated layers is fundamental to the understanding of folding of single or multiple stiff layers in a medium because the stiff layers largely control the fold forms and in many respects the stiff layers behave as though they were isolated. The theory of buckling of isolated layers with small wavelength/thickness ratios has been discussed at some length for viscous materials by Ramberg (I, 1970, p. 209; I, 1961) and for elastic materials by Biot (I, 1965a), and our solution is patterned after theirs.

Solution

The method of solution is to determine conditions under which the linearized equilibrium equations [eqs. (II, 31)] will allow sinusoidal solutions which satisfy the boundary conditions [eqs. (II, 32)]. The linearized equilibrium equations are:

$$A(\partial^2 u_o/\partial x^2) + (B + S_{\bar{x}\bar{x}} + G)(\partial^2 v_o/\partial x \partial y) + (G + S_{\bar{y}\bar{y}})(\partial^2 u_o/\partial y^2) = 0 \qquad (1a)$$

$$A(\partial^2 v_o/\partial y^2) + (B + S_{\bar{y}\bar{y}} + G)(\partial^2 u_o/\partial x \partial y) + f(\partial^2 v_o/\partial x^2) = 0 \qquad (1b)$$

$$\text{where: } f = (G + S_{\bar{x}\bar{x}}) \qquad (1c)$$

The quantity $(G + S_{\bar{x}\bar{x}})$ is given a special symbol because it plays a critical role in our analyses later on. We have assumed that the initial shear stresses are zero and that the initial normal stresses, $S_{\bar{x}\bar{x}}$ and $S_{\bar{y}\bar{y}}$, are homogeneous, that is, the same everywhere in the body. The boundary conditions in the absence of initial shear stresses are:

$$\tau_{\bar{x}x} = S_{\bar{x}\bar{x}} + A(\partial u_o/\partial x) + B(\partial v_o/\partial y) \qquad (2a)$$

$$\tau_{\bar{x}y} + S_{\bar{x}\bar{x}}(\partial v_o/\partial x) + G[(\partial v_o/\partial x) + (\partial u_o/\partial y)] \qquad (2b)$$

$$\tau_{\bar{y}x} = S_{\bar{y}\bar{y}}(\partial u_o/\partial y) + G[(\partial v_o/\partial x) + (\partial u_o/\partial y)] \qquad (2c)$$

$$\tau_{\bar{y}y} = S_{\bar{y}\bar{y}} + A(\partial v_o/\partial y) + B(\partial u_o/\partial x) \qquad (2d)$$

$$\text{where: } A = 2G + \lambda,\ B = \lambda,\ G = E/[2(1+\nu)] \text{ and } \lambda = E\nu/[(1+\nu)(1-2\nu)] \qquad (2e)$$

in which G is modulus of rigidity, E is Young's modulus, ν is Poisson's ratio and G and λ are Lamé's constants. The variables u_o and v_o in eqs. 1 and 2 are the *incremental* displacements. The total displacements are equal to the incremental plus the initial displacements, as explained in Part II. In deriving eqs. 1 and 2 we have assumed that the material is compressible, but the equations are valid for nearly incompressible materials; we may make ν arbitrarily close to 0.5.

Sinusoidal solutions to eqs. 1 are facilitated by writing eqs. 1 as a single, fourth-order equation in v_o. Solving eq. 1b for $\partial^2 u_o/\partial x \partial y$:

$$\partial^2 u_o/\partial x \partial y = -[f/(B + S_{\bar{y}\bar{y}} + G)](\partial^2 v_o/\partial x^2) - [A/(B + S_{\bar{y}\bar{y}} + G)](\partial^2 v_o/\partial y^2) \qquad (3)$$

Differentiating eq. 1a with respect to x and y, substituting eq. 3 into the result, and rearranging:

$$(\partial^4 v_o/\partial y^4) + 2\mathbf{A}(\partial^4 v_o/\partial x^2 \partial y^2) + \mathbf{B}(\partial^4 v_o/\partial x^4) = 0 \qquad (4a)$$

where: $\mathbf{A} = [A^2 - (B + S_{\bar{x}\bar{x}} + G)(B + S_{yy} + G) + (S_{\bar{y}\bar{y}} + G)f]/[2A(S_{\bar{y}\bar{y}} + G)]$

(4b)

and: $\mathbf{B} = f/(S_{\bar{y}\bar{y}} + G)$

By a similar procedure we can write a fourth-order equation in u_o by solving eq. 1a for $(\partial^2 v_o/\partial x \partial y)$ and substituting the result into eq. 1b:

$$(\partial^4 u_o/\partial y^4) + 2\mathbf{A}(\partial^4 u_o/\partial x^2 \partial y^2) + \mathbf{B}(\partial^4 u_o/\partial x^4) = 0 \qquad (5)$$

Now we find sinusoidal solutions to eq. 4a by substitution of the expression:

$v_o = F(ly)\sin(lx)$

where: $l = 2\pi/L$ (6)

L is wavelength and $F(ly)$ is some function of y. The result, after a few manipulations, is:

$$v_o = [C_1 \exp(m_1 ly) + C_2 \exp(-m_1 ly) + C_3 \exp(m_2 ly) + C_4 \exp(-m_2 ly)]\sin(lx) \qquad (7a)$$

or equivalently:

$$v_o = [C_5 \sinh(m_1 ly) + C_6 \cosh(m_1 ly) + C_7 \sinh(m_2 ly) + C_8 \cosh(m_2 ly)]\sin(lx) \qquad (7b)$$

where: $m_1 = +\sqrt{\mathbf{A} + \sqrt{\mathbf{A}^2 - \mathbf{B}}}$

(8)

and: $m_2 = +\sqrt{\mathbf{A} - \sqrt{\mathbf{A}^2 - \mathbf{B}}}$

Similarly, eq. 5 can be solved for u_0:

$$u_o = [C_1 M_1 \exp(m_1 ly) - C_2 M_1 \exp(-m_1 ly) + C_3 M_2 \exp(m_2 ly) - C_4 M_2 \exp(-m_2 ly)]\cos(lx) \qquad (9a)$$

or:

$$u_o = [C_5 M_1 \cosh(m_1 ly) + C_6 M_1 \sinh(m_1 ly) + C_7 M_2 \cosh(m_2 ly) + C_8 M_2 \sinh(m_2 ly)]\cos(lx) \qquad (9b)$$

where: $M_1 = [m_1(B + S_{\bar{x}\bar{x}} + G)]/[A - (S_{\bar{y}\bar{y}} + G)m_1^2]$

(10)

$M_2 = [m_2(B + S_{\bar{x}\bar{x}} + G)]/[A - (S_{\bar{y}\bar{y}} + G)m_2^2]$

The coefficients M_1 and M_2 and the arbitrary constants, C_1 ... C_8 in eqs. 9 were derived by substituting the solutions for u_o and v_o into equilibrium equations 1a, or 1b, and comparing arbitrary constants. The procedure is somewhat tedious, but substitution of eqs. 7a and 9a or 7b and 9b into eqs. 1 verifies that eqs. 7 and 9 are, indeed, solutions to the differential equations.

Much of the work of deriving solutions for sinusoidal folds in elastic multilayers now is complete, except for actual numerical evaluation of the results, because all relevant solutions are contained in eqs. 7 and 9. In the following pages we merely use eqs. 7 and 9 and appropriate boundary conditions to solve a variety of problems.

Let us consider buckling of an isolated layer. Figs. 1A—C show deformed grid lines in three ideal layers buckled theoretically. The lengths shown are those after deformation as a result of initial axial compression that produced the buckles. The original lengths of the layers were somewhat larger than shown in Figs. 1. In order to derive a solution for the form of the buckled layers and the critical axial compression required for buckling, we use eqs. 7b and 9b and boundary conditions 2c and 2d. The orientations of positive boundary stresses are shown in Fig. 1A. The boundary conditions are (at $y = T/2$):

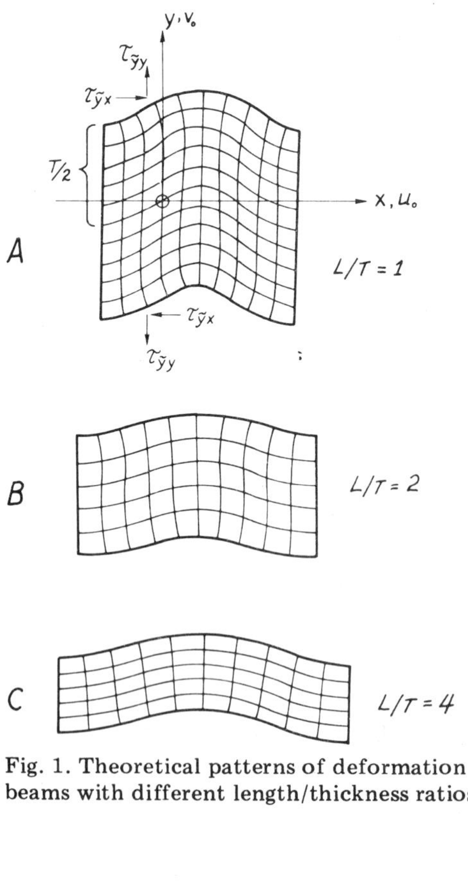

Fig. 1. Theoretical patterns of deformation of rectangular grid lines drawn on deep beams with different length/thickness ratios before buckling but after initial shortening.

$$\tau_{\tilde{y}y} = S_{\tilde{y}\tilde{y}} = A(\partial v_o/\partial y) + B(\partial u_o/\partial x) + S_{\tilde{y}\tilde{y}} \tag{11a}$$

$$\tau_{\tilde{y}x} = 0 = G[(\partial u_o/\partial y) + (\partial v_o/\partial x)] + S_{\tilde{y}\tilde{y}}(\partial u_o/\partial y) \tag{11b}$$

that is, the upper and lower surfaces are subjected to normal stress (if $S_{\tilde{y}\tilde{y}} \neq 0$) but to zero shear stress.

Substituting expressions 7b and 9b for the displacements into eq. 11b and rearranging:

$$C_5 = C_7 = 0 \qquad C_6 = -C_8 C$$

where: $C = (1 + m_2M_2)\cosh(m_2lT/2)/[(1 + m_1M_1)\cosh(m_1lT/2)]$ (12)

If we let the amplitude of the deflections at $x = L/4$ and $y = 0$ be v_o^o, then eq. 7b indicates that:

$$v_o^o = C_6 + C_8 \tag{13}$$

which evaluates the constants:

$$C_6 = v_o^o[C/(C-1)] \qquad C_8 = -v_o^o[1/(C-1)] \tag{14}$$

and expressions 7b and 9b become:

$$v_o = [v_o^o/(C-1)][C\cosh(m_1ly) - \cosh(m_2ly)]\sin(lx) \tag{15a}$$

$$u_o = [v_o^o/(C-1)][CM_1\sinh(m_1ly) - M_2\sinh(m_2ly)]\cos(lx) \tag{15b}$$

Eqs. 15 describe the distribution of displacements throughout the buckled layer, but they are valid only if the initial stresses are sufficiently large to allow buckling. We use boundary condition 11a to determine the magnitudes of the initial stresses. Substituting eqs. 15 into eq. 11a:

$$[v_o^o l/(C-1)][A\{Cm_1\cosh(m_1lT/2) - m_2\cosh(m_2lT/2)\}$$
$$- B\{CM_1\sinh(m_1lT/2) - M_2\sinh(m_2lT/2)\}]\sin(lx) = 0$$

and in order for there to be a solution where v_o^o and $\sin(lx)$ are not zero, that is, a non-trivial solution, the term in brackets must be zero, or:

$$A[Cm_1\cosh(m_1lT/2) - m_2\cosh(m_2lT/2)] = B[CM_1\sinh(m_1lT/2)$$
$$- M_2\sinh(m_2lT/2)] \tag{16}$$

A way to solve eq. 16 is to normalize the elasticity moduli and stresses with the shear modulus and to find, by iterative methods and a digital computer, ratios of wavelength to thickness, L/T, that satisfy eq. 16 for each value of initial stress. Fig. 2 shows the relation between normalized axial stress, $S_{\tilde{x}\tilde{x}}/G$, and L/T for three values of confining stress, $S_{\tilde{y}\tilde{y}}/G$, derived by this method. Also plotted in Fig. 2 is the relation between the Euler buckling

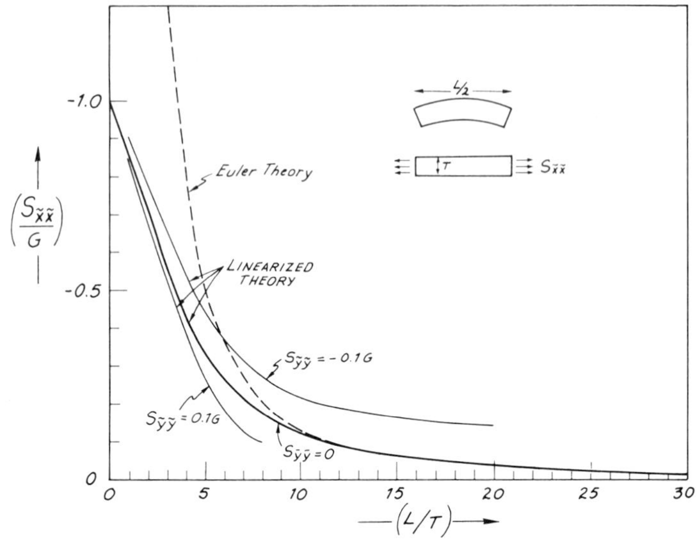

Fig. 2. Relations between normalized critical axial stress, $S_{\tilde{x}\tilde{x}}/G$, in plane strain, including effect of normalized confining stress, $S_{\tilde{y}\tilde{y}}/G$, according to linearized theory and according to Euler's theory.

stress and wavelength according to the elementary theory (e.g., Johnson, I, 1970, p. 90):

$$(S_{\tilde{x}\tilde{x}}/G)_{\text{crit}} = -(2/3)[(1+\nu)/(1-\nu^2)]\pi^2(T/L)^2 \tag{17}$$

where ν is Poisson's ratio, assumed to be 0.49 for the computations.

Comparison of the curves predicted by the Euler and linearized theories indicates a close correlation except where the wavelength of the layer is less than about five times the thickness. According to the Euler theory, the wavelength approaches zero only if the magnitude of the axial load approaches infinity, whereas the linearized theory indicates that the wavelength approaches zero if the magnitude of the axial compression approaches the shear modulus, G, that is, if the parameter, f, eq. 1c, approaches zero. Biot's (I, 1965a) theory makes predictions qualitatively similar to ours.

Now, solutions to eq. 16 provide magnitudes of axial compression required to buckle isolated layers of various length/thickness ratios after initial stresses are applied but just before buckling begins. These values are then substituted into eqs. 15 in order to compute forms of deflected layers. Figs. 1A—C show theoretical forms and deformed grid patterns on layers with L/T ratios of 1, 2 and 4. The grid pattern in the layer that had a length/thickness ratio of 4 just before buckling, Fig. 1C, shows deformation similar to that of pure bending as assumed in elementary beam theory, so it is not surprising that elementary beam theory provides a reasonable estimate of the critical axial compression (Fig. 2). The grid lines that were originally vertical are largely rotated. The layer with a length/thickness ratio of unity, Fig. 1A, shows the phenomenon of simple shear, $\partial u_o/\partial y \approx 0$, throughout most of the thickness of the layer, as explained in the Discussion of Part II. As indicated there, if the magnitude of the axial compression approaches the magnitude of the incremental shear modulus, the material tends to deform in simple shear. Thus, the thick layer shown in Fig. 1A buckles not by bending, but mostly by shear. The exception is near the bounding surfaces where, of course, the shear stress approaches zero.

Experiments

Several experiments were made with deep beams of photoelastic rubber and gelatin confined laterally to approximate the conditions of plane strain assumed in the theoretical analyses. Grid patterns were drawn on the deep beams to visualize shear strains and rotations.

The use of gelatin as a photoelastic material is well known (e.g., Hatschek, 1933; Ward and Saunders, 1958) so we will not discuss the properties of gelatin in detail. The gelatin mixture we use consists of about 400 g of dry gelatin (Kind and Knox Gelatin Co., type 2527), 1000 g of deionized water and about 500 ml of glycerin. Typically the mixture is heated in a water bath at 50°C for five or six hours until the mixture clarifies and then is poured into a special mold used to make the layers. The photoelastic rubber is type PSM-4, supplied by Photoelastic Inc., Pennsylvania.

We will report ratios of elasticity moduli of rubber and gelatin as we discuss various experiments in the following pages, but the meaning of the ratios is not completely clear. The moduli are measured by extending a strip of gelatin or rubber under various loads and measuring changes in distances between points on the strips. Unfortunately, nonlinear effects cause the modulus in tension to be somewhat higher than that in compression (e.g., see Hatschek, 1933). Also, the properties of gelatin are time-dependent and some of the strain is irrecoverable. For example, a strip of gelatin was strained longitudinally $3.19 \cdot 10^{-2}$ under a mean axial stress of about $4.03 \cdot 10^4$ dyn/cm^2, and there was no measurable permanent strain. However, when it was strained $4.51 \cdot 10^{-2}$ under a mean stress of $5.81 \cdot 10^4$ dyn/cm^2 and then the load removed, there remained a permanent strain of $0.16 \cdot 10^{-2}$, or about 3% of the total strain. The new, "unstressed" length of the strip was measured and then the strip was strained $6.50 \cdot 10^{-2}$ under a mean axial stress of about $8.47 \cdot 10^4$ dyn/cm^2. The load was maintained for three minutes and the strain remeasured; the amount of time-dependent strain was $0.32 \cdot 10^{-2}$, or about 5% of the strain measured immediately after loading. Clearly, the gelatin does have time-dependent properties, but we ignore them when we report ranges of elasticity moduli. The apparent Young's modulus for the strip described above ranged from 1.26 to $1.30 \cdot 10^6$ dyn/cm^2.

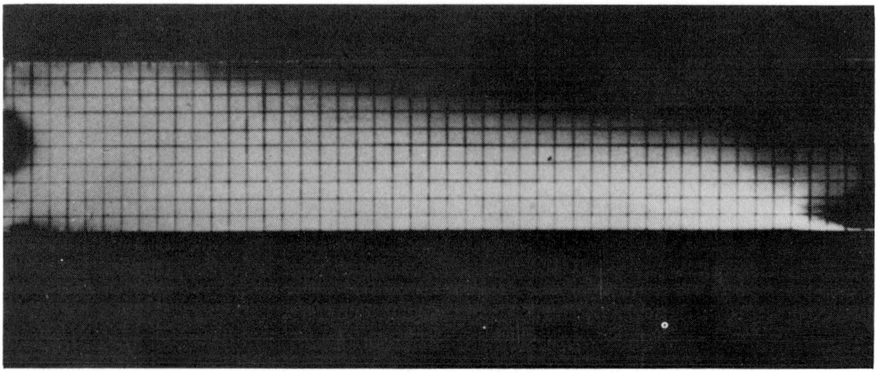

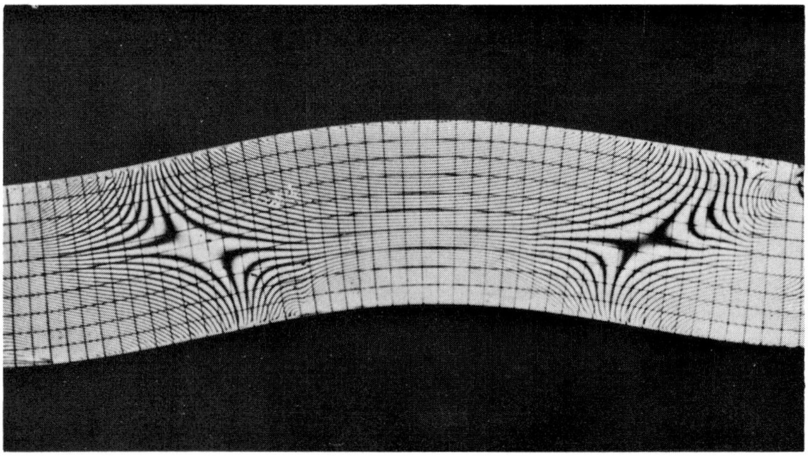

B

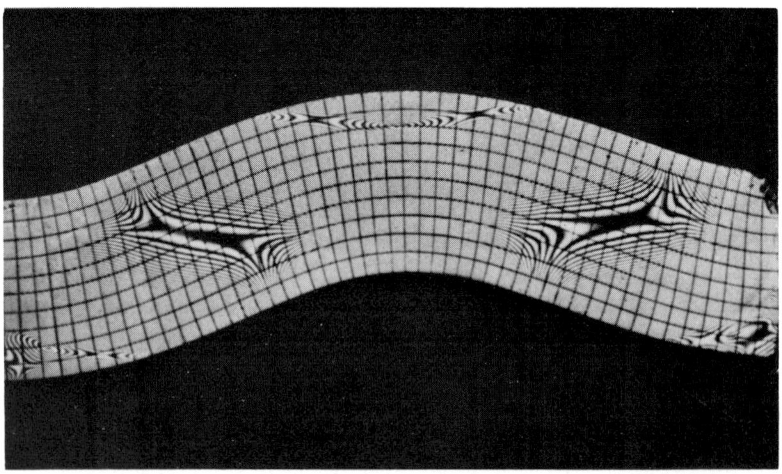

C

Fig. 3. Buckled deep beam of photoelastic rubber confined front and back by plexiglass and laterally with rigid platens. Unconfined top and bottom. Grid pattern drawn on beam before shortening.

We have detected no time-dependent or permanent strain in the photoelastic rubber, even where strips were tightly bent at hinges of chevron folds.

Fig. 3 shows a deep beam of photoelastic rubber buckled by axial shortening. The initial wavelength/thickness ratio, L/T, was about 4.5 at the time of buckling, and the deformation of the grid lines drawn on the deep beam closely resembles the theoretical pattern of deformed grid lines for a layer with an L/T ratio of 4.0, shown in Fig. 1C. The regular, narrow lines shown in Fig. 3 are grid lines and the other lines are isochromatics. The deep beam shown in Fig. 3 did not have "built-in" ends, so shortly after the photo shown in Fig. 3C was taken, a "pinch" formed at the base of the beam, the ends of the beam rotated away from the sides, the beam suddenly flipped into one-half a wave, an extension crack formed at the top and the beam failed.

A set of experiments using four deep beams of gelatin was designed to study effects of thickness on the buckling mode and the formation of the "pinch" recognized in the rubber. As in the experiment shown in Fig. 3, the deep beams were confined laterally with plexiglass sheets about 1.3 cm in thickness to simulate plane-strain conditions, and the four deep beams were simultaneously shortened axially by advancing a loading platen with screws. Each end of the beams was "built-in" to a strip of gelatin reinforced with a glass rod so that the beams were essentially cantilevered at each end. The gelatin had an apparent Young's modulus of about $2.42—2.44 \cdot 10^6$ dyn/cm^2. The thicknesses of the four beams before shortening were about 3.8, 1.9, 2.5 and 5.1 cm, respectively, from bottom to top as shown in Fig. 4A. A grid pattern and a circle were drawn on each of the deep beams before loading.

The thinnest beam, 1.9 cm thick, buckled first and deflected laterally most strongly as shortening increased (Figs. 4A—N). It eventually deflected so strongly that it interferred with another beam (Fig. 4G). The inner and outer surfaces of the thinnest beam remained smooth curves, even for large deflections (Fig. 4M).

At first, the second thinnest beam, originally of 2.5 cm thickness, deflected much the same way as the thinnest one (Figs. 4A—G), but suddenly a "pinch" formed in the core of the bend at the upper edge of the layer, as shown in Fig. 4H, and the pinch deepened (Fig. 4I) until the beam failed (Fig. 4J), as noted in the deep beam of rubber.

The second thickest beam, originally of 3.8 cm thickness, was disturbed by the thinnest beam, but eventually (Fig. 4L) it separated from the thin beam and formed a pinch in the core of the bend, and the pinch deepened until the beam failed (Fig. 4N).

The thickest beam, originally 5.1 cm thick, shortened and thickened considerably before there was any sign of buckling. It was about 5.9 cm thick when buckling started (Fig. 4G). The first buckling (Fig. 4H) produced broadly curved, outer surfaces, but with continued shortening the curvature increased markedly in the core of the bend (Fig. 4J). Then a pinch formed at the core (Fig. 4K) and greatly deepened (Figs. 4K—N) before the beam failed.

Analysis of experiments

Lines that were originally vertical, normal to upper and lower surfaces of the deep beams, rotated and distorted different amounts with deflection, depending on the thickness of the beam, as predicted theoretically (Figs. 1A—C). The lines on the thinnest beam rotated and remained quite straight as deflection progressed. The lines became slightly S-shaped on the limbs, particularly noticeable at large deflections (Fig. 4M), but they remained straight near the hinge. Also, the spacings between the lines decreased roughly linearly toward the core and increased roughly linearly toward the outer arc of the flexure,

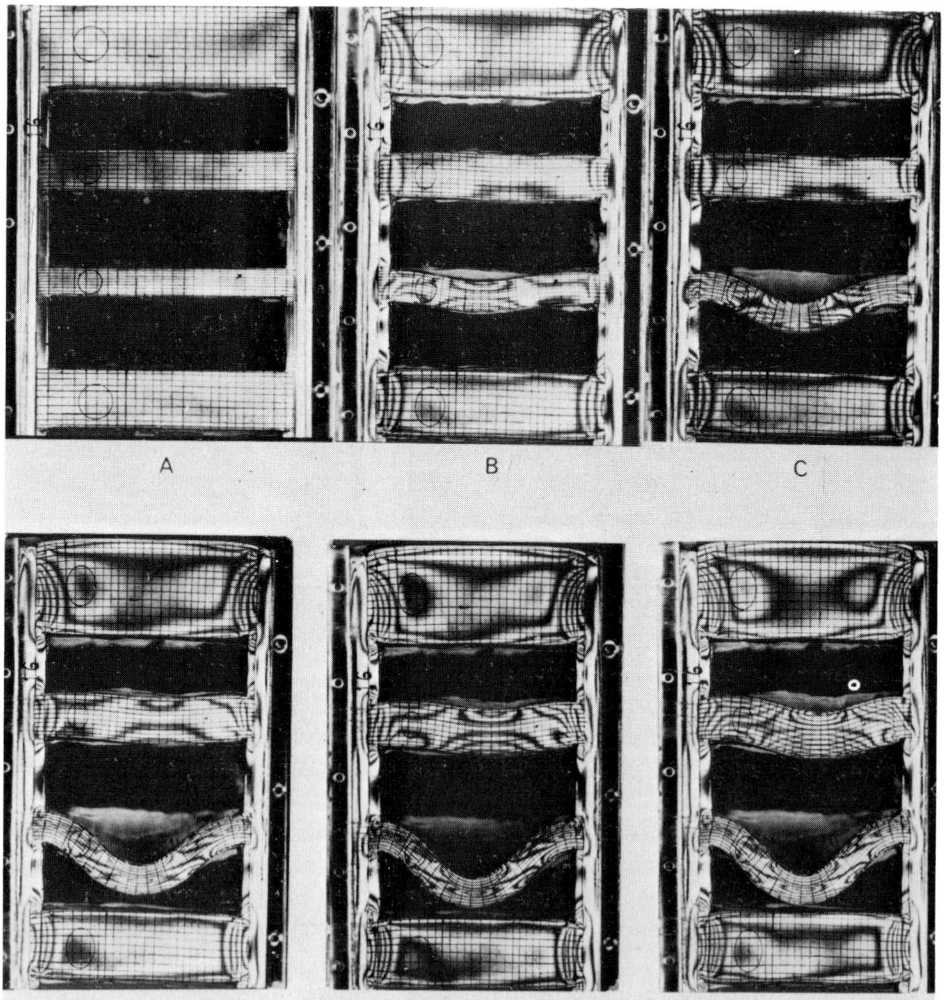

A B C

D E F

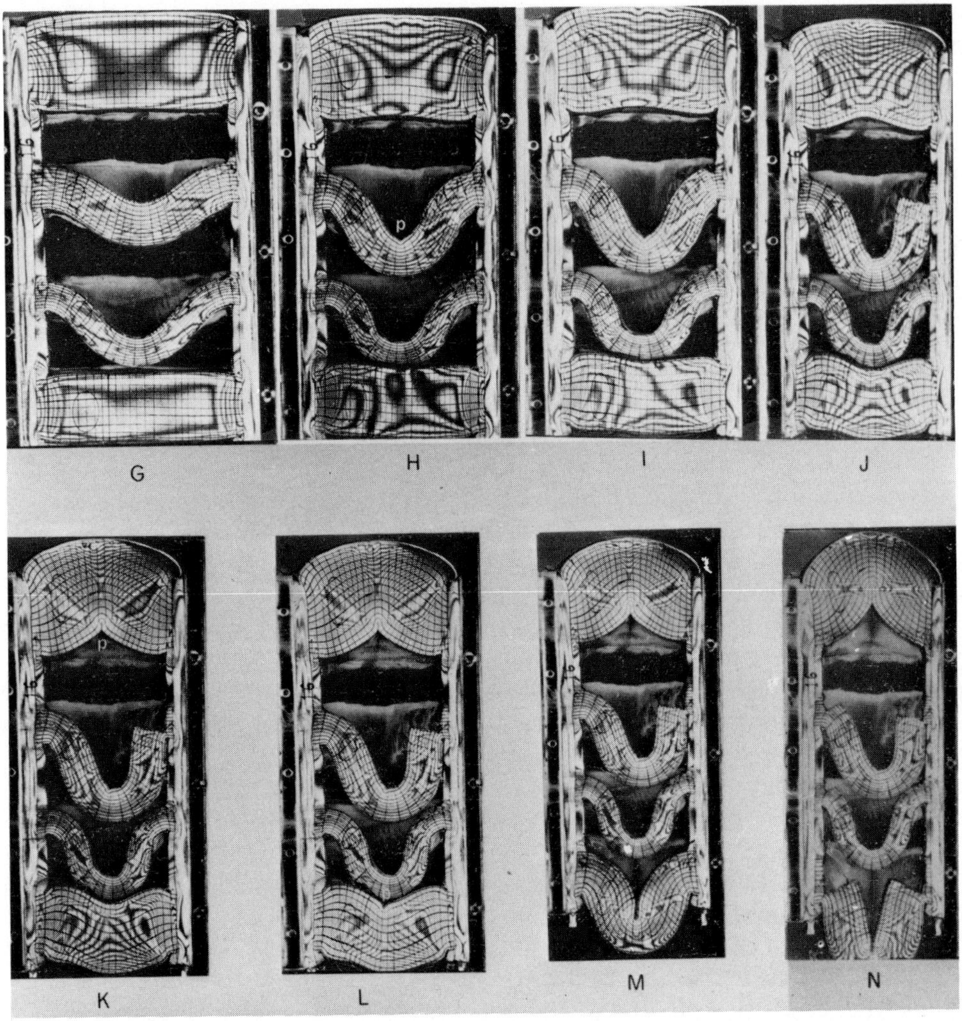

Fig. 4. Simultaneous buckling of four deep beams of gelatin confined as rubber beam shown in Fig. 3. Original thicknesses were 3.8, 1.9, 2.5 and 5.1 cm, respectively, from bottom to top in A.

indicating that the "fiber strains" were roughly as assumed in elementary beam theory.

The lines that were originally vertical, normal to the upper and lower surfaces, in the thickest beam remained nearly vertical for small to moderate deflections in the central half-length of the beam (Figs. 4H—J), closely resembling the theoretical pattern shown in Figs. 1A and 1B. Along the boundaries, of course, the lines remained normal to the stress-free surface and rotated as the surfaces rotated, as is required for equilibrium of shear stresses. Thus, the lines that were originally vertical rotated near the boundaries of the thickest beam much as the lines did throughout the entire depth of the thinnest beam. Away from the boundaries, toward the center, however, the slope $\partial u_0/\partial y$, of the lines remained constant.

One phenomenon which is not predicted explicitly by the solution for the buckling of a layer is the formation of the "pinch". Theoretically, a pinch might form if the parameter, f, eq. 1c, which is the coefficient of the curvature term, $\partial^2 v_o/\partial x^2$, in equilibrium equation 1b, approaches zero, because if the coefficient approaches zero the curvature becomes indeterminant. As shown in Fig. 2, the wavelength approaches zero as the magnitude of the initial compression approaches the shear modulus. These theoretical observations prompt us to suggest that a pinch can be a result of second-order effects not considered in the linearized theory, in particular that a pinch is a result of incremental compression near the inner arc and tension near the outer arc, superimposed on the high initial compression required to initiate buckling. Thus, in effect, the product of the incremental axial stress, $s_{\bar{x}\bar{x}}$, and the incremental slope, $\partial v_o/\partial x$, is not always negligible, as we assumed in order to linearize the theory in Part II, so that the parameter f should be modified to a parameter f^*, where:

$$f^* = S_{\bar{x}\bar{x}} + s_{\bar{x}\bar{x}} + G$$

We recommend a simple experiment to illustrate the importance of the axial compression in determining whether a pinch will form at the edge of an elastic beam or plate. A piece of moderately stiff, relatively incompressible rubber about 1 cm by 1.5 cm by several centimeters in length is satisfactory. First, bend the strip until the ends touch, by applying bending moments at the ends of the strip with the fingers. A pinch does not form; rather, the inner and outer surfaces remain smooth. Now, however, take the ends of the strip firmly in each hand and shorten the strip as much as possible and simultaneously allow the strip to deflect laterally. A pinch forms at the core of the bend, at least in the rubber with which we have experimented.

We can test the concept, using equations already derived, that the pinches in the deep beams are second-order effects, by computing the magnitude of incremental axial compression at the cores of buckled layers. The incremental axial stress is:

$$s_{\bar{x}\bar{x}} = A(\partial u_o/\partial x) + B(\partial v_o/\partial y); \text{ at } y = -T/2,\ x = L/4$$

Substituting eqs. 15 for the incremental displacements into this equation and rearranging:

$$(T/v_o^o)(s_{\bar{x}\bar{x}}/G) = (T/L)[2 + (\lambda/G)][2\pi/(C-1)][C(M_1 - m_1)\sinh(m_1 lT/2) - (M_2 - m_2)\sinh(m_2 lT/2)] \tag{18}$$

Values of $(T/v_o^o)(s_{\bar{x}\bar{x}}/G)$ as a function of L/T are shown in Fig. 5. Let us consider the deep beam shown in Fig. 3 to illustrate the use of Fig. 5. The wavelength/thickness ratio, Fig. 3B, was about $L/T \approx 4.6$ at the time of buckling.

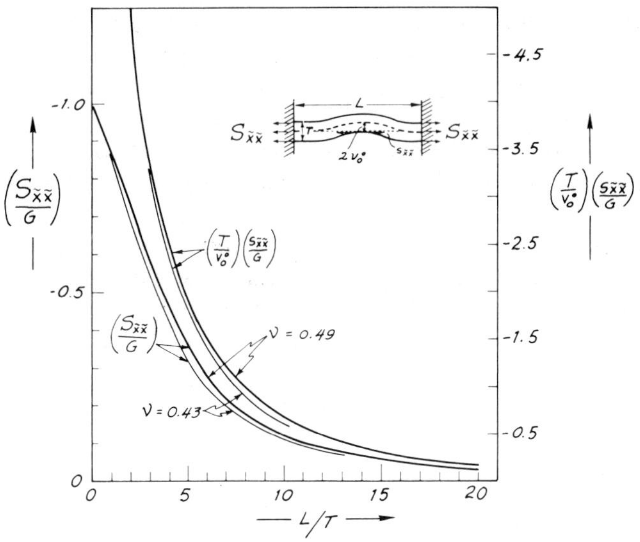

Fig. 5. Theoretical relations between normalized initial axial stress, $S_{\tilde{x}\tilde{x}}$, and wavelength, L, and between incremental axial stress, $s_{\tilde{x}\tilde{x}}$, at midlength of inner arc and wavelength of isolated layer. One pair of curves for nearly incompressible materials, $\nu = 0.49$, and the other pair for materials in which $\nu = 0.43$.

According to Fig. 5, the critical axial compression is approximately:

$(S_{\tilde{x}\tilde{x}}/G) \approx -0.41$

Also, according to Fig. 5 the value of incremental compression was, approximately:

$(T/v_o^o)(s_{\tilde{x}\tilde{x}}/G) \approx -2.2$

so that: $(s_{\tilde{x}\tilde{x}}/G) \approx -2.2(v_o^o/T)$

Thus, in order for $f^* = S_{\tilde{x}\tilde{x}} + s_{\tilde{x}\tilde{x}} + G \approx 0$:

$v_o^o/T \approx 0.27$

The ratio of the amplitude to the thickness at the time of pinching of the deep beam of rubber, according to Fig. 3C, which was taken immediately prior to pinching, was about 0.26, so there is close agreement between the theoretical value of 0.27 and the experimental value of 0.26 for the rubber beam.

Similar calculations for the gelatin beams shown in Fig. 4 indicate less close agreement between theoretical and actual values of v_o^o/T at time of pinching, perhaps in part because of some permanent deformation of the gelatin. In any case, for the deepest beam, the L/T ratio was initially about 1.7, and the theoretical critical value of v_o^o/T indicated by Fig. 5 is 0.04. The actual value of v_0°/T at middepth immediately before pinching was about 0.09. For the second beam from the top in Fig. 4, L/T was 4.5 and the critical value of v_0°/T is 0.31. The actual value of v_0°/T was about 0.46. Finally, for the thinnest layer, L/T was about 7 and v_0°/T theoretically must be about 0.72 for pinching, yet v_0°/T actually was slightly greater than 1.2 *without* pinching (Fig. 4N). The amplitude of the thin layer, of course, greatly exceeded the limits of validity of the linearized theory.

Thus a pinch apparently forms as a result of high, initial axial compression or a result of a combination of initial axial compression and a second-order effect of the incremental axial compression induced at the cores of the buckles. The criterion:

$$f = S_{\bar{x}\bar{x}} + G \approx 0$$

or: $f^* = S_{\bar{x}\bar{x}} + s_{\bar{x}\bar{x}} + G \approx 0$

closely approximates the state of stress at the time of pinching.

GENERALIZED SOLUTIONS FOR MULTILAYERED BODIES

The solution for the buckling of a single, isolated layer discussed in preceding pages is fundamental to an understanding of folding of multilayers. Indeed, it is one basis for our understanding of chevron and concentric-like folding of elastic multilayers, as we will discuss in the following pages. Now we will consider folding of a single layer in a soft medium and folding of multilayers, using the linearized theory. The derivations are tedious so, in order to keep them brief, we will develop the solution for the general problem involving an indefinite number of layers. The method of determining wavelengths is similar to that developed by Ramberg (I, 1961; I, 1970) for viscous materials. After indicating the method of solution we will present some numerical evaluations of the solutions and compare them with those of Ramberg, with some experiments and with elementary folding theory. We will describe some experiments using multilayers of rubber and gelatin which show transitions from sinusoidal to concentric-like to chevron folding. Finally, we will show how the nonlinear theory of elasticity explains the transitions.

Let us designate the layers with j and the interfaces between layers with i (Fig. 6). Each layer is characterized by shear modulus, G_j, Poisson's ratio, ν_j, and thickness, t_j. The initial stress $S_{\bar{y}\bar{y}}$ is the same in each layer, as required for equilibrium. The initial axial stress, $(S_{\bar{x}\bar{x}})_j$, in each layer can be calculated by assuming that the initial axial strain in the x-direction, $\partial U/\partial x$, is the same in each layer and that the materials are compressible and linear. Then, stress—strain equations II, 28 indicate that:

$$(S_{\bar{x}\bar{x}})_j = A_j(\partial U/\partial x) + B_j(\partial V/\partial y)_j$$

$$S_{\bar{y}\bar{y}} = A_j(\partial v/\partial y)_j + B_j(\partial U/\partial x)$$

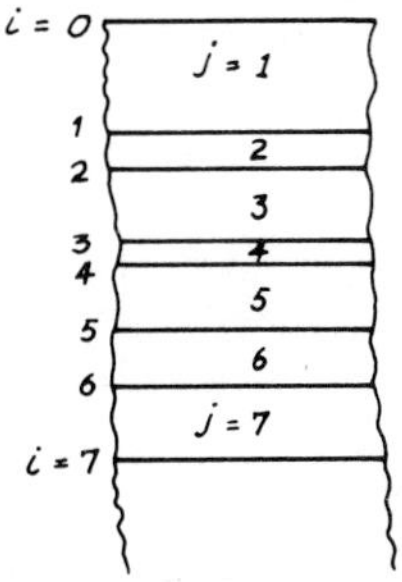

Fig. 6. Indices for identifying interfaces, i, and layers, j, in multilayered body.

Solving the second equation for $(\partial V/\partial y)_j$ and substituting into the first:

$$(S_{\bar{x}\bar{x}})_j = [(A^2_j - B^2_j)/A_j](\partial U/\partial x) + (B_j/A_j)S_{\bar{y}\bar{y}} \tag{19}$$

Now, if each of the layers is bonded to its neighbors at interfaces, the displacements, v_{0j} and u_{0j} and the boundary stresses, $\tau_{\bar{y}y}$ and $\tau_{\bar{y}x}$, must be the same at each interface. The expressions for the displacements and boundary stresses can be derived with eqs. 7b, 9b, 2c and 2d, by setting $y_i = \pm t_j/2$ for each layer.

$$u_{0j} = [C_{5j}M_{1j}\cosh(m_{1j}l\gamma_j t_j/2) + C_{6j}M_{1j}\sinh(m_{1j}l\gamma_j t_j/2) + C_{7j}M_{2j}\cosh(m_{2j}l\gamma_j t_j/2) + C_{8j}M_{2j}\sinh(m_{2j}l\gamma_j t_j/2)]\cos(lx) \tag{20a}$$

$$v_{0j} = [C_{5j}\sinh(m_{1j}l\gamma_j t_j/2) + C_{6j}\cosh(m_{1j}l\gamma_j t_j/2) + C_{7j}\sinh(m_{2j}l\gamma_j t_j/2) + C_{8j}\cosh(m_{2j}l\gamma_j t_j/2)]\sin(lx) \tag{20b}$$

$$(\tau_{\bar{y}x})_j = (G_j + S_{\bar{y}\bar{y}})l[C_{5j}M_{1j}m_{1j}\sinh(m_{1j}l\gamma_j t_j/2) + C_{6j}M_{1j}m_{1j}\cosh(m_{1j}l\gamma_j t_j/2) + C_{7j}M_{2j}m_{2j}\sinh(m_{2j}l\gamma_j t_j/2) + C_{8j}M_{2j}m_{2j}\cosh(m_{2j}l\gamma_j t_j/2)]\cos(lx) + G_j l[v_{0j}\,\mathrm{ctn}(lx)] \tag{20c}$$

$$(\tau_{\bar{y}y})_j = S_{\bar{y}\bar{y}} + A_j l[C_{5j}m_{1j}\cosh(m_{1j}l\gamma_j t_j/2) + C_{6j}m_{1j}\sinh(m_{1j}l\gamma_j t_j/2) + C_{7j}m_{2j}\cosh(m_{2j}l\gamma_j t_j/2) + C_{8j}m_{2j}\sinh(m_{2j}l\gamma_j t_j/2)]\sin(lx) - B_j l[u_{0j}\tan(lx)] \tag{20d}$$

where v_{0j} and u_{0j} in eqs. 20c and 20d are given by eqs. 20a and 20b, and where:

$$\gamma_j = +1;\ \text{for tops of layers} \qquad \gamma_j = -1;\ \text{for bottoms of layers} \tag{21}$$

The coefficients, m_{1j}, m_{2j}, M_{1j}, and M_{2j} can be derived from eqs. 8 and 10 by subscripting variables and constants:

$$m_{1j} = +\sqrt{\mathrm{A}_j + \sqrt{\mathrm{A}_j^2 - \mathrm{B}_j}}$$
$$m_{2j} = +\sqrt{\mathrm{A}_j - \sqrt{\mathrm{A}_j^2 - \mathrm{B}_j}} \qquad (22a)$$

$$\mathrm{A}_j = [A_j^2 - \{B_j + (S_{\bar{x}\bar{x}})_j + G_j\}(B_j + S_{\bar{y}\bar{y}} + G_j) + (S_{\bar{y}\bar{y}} + G_j)\{(S_{\bar{x}\bar{x}})_j + G_j\}]/[2A_j(S_{\bar{y}\bar{y}} + G_j)] \qquad (22b)$$

$$\mathrm{B}_j = [(S_{\bar{x}\bar{x}})_j + G_j]/(S_{\bar{y}\bar{y}} + G_j)$$

$$M_{1j} = [m_{1j}\{B_j + (S_{\bar{x}\bar{x}})_j + G_j\}]/[A_j - (S_{\bar{y}\bar{y}} + G_j)m_{1j}^2]$$
$$M_{2j} = [m_{2j}\{B_j + (S_{\bar{x}\bar{x}})_j + G_j\}]/[A_j - (S_{\bar{y}\bar{y}} + G_j)m_{2j}^2] \qquad (22c)$$

Thus we have derived expressions for the boundary stresses and the displacements for the top and bottom of each layer. The next step is to equate boundary values for each interface. Let us consider the first three interfaces, $i = 0$, 1 and 2 (Fig. 6). The interface 0, the top of the uppermost layer, is free so the displacements are unknown but the boundary stresses are known and the boundary conditions there are:

$$y_1 = +t_1/2; \quad (\tau_{\bar{y}x})_1 = 0; \quad (\tau_{\bar{y}y})_1 = S_{\bar{y}\bar{y}}$$

and according to eqs. 20c and 20d:

$$a_{11}C_{51} + a_{12}C_{61} + a_{13}C_{71} + a_{14}C_{81} = 0$$
$$a_{21}C_{51} + a_{22}C_{61} + a_{23}C_{71} + a_{24}C_{81} = 0 \qquad (23)$$

where, for example:

$$a_{11} = (G_1 + S_{\bar{y}\bar{y}})M_{11}m_{11}\sinh(m_{11}lt_1/2) + G_1\sinh(m_{11}lt_1/2)$$

......

......

$$a_{24} = A_1 m_{21}\sinh(m_{21}lt_1/2) - B_1 M_{21}\sinh(m_{21}lt_1/2)$$

For interface 1, the bottom of the medium and the top of the first stiff layer (Fig. 6), both the boundary stresses and the boundary displacements must match, so the boundary conditions are:

at $y_1 = -t_1/2$ and $y_2 = +t_2/2$; $\gamma_1 = -1$, $\gamma_2 = +1$

$$\begin{aligned} u_{01} - u_{02} = 0 = {} & a_{31}C_{51} + a_{32}C_{61} + a_{33}C_{71} + a_{34}C_{81} \\ & + a_{35}C_{52} + a_{36}C_{62} + a_{37}C_{72} + a_{38}C_{82} = 0 \\ v_{01} - v_{02} = 0 = {} & a_{41}C_{51} \cdots\cdots \cdots\cdots + a_{48}C_{82} = 0 \\ (\tau_{\bar{y}x})_1 - (\tau_{\bar{y}x})_2 = 0 = {} & a_{51}C_{51} \cdots\cdots \cdots\cdots + a_{58}C_{82} = 0 \\ (\tau_{\bar{y}y})_1 - (\tau_{\bar{y}y})_2 = 0 = {} & a_{61}C_{51} \cdots\cdots \cdots\cdots + a_{68}C_{82} = 0 \end{aligned} \qquad (24)$$

The same procedure is repeated for interface 2 and so on throughout the entire multilayer and medium, from the top interface to the bottom interface. For the first three interfaces the matrix of coefficients is, from eqs. 23 and 24:

$$\begin{array}{ll}
\text{Interface 0} & \left\{ \begin{array}{l} a_{11}\ a_{12}\ a_{13}\ a_{14}\ 0\quad 0\quad 0\quad 0\quad 0 \ldots \\ a_{21}\ a_{22}\ a_{23}\ a_{24}\ 0 \ldots\ldots\ldots \end{array} \right. \\
\text{Interface 1} & \left\{ \begin{array}{l} a_{31}\ a_{32}\ a_{33}\ a_{34}\ a_{35}\ a_{36}\ a_{37}\ a_{38}\quad 0\quad 0 \ldots\ldots \\ a_{41} \ldots\ldots\ldots\ldots\ldots\ldots\ldots a_{48}\quad 0 \ldots\ldots\ldots \\ a_{51} \ldots\ldots\ldots\ldots\ldots\ldots\ldots a_{58}\quad 0 \ldots\ldots\ldots \\ a_{61}\ a_{62}\ a_{63}\ a_{64}\ a_{65}\ a_{66}\ a_{67}\ a_{68}\quad 0\quad 0 \ldots\ldots \end{array} \right. \\
\text{Interface 2} & \left\{ \begin{array}{l} 0\quad 0\quad 0\quad 0\quad a_{75}\ a_{76}\ a_{77}\ a_{78}\quad a_{79}\ a_{7,10}\ a_{7,11}\ a_{7,12}\ 0\ 0\ 0 \\ 0 \ldots\ldots\ldots 0\quad a_{85} \ldots\ldots\ldots\ldots\ldots\ldots\ldots\ldots a_{8,12} \ldots\ldots \\ 0 \ldots\ldots\ldots 0\quad a_{95} \ldots\ldots\ldots\ldots\ldots\ldots\ldots\ldots a_{9,12} \ldots\ldots \\ 0 \ldots\ldots\ldots 0\quad a_{10,5} \ldots\ldots\ldots\ldots\ldots\ldots\ldots\ldots a_{10,12} \ldots\ldots \end{array} \right. \\
\text{Interface 3} & \left\{ \ 0 \ldots\ldots\ldots\ldots 0 \ldots\ldots\ldots\ldots a_{11,9} \ldots\ldots\ldots\ldots\ldots \right.
\end{array} \qquad (25)$$

Eqs. 23 and 24 and similar equations for other interfaces represent a system of linear equations defining relations amongst the arbitrary constants, C_{51}, C_{61}, C_{71}, C_{81} ... C_{5j}, C_{6j}, C_{7j}, C_{8j}. The equations are homogeneous, so there is no solution unless the determinant of the coefficients of the arbitrary constants, that is, the determinant of matrix 25, vanishes. If the determinant vanishes, the incremental displacements associated with sinusoidal folding are possible, thus it is this condition that is of interest in folding theory. The way we solve eqs. 25 is to adjust values of normalized axial stress, $(S_{\bar{x}\bar{x}})_j/G_j$, in stiff layers, that is, adjust the initial axial strain, for each value of wavelength/thickness ratio, L/t_j, in stiff layers until the determinant is zero. Each pair of values of initial stress and wavelength define a point on a curve, and the minimum of the curve represents the wavelength for which the axial compression is minimized, as in elementary theory of folding (e.g., Johnson, I, 1970, p. 158). For example, Fig. 7 shows relations between axial stress and L/T for a single stiff layer between two soft media for various modulus ratios of soft to stiff layers, assuming Poisson's ratio is 0.49 for both materials. The wavelength/thickness ratio, L_m/T for which the axial compression is minimized is shown for three of the curves in Fig. 7, as determined by an iterative procedure that locates minima. Fig. 7 shows that the L/T ratio decreases with increasing ratio of moduli, G_3/G_1, of soft to stiff materials, as in the elementary theory. Further, it shows that the minimum becomes less well-defined as the ratio increases and that, in the limit where the ratio becomes unity, there is no minimum.

For a problem involving many layers, one could write the entire matrix for the problem and solve for its determinant by normal methods. However, the values in the matrix 25 cluster about the diagonal, so we solve it in blocks by usual pivot methods (e.g., Gere and Weaver, 1965). Thus, we solve it to line 9, coefficient a_{98} in eqs. 25, then replace coefficients a_{11} ... a_{14} with coefficients a_{98} ... $a_{9,12}$, after modification by the pivot method, and replace coefficients a_{21} ... a_{24} by modified coefficients $a_{10,8}$ ... $a_{10,12}$, recompute the

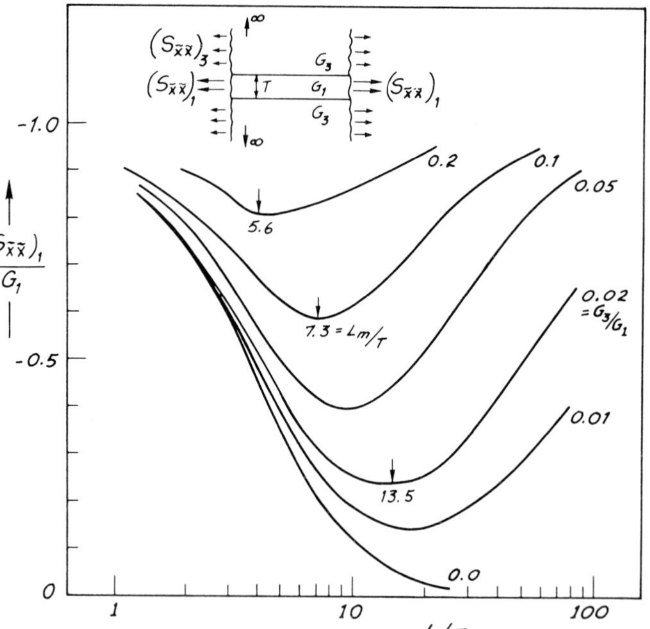

Fig. 7. Examples of curves relating normalized initial axial stress in stiff layer and wavelength/thickness ratio of stiff layer for various modulus ratios of stiff and soft materials. Minima define wavelengths, L_m/T, that require least axial compression. Poisson's ratio assumed to be 0.49.

rest of the coefficients in the block, eqs. 25, and proceed with the pivoting. In this way we are able to solve any size of matrix with a small computer, using only the amount of storage required for an 8×10 matrix. Of course, the larger the matrix, the longer it takes to solve the problem. For problems where the pattern of layers is symmetric, one can stop the calculations at the top interface of the middle layer and save considerable computer time.

FOLDING OF A HOMOGENEOUS STIFF LAYER IN A SOFT MEDIUM

One of the classic problems in structural geology is the folding of a single stiff layer in a soft medium. The viscosity analog of the problem was solved by Ramberg (I, 1961; I, 1970), and our solution for elastic materials closely follows Ramberg's. The problem has received an inordinant amount of attention in geological literature (see Part I), considering the relative scarcity of single-layer folds in nature. Nevertheless, we contribute results of one more solution here in Figs. 7 and 8.

Relations between wavelength and axial compression for various modulus ratios, and values of wavelength ratios, L_m/T, for which axial compression is minimized for several modulus ratios are shown in Fig. 7. Fig. 8 shows relations between L_m/T and the modulus ratio for a wide range of modulus ratios. Both Figs. 7 and 8 show that, if the moduli of the layer and medium are nearly the same, the wavelength is small, approaching zero as the moduli become the same. Fig. 7 shows that the magnitude of the axial compression required to buckle the stiff layer approaches the magnitude of the shear modulus as the wavelength decreases.

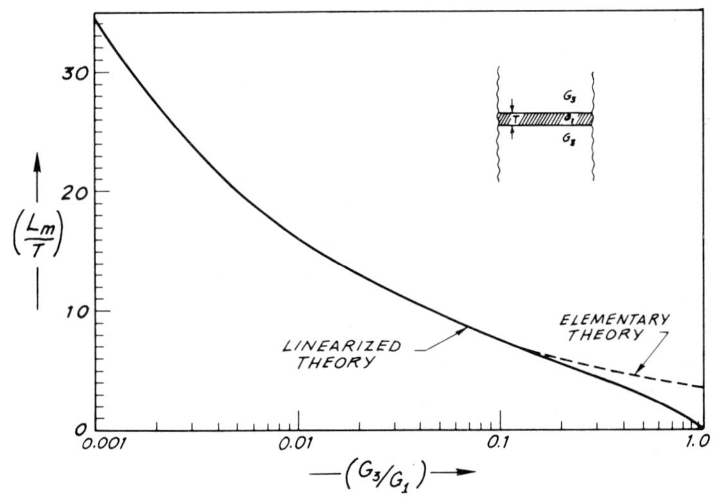

Fig. 8. Relation between wavelength/thickness ratio and modulus ratio for single stiff layer in soft media as derived from the linearized theory and the elementary theory. Poisson's ratio assumed to be 0.49.

Also shown in Fig. 8 is the relation between wavelength and modulus ratio according to the elementary theory (e.g., Currie et al., I, 1962):

$$L_m/T = 2\pi[G_1(1-\nu_3^2)(1+\nu_1)/\{G_3(1-\nu_1^2)(1+\nu_3)6\}]^{1/3} \tag{26}$$

The correlation between the predictions of the linearized theory and the elementary theory is striking. Apparently, for modulus ratios, G_3/G_1, lower than about 1/5, or even 1/3, the two theories provide nearly the same estimate of wavelength/thickness ratio. Thus, apparently unlike the elementary theory of folding of viscous materials (Sherwin and Chapple, I, 1968), the elementary theory of folding of an elastic layer in a soft elastic medium is quite accurate, even for low contrasts in properties. The correlation is surprising. The linearized theory assumes bonding between layer and medium whereas the elementary theory assumes free slip. The linearized theory automatically incorporates effects of shear as well as bending within the stiff layer and incorporates the change in length of the layer during the initial loading whereas the elementary theory considers effects of bending only. There must be compensating errors in the elementary theory which we do not understand. In any case, if one is interested primarily in wavelength/thickness ratios for a single stiff layer in a soft medium, one might as well use the elementary theory and save considerable effort!

Based on experimental observation and theoretical analysis of "pinching" in isolated layers, discussed in earlier pages, we would suggest that, if the parameter f or f^* approaches zero, the stiff layer will "yield" in cores of folds, much as a plastic material, except that the strains would be recoverable. Thus, we suggest that the same phenomenon that leads to pinching in an isolated stiff layer leads to chevron-like folding in a stiff layer embedded in a soft medium, where the curvature is concentrated in the hinges and where the limbs are relatively straight. The magnitudes of the parameter $(T/v_0^0)(s_{\bar{x}\bar{x}}/G_1)$ in cores of folds in stiff layers have been computed much as they were for an isolated layer (e.g., eq. 18), and the magnitudes of the parameter as a function of (L_m/T) are shown in Fig. 9. As for an isolated layer, the values are in terms of units of amplitude relative to thickness of the stiff layer. Also shown in Fig. 9 are the minimum values of initial axial compression in the stiff layers required for folding. Figs. 8 and 9 show that, for wavelengths less than about ten times the thickness, $L_m/T \lesssim 10$, that is, for modulus ratios greater than about 0.05, $G_3/G_1 \gtrsim 1/20$, the factor f^*:

$$f^* = S_{\bar{x}\bar{x}} + s_{\bar{x}\bar{x}} + G$$

can approach zero even though the amplitude of the fold is less than the thickness of the layer, $v_0^0/T \lesssim 1.0$. This conclusion can be verified as follows. According to Fig. 9, if the ratio L_m/T is 10, the ratio $S_{\bar{x}\bar{x}}/G_1$ is about -0.385 and the ratio $(T/v_0^0)(s_{\bar{x}\bar{x}}/G_1)$ is about -0.6. Thus, if $v_0^0/T = 1$:

$$f^*/G_1 = 1 + S_{\bar{x}\bar{x}}/G_1 + s_{\bar{x}\bar{x}}/G_1$$

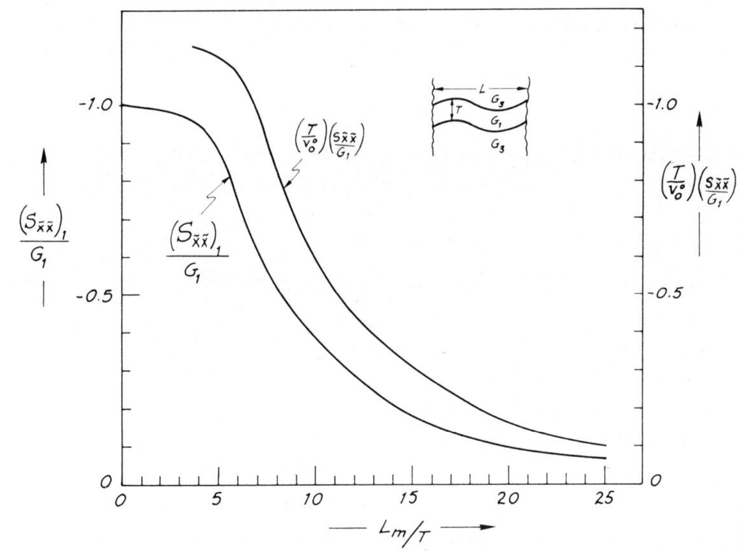

Fig. 9. Relation between minimum axial compression, $S_{\bar{x}\bar{x}}/G_1$, and corresponding wavelength ratio, L_m/T, and relation between incremental axial compression in core of stiff layer and wavelength ratio according to linearized theory.

so that: $f^*/G_1 \approx 1 - 0.6 - 0.385 \approx 0.015$

which is nearly zero. According to Fig. 8, the ratio $L_m/T = 10$ corresponds with the modulus ratio of about $G_3/G_1 = 1/20$.

Single-layer folds were produced in a strip of stiff elastic rubber embedded in soft gelatin. The apparent modulus ratio, G_3/G_1, was about 0.045. According to Fig. 8, the wavelength/thickness ratio of dominant folds should be about 10 and according to Fig. 9:

$(S_{\bar{x}\bar{x}})_1/G_1 \approx -0.38$ and $(s_{\bar{x}\bar{x}})_1/G_1 \approx -0.6(v_0^0/T)$

so that for the parameter f^* to be zero:

$v_0^0/T \approx 1$

According to the experiment, the wavelength/thickness ratio was about 11 rather than 10 as predicted theoretically, so that the effective modulus ratio was about 0.033 rather than 0.045 as measured in tension tests. Thus, the critical ratio of v_0^0/T should be about 1.27. The experimental folds maintained a smoothly curved, sinusoidal-like form for amplitude/thickness ratios less than about 1.5. When the folds had higher amplitude ratios, they clearly had straightened limbs but the hinges remained rounded. With slightly further shortening, however, the stiff layer separated from the gelatin at cores of folds, the stiff layers pinched at cores and the limbs straightened further, producing a perfect chevron pattern. Therefore, chevron folds can be produced even in a single, isolated elastic layer in relatively stiff media.

FOLDING OF ELASTIC MULTILAYERS

Experimental

Several experiments with multilayers of black rubber and photoelastic gelatin were described in Part I. We will now describe a series of experiments with multilayers of gelatin and photoelastic rubber, the properties of which were described in earlier pages. The strips of rubber were 0.25 cm thick, 1.3 cm wide and 25 cm long. The spacings between the rubber strips were different for different experiments, the thickness of gelatin interbeds ranging from one to nine times the thickness of the rubber strips. The gelatin was poured in two stages. First, the rubber strips were placed in a special mold that held the strips straight and at the proper spacing, and gelatin was poured around the parts of the strips that extended into an open space of the mold. After the gelatin set, the rubber strips and gelatin were removed together, placed in a form, and the remainder of the gelatin interbeds and media poured. After the multilayer cured for 18—24 hours at room temperature, it was removed and placed in a five-sided box constructed of thick plexiglass sheets for the experiment. A platen was advanced by screws into the open side of the box so that the multilayer essentially was subjected to uniaxial loading under plane-strain conditions. The media themselves were unconfined and free to expand in one dimension except for friction between the gelatin and the plexiglass plates.

Figs. 10A—C show three of the multilayers. One consisted of 16 stiff rubber strips and 15 soft gelatin strips of equal thickness of about 0.25 cm, forming a composite about 8.1 cm thick and 25 cm long. Each medium was 7.4 cm thick (Fig. 10A1). The apparent modulus ratio, E_1/E_2, of stiff to soft layers was about 22. A second (Fig. 10B1) had essentially the same overall dimensions, but the thickness ratio, t_1/t_2, of stiff to soft layers was about 1/3 and the apparent modulus ratio was about 44. The third (Fig. 10Cl) had the same geometry as the second, but the gelatin interbeds were softer than the gelatin medium, and the modulus ratios were $E_1/E_3 \approx 28$ and $E_1/E_2 \approx 67$, where the subscript 1 refers to the rubber, 2 to the interlayers and 3 to the media.

Lines were drawn on the gelatin normal to layering before each experiment so that the distribution of $\partial u_0/\partial y$ and $\partial v_0/\partial x$, that is, the incremental shear strain and rotation, could be visualized. The apparent curvature of one end of the multilayer shown in Figs. 10A is merely the edge of the polarizing plate. The multilayers had plane ends. Also, the tendency for the fold patterns to be more intense near one end of the models than the other is a reflection of friction between the plexiglass sheets and the models. The problem was partly corrected in later experiments, such as that shown in Figs. 10C, by using silicone grease instead of silicone oil as lubricant.

The three experiments shown in Figs. 10A—C are interesting for their contrasts as well as their similarities of fold pattern. The fold pattern in two

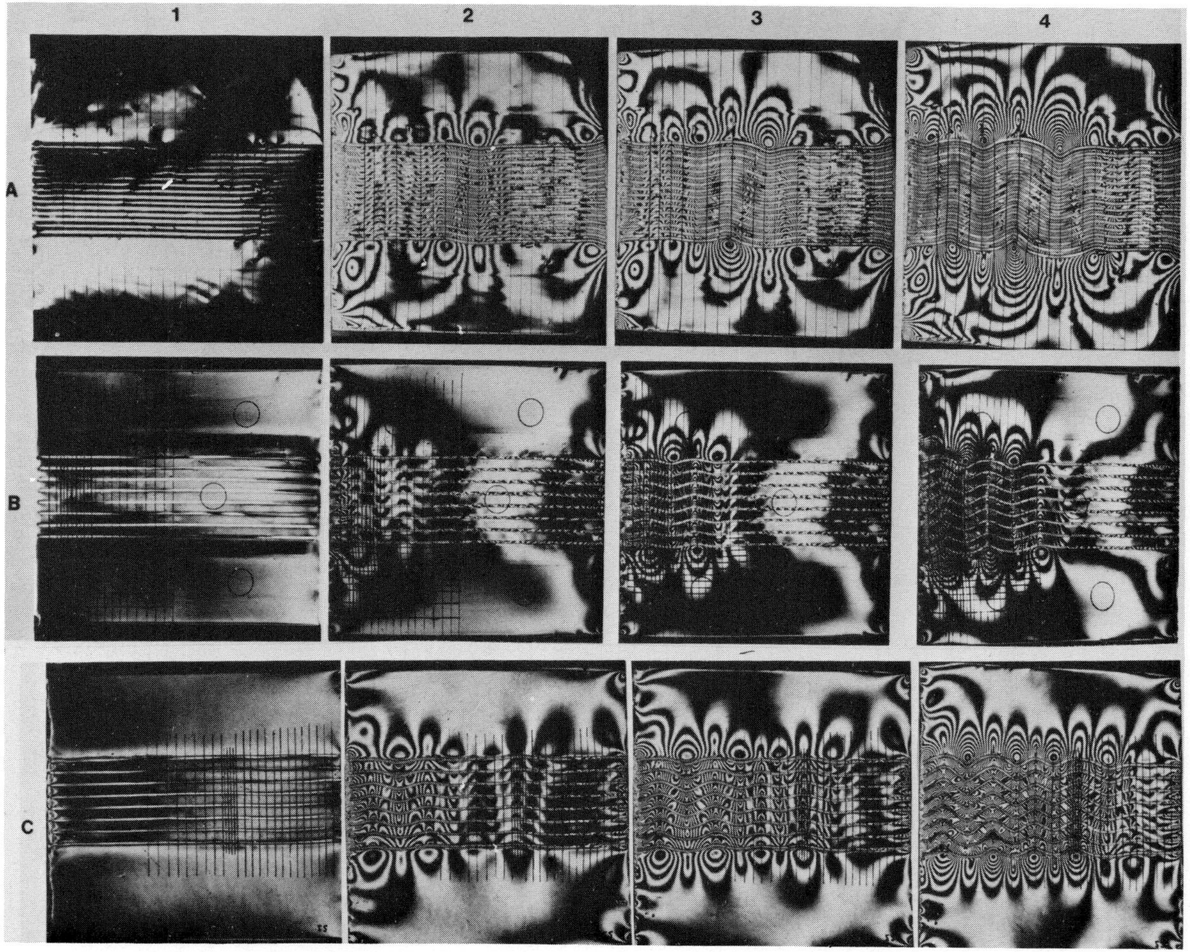

Fig. 10. Folding of multilayers of soft gelatin and stiff photoelastic rubber in medium of gelatin subjected to axial shortening. A. Thickness ratio t_2/t_1 = 1; modulus ratios, $E_1/E_2 = E_1/E_3$ = 22. B. Thickness ratio t_2/t_1 = 3; modulus ratios, $E_1/E_2 = E_1/E_3$ = 44. C. Thickness ratio t_2/t_1 = 3, modulus ratios, E_1/E_2 = 67, $E_1/E_3 \approx$ 28. Soft interlayers, firm media, stiff layers.

experiments began as waves with roughly constant amplitudes and wavelengths throughout the stack of layers (Figs. 10A2 and 10B2). The patterns might be called sinusoidal in this respect. The peculiar, kink-like pattern at midlength in the multilayer shown in Fig. 10A2 apparently was a result of an accidental alignment of air bubbles in the gelatin layers which are visible in Fig. 10A1. The sinusoidal-like pattern at midlength in the model shown in Fig. 10A2 eventually was transformed into a concentric-like pattern (Fig. 10A3) and then into a combination of concentric-like folds near the crests and chevron-like folds in the cores of the concentric-like folds (Fig. 10A4). The folds of chevron-like form had rounded hinges but nearly straight limbs. Thus the sequence of fold development shown in Figs. 10A is the same as that shown in three experiments reported in fig. 19 of Part I.

The fold pattern in the multilayer shown in Fig. 10B, with a thickness ratio of $t_1/t_2 = 1/3$, shows the same sequence but the concentric-like pattern is more subdued; the angle between the imaginary radii bounding the concentric-like arcs is smaller. The angle between imaginary radii is about 15° for a fold shown in Fig. 10B2, whereas it is about 30° for the fold shown at midlength in Fig. 10A2.

The fold pattern shown in Fig. 10C is somewhat different from the others during early stages of folding but becomes similar to that shown in Fig. 10A during final stages. The first-formed waves (Fig. 10C2) had higher amplitudes at midheight than near the upper and lower boundaries of the multilayers. Also, the wavelengths were slightly less in the rubber strip at the crests of the folds at each interface with the media than in the second rubber strip from the interface. The wavelengths can readily be measured by recognizing the photoelastic pattern that occurs at inflection points in the rubber strips. As overall shortening proceeded, the folds assumed a concentric-like pattern (Fig. 10C3) and then a chevron-like pattern (Fig. 10C4).

One striking feature of the folds shown in Figs. 10, in fact in all the folds we have produced in multilayers, is that lines originally drawn vertically, normal to layers, remain, *on the average*, vertical. Indeed, the lines rotate in stiff layers and are somewhat curved in the soft layers, but the average, taken over pairs of stiff and soft layers, produces a line which remains essentially vertical, regardless of amplitude of folding. Close examination of Fig. 10A3 is required to detect the distortion of the lines locally as the lines cross from stiff to soft layers. The rotations are most apparent near the upper and lower boundaries of the multilayer.

The last experiments we will describe were performed by Strömgård (Roberts and Strömgård, I, 1972), using rubber and gelatin. Each multilayer consisted of thick media of gelatin, two stiff rubber strips and a gelatin interbed, roughly twice as thick as the rubber strip (Fig. 11). In one experiment, the interbed, 2, (Fig. 11), was about half as stiff as the medium and in the second, Fig. 11B, the interbed was twice as stiff as the medium. The multilayers were deformed by axial shortening, as in the experiments already described. As shown in Fig. 11A, the multilayer with the softer interbed maintained a

A

B

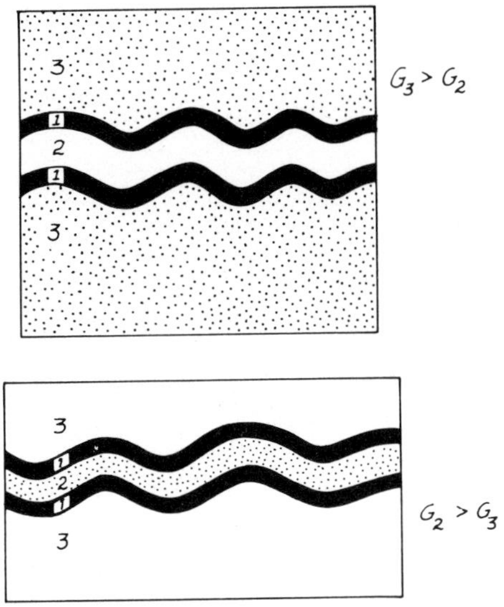

Fig. 11. Sketches of folds in experimental multilayers produced by Strömgård. A. Sinusoidal-like pattern of folds in multilayer with soft interlayer, firm media. t_2/t_1 = 2, G_1/G_2 = 144, G_1/G_3 = 76. B. Concentric-like pattern in multilayer with firm interlayer, soft media. t_2/t_1 = 2, G_1/G_2 = 76, G_1/G_3 = 144.

sinusoidal-like form for relatively large deflections whereas the multilayer with the stiffer interbed, Fig. 11B, developed a concentric-like form. This experiment is the only one with which we are familiar where sinusoidal-like waveforms were maintained during folding to high amplitudes in elastic multilayers containing interbeds the thickness of which is less than the wavelength in the stiff layers. In all other experiments the waveforms assume concentric-like shapes. A suitable theory should explain not only why most multilayers fold into concentric-like forms but why the multilayer with the soft interbed (Fig. 11A) maintained sinusoidal-like forms, even for high amplitude.

Theoretical wavelengths of sinusoidal folds

We have solved several problems of sinusoidal folding of multilayers confined by elastic or rigid media, using the theory presented in earlier pages (eqs. 19—25), but the number of variables is so great that the results are difficult to present in useful form. Thus, we will restrict our attention here to comparison of theoretical results with those of one of our experiments and with two experimental and theoretical results by Ramberg and Strömgård (I, 1971), who used a theory of folding of viscous layers.

Ramberg and Strömgård (I, 1971) performed some experiments with multilayers of rubber and gelatin and compared the wavelengths developed experimentally with those predicted theoretically with Ramberg's (I, 1970)

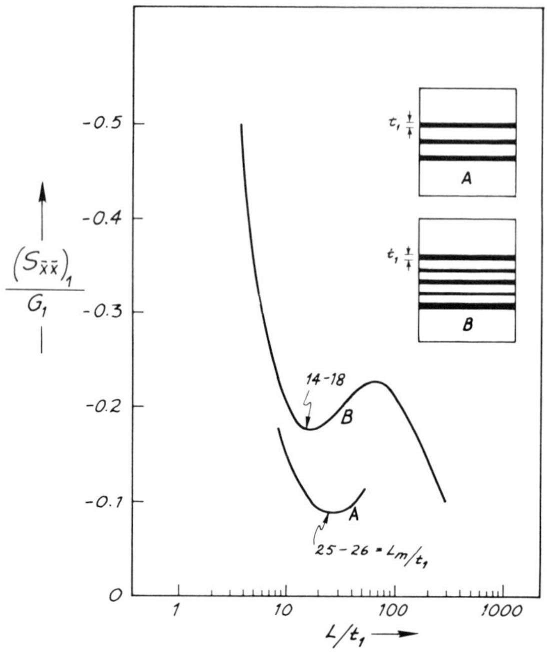

Fig. 12. Relations between initial axial compression and wavelengths for two experimental multilayers.

theory. One experimental multilayer contained five layers, three stiff and two soft, with a modulus ratio G_1/G_2 of about 126, confined by two thick, soft media of gelatin. Ramberg and Strömgård computed a wavelength/thickness ratio of about 25 for the stiff layers and measured a value of 22. Our theory, Fig. 12 curve A, assuming Poisson's ratio of 0.43 for the gelatin and 0.46 for the rubber (K.E. Strömgård, personal communication, 1971), indicates a wavelength/thickness ratio of between 25 and 26. One experimental multilayer was designed to test the ability of Ramberg's theory to predict two simultaneous wavelengths in a multilayer (op. cit., fig. 1). Ramberg computed a wavelength for general buckling of the multilayer to be about 16 times the thickness of the stiff layers and measured a wavelength of 17. Our theory, Fig. 12, curve B, indicates a wavelength of between 14 and 18 times the thickness. Thus, there appears to be close correlation between the wavelength/thickness values we would predict and Ramberg would predict, using quite different theories.

We computed initial wavelengths for the experimental folds shown in Fig. 10C to be, L/t_1 = 21—22. The value measured in Fig. 10C2 is between 21.2 and 21.5.

The comparisons between computed and actual wavelengths of folds in the experimental multilayers are encouraging, indeed. However, if one is interested primarily in estimates of wavelengths for multilayers confined by soft media, the lengthy computations probably are generally unnecessary. As indicated in Fig. 8, predictions of wavelengths with the elementary theory of folding of single layers closely correlate with those with the linearized theory, so we should imagine that the elementary theory also will predict accurately wavelengths of folds in multilayers. One elementary theory of buckling of multilayers assumes that all stiff layers buckle with the same wavelength and curvature, that the contacts between stiff layers are frictionless, and that there is a soft medium on each side of the multilayer. In this case (e.g., Currie et al., I, 1962; Johnson, I, 1970, p. 152), the wavelength, L_m, requiring the minimum axial compression is:

$$L_m = 2\pi \left[\left(\sum_{n=1}^{N} B_n t_n^3 \right) / 6B_0 \right]^{1/3} \qquad (27)$$

where Σ indicates summation for all N layers, B_n is a modulus of a stiff layer and B_0 of the medium:

$$B_n = 2G_n(1 + \nu_n)/(1 - \nu_n^2)$$

Let us assume that the interbeds are sufficiently soft to behave essentially as lubricants between stiff layers so that N in eq. 27 is the number of *stiff* layers. Then, using eq. 27, we compute an L_m/t_1 ratio of about 25 for experimental model A, Fig. 12, whereas the experimental data indicated 22, Ramberg's theory, 25 and the linearized theory, between 25 and 26. For experimental model B, Fig. 12, we compute a ratio of 17.4 with eq. 27, the experimental data indicated 17, Ramberg's theory 16, and the linearized theory between 14 and 18. Finally, for the multilayer shown in Fig. 10C, we compute an L_m/t_1 ratio of 21 with eq. 27, the experimental data indicate 21.2—21.5 and the linearized theory indicates between 21 and 22. The wavelengths for these examples, therefore, can be estimated closely using the elementary theory, saving considerable time and effort.

Thus, we suggest that if one is interested primarily in estimating wavelengths of sinusoidal folds in elastic multilayers confined by soft media, one should proceed as follows. First, compute the wavelength of folds for a single stiff layer, assuming the layer folds independently in the soft interbeds (using eq. 26). If the thickness of soft interbeds is greater than the wavelength thus computed, the theoretical wavelength should accurately predict wavelengths of folds in stiff layers. If, however, the thickness of soft interbeds is less than the wavelength computed with eq. 26, the multilayer will fold as a unit and eq. 27 is the appropriate equation for predicting wavelengths, whether the stiff layers are of the same or of different thicknesses and stiffnesses. Finally, if the multilayer is confined above and below by very stiff or even by rigid media, the wavelength of sinusoidal folds can be computed approximately as the Biot wavelength (Johnson, I, 1970, p. 243; I, 1969):

$$L_B = (4\pi T)^{1/2}[(B_h/B_v)\{t_1^3/12(t_1 + t_2)\}]^{1/4} \qquad (28)$$

where: $B_h = B_1 + B_2[\{(t_1 + t_2)/t_1\}^3 - 1]$

and: $B_v = (t_1 + t_2)B_1B_2/(t_1B_2 + t_2B_1)$

in which 1 refers to stiff and 2 to soft layers, t is thickness of individual layers and T is thickness of the multilayer.

Variations of wavelengths within multilayers

Although elementary folding theory, Ramberg's theory for viscous materials and the linearized theory for elastic materials all predict closely the wavelengths of sinusoidal folds, none of the theories predicts explicitly the development of concentric-like folds in multilayers. The experimental folds (Figs. 10 and fig. 19, Part I) in multilayers of rubber and gelatin become concentric-like even for small amplitudes. Now we will suggest an explanation for the concentric-like form.

The linearized theory can be used to explain concentric-like forms if we assume that the forms can be characterized by wavelengths that change from place to place within a multilayer. According to this analysis, the wavelengths become modified by processes which are higher-order effects of the state of stress induced by the folding itself. Two different, higher-order effects can cause wavelengths in stiff layers at cores of folds in multilayered bodies to decrease and wavelengths in stiff layers at crests to increase, producing overall waveforms that appear to be concentric. First, assume that the media have different elastic properties from the soft interbeds of the multilayer shown schematically in Fig. 13A. If there is no medium, or if the media are softer than the interbeds between stiff layers, the shear stress, τ_3, at the interface between the base of a stiff layer (1) and the medium (3), is less than the shear stress, τ_2, at the interface between the interbed (2) and the top of the stiff layer (1, Fig. 13A). The average gradient in the shear stresses, τ_2 and τ_3, from bottom to top of the lowermost stiff layer (Fig. 13A) is:

$$(\tau_2 - \tau_3)/t_1$$

Further, for equilibrium of stresses, there must be a gradient in the average axial stress, σ (Fig. 13A), along the stiff layer:

$$\frac{\partial\sigma}{\partial x} + \frac{\tau_2 - \tau_3}{t_1} \approx 0$$

Because $\tau_2 > \tau_3$ for the lowermost stiff layer, σ must be a larger compressive stress on the left-hand side of the small element than at the right-hand side, as shown in Fig. 13A. Accordingly, the axial compression in the lowermost stiff layer increases in magnitude from right to left, from a minimum at the crest of the synclinal fold to a maximum at the core of the anticlinal fold, shown in Fig. 13A. Conversely, the gradient in axial compression from the synclinal fold to the anticlinal fold is opposite for the uppermost stiff layer

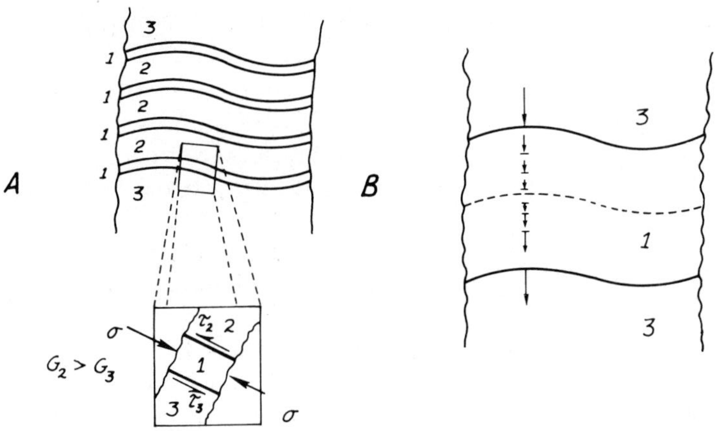

Fig. 13. Schematic diagrams of incremental states of stress that cause sinusoidal pattern to be transformed into concentric-like pattern. A. Gradient in shear stress from stiffer interlayer, 2, to softer medium, 3, across stiff layer, 1, induces gradient in incremental axial stress along stiff layer. B. Interaction between a homogeneous layer, 1, in medium, 3, induces incremental tension at cores of folds and compression at crests.

shown in Fig. 13A because the gradient in shear stress is reversed there. Finally, we can qualitatively combine this knowledge of the increase in axial compression in stiff layers at cores of folds in multilayers with the knowledge that the buckling wavelength of an isolated layer decreases as the axial compression increases (Fig. 2). For the same reason the decrease in axial compression in the stiff layers at the crests of folds in multilayers causes the wavelengths to increase there. Thus, if the medium is softer than the interbeds, the stiff layers near cores of folds will tend to decrease in wavelength, and the layers near the crests of folds will tend to increase in wavelength.

The opposite effect should occur if the media are stiffer than the interbeds; the wavelength in stiff layers at the top and bottom of the multilayer should increase at cores of folds and decrease at crests. We described exactly this phenomenon in the experimental multilayer shown in Fig. 10C2.

The second, higher-order process we will discuss occurs only if there is a confining medium. Let us consider a folded, homogeneous layer, 1, bonded to a medium, 3 (Fig. 13B). The normal stress in the y-direction is zero at mid-height in the layer by symmetry. Further, it becomes a larger tensile stress toward the core of an anticline and a larger compressive stress toward the crest of an anticline (Fig. 13B). Now, suppose that the homogeneous layer is replaced by a multilayer. The same general conclusions apply (Johnson, I, 1969) so that stiff layers near the cores of anticlines are subjected to incremental tension and those near crests to incremental compression in the y-direction. Let us conceptually remove a stiff layer and analyze the higher-order effect of the incremental stress, $s_{\bar{y}\bar{y}}$, on the wavelength of the stiff layer. We assume that the incremental stress enters the analysis the same way as the initial stress, $S_{\bar{y}\bar{y}}$, and it is in this sense that we are now considering a higher-order effect ignored in the linearized theory. As shown in Fig. 2, if there is significant tension, $s_{\bar{y}\bar{y}}$ ($S_{\bar{y}\bar{y}}$), the wavelength of a single layer is decreased and if there is significant compression the wavelength is increased. This correlates with the observation that wavelengths decrease near cores of folds and increase near crests. For example, if the normalized initial axial stress were —0.25, and the confining stress, $S_{\bar{y}\bar{y}}$, were zero, the critical wavelength/thickness ratio, L/T, for an unconfined layer would be about 6.2 (Fig. 2). But if the normalized confining stress were tensile, and $S_{\bar{y}\bar{y}}/G_1 = +0.1$, L/T would be decreased to 5.1 and if $S_{\bar{y}\bar{y}}/G_1 = -0.1$, that is, compressive, L/T would be increased to about 7.3.

In this way we can combine our understanding of the buckling of an isolated layer with our understanding of the state of stress within a multilayer folded sinusoidally to understand why stiff layers within a multilayer tend to buckle with different wavelengths at different places within the multilayer. And thus we can explain why multilayers tend to fold into concentric-like forms which are characterized by groups of stiff layers in which wavelengths decrease toward cores and increase towards crests of folds (Figs. 10).

AN EXPLANATION FOR THE TRANSITION FROM SINUSOIDAL TO CONCENTRIC-LIKE TO CHEVRON FOLDS

The experiments reported here and in Part I show the development of folds in relatively simple experimental materials which closely resemble ideal sinusoidal, concentric and chevron folds, and the theory provides the elements required to understand the transitions from one fold form to another. Now we will combine the experimental and theoretical observations in order to explain the transitions.

We begin with a multilayered sequence of alternating stiff and soft layers, originally horizontal, embedded in confining media, and subjected to axial shortening. Each layer remains straight but shortens and thickens as the multilayer is shortened. The shortening is homogeneous and may be finite. The confining media are unconfined themselves. The stiff and soft layers are subjected to axial compression of different amounts, related to the thicknesses and rheological properties of the layers. The stiffer layers carry higher axial compression than the softer layers. This is the state of initial stress of the multilayer and the media.

There is a certain, critical, axial compression at which the initially straight form is an unstable configuration. A series of sinusoidal waves satisfies equilibrium as well as the boundary conditions and is, therefore, a possible alternative configuration. The nonuniform stresses and strains associated with the sinusoidal waves are the incremental stresses and strains. At first, sinusoidal-like waves of various lengths develop and are overprinted until an optimum pattern is developed. The factors that control the wavelengths in multilayers are many but can be understood in a general way. The relation between axial stress and wavelength of stiff layers in a multilayer can be separated conceptually into resistance of the stiff layers and resistance of the medium, although the relations cannot be expressed in terms of simple variables. Instead of axial load, as in elementary theory, the critical parameter is f_1, where (eq. 1c):

$$f_1 = (S_{\bar{x}\bar{x}})_1 + G_1$$

which depends on the relative magnitude of the incremental shear modulus, G_1, of stiff layers and the magnitude and sign of the initial axial stress $(S_{\bar{x}\bar{x}})_1$, which is negative if compressive and positive if tensile. As for a single layer according to the elementary theory, the linearized theory indicates that there is a wavelength, L_m, that requires the minimum axial compression, that is, a *maximum* f_1. The resistance of each stiff layer depends upon thickness, t_1, modulus, G_1, Poisson's ratio, ν_1, and state of initial stress, $(S_{\bar{x}\bar{x}})_1$ and $S_{\bar{y}\bar{y}}$, such that, the larger are t_1, G_1, ν_1 and the more compressive is $S_{\bar{y}\bar{y}}$, the larger is the wavelength in the stiff layers (Fig. 2). A factor not considered in the elementary theory is that, the larger the axial compression, the smaller the effective modulus, f, of the stiff layers, and hence, the smaller the wavelength of the stiff layer. Indeed, the linearized theory indicates that, if the axial compression becomes sufficiently large, the factor f approaches zero and the wavelength approaches zero (Fig. 2). An increase in initial axial compression, therefore, lowers the resistance of the stiff layer to buckling and if other things are equal, decreases the wavelength, L_m.

The resistance of the medium is somewhat more complex in the linearized folding theory than in the elementary folding theory; in part because the medium in effect can be between the stiff layers as well as bounding the multilayer. However, similarities remain. In both the elementary theory (e.g., Johnson, I, 1970, p. 152) and in the linearized theory, the larger the number of layers in the multilayer, the softer is the effective behavior of the media, because the resistance of the media is shared by all the layers. Thus, the thicker the stack in a multilayer, the larger the wavelengths of folds, assuming the stiff layers are sufficiently closely spaced so that they strongly interact. Further, the stiffer the media, the more will be the tendency for amplitudes of folds to be larger at middepth in the multilayer than at boundaries with the media (Fig. 10C2) and the shorter will be the wavelengths. Finally, the smaller the thickness ratio, t_1/t_2, and the modulus ratio, G_1/G_2, of stiff layers and soft interbeds, the shorter will be the wavelength in the stiff layers.

The wavelengths of subsequent folds are largely determined by these early, sinusoidal waves, but the waves become modified by two processes which are results of second-order stresses induced by growth of the sinusoidal folds. One process is change in axial compression of stiff layers near upper and lower edges of a multilayered body as a result of differences in moduli between interbeds and media. If the media are softer than the interbeds, the wavelengths decrease in cores of folds and increase in crests, tending toward a concentric-like form in the multilayer. If the media are stiffer than the interbeds, on the other hand, the wavelengths increase in cores of folds and decrease in crests. The other process is change in confining stress, acting normally to layers, as fold amplitudes increase. Stiff layers in areas of decreased confining compression, in areas above midheight in synclines and below midheight in anticlines, decrease in wavelength. Stiff layers in areas of increased confining compression, however, increase in wavelength. Both processes make sinusoidal patterns which we compute with the linearized theory unstable and short-lived, causing wavelengths to change from place to place in the fold pattern rather than to remain constant as we would predict with the linearized theory alone. Further, just as the linear theory of elasticity predicts a pattern of uniform shortening and thickening in response to axial loading, a pattern which the linearized theory indicates to be unstable if loads are sufficiently high, the linearized theory predicts a pattern of sinusoidal folds where wavelengths are constant throughout, a pattern which the higher-order theory indicates to be unstable if displacements are sufficiently high.

We conclude that sinusoidal-like patterns with constant wavelengths are nearly always unstable in multilayers. Growth of the folds will cause them to be overprinted by a concentric-like pattern, characterized by variable wavelengths. An exception is shown in Strömgård's experiment (Fig. 11A), where there are only a few layers and where the higher-order effects cancel each other, that is, where the medium is sufficiently stiffer than the interbeds to counteract the tendency for the incremental normal stress, $s_{\bar{y}\bar{y}}$, induced by folding, to modify the wavelengths. As shown in Fig. 10C, the concentric-like pattern develops inside a thick multilayer bounded by stiff media because the effect of the shear gradient is restricted to the vicinity of the upper and lower boundaries of the multilayer, whereas the effect of the gradient of normal confining stresses is throughout most of the multilayer. The concentric-like form, therefore, is typical of folds in multilayers where thicknesses of interbeds are less than the wavelengths of folds in the stiff layers.

Chevron-like folds also can be explained in terms of higher-order effects. Such folds have tightly bent or even sharp hinges and relatively straight limbs, and they appear to be analogous to pinches in single, isolated layers. We imagine that they form as follows: when the combination of the initial axial stress, $S_{\bar{x}\bar{x}}$, the incremental axial stress, $s_{\bar{x}\bar{x}}$, and the shear modulus, G_1, of a stiff layer approaches zero locally, as expressed in the parameter f^*, $f_1{}^* = G_1 + (S_{\bar{x}\bar{x}})_1 + (s_{\bar{x}\bar{x}})_1$, the elastic material behaves as though it loses further resistance to shear stress or bending. In effect, the layer "yields" elastically in a small area and a hinge forms. The yielding, however, is completely recoverable if the material is an ideal elastic substance. Further, as the hinge forms, the material in the hinge cannot support high bending moments so that the limbs between hinges tend to remain straight during growth of the folds. In this respect a stiff layer behaves similarly to a short strip of cardboard buckled by loads applied by one's fingers. The yielding and the straightening produce a chevron form. According to this concept of the formation of chevron folds in elastic materials, the shorter the wavelength

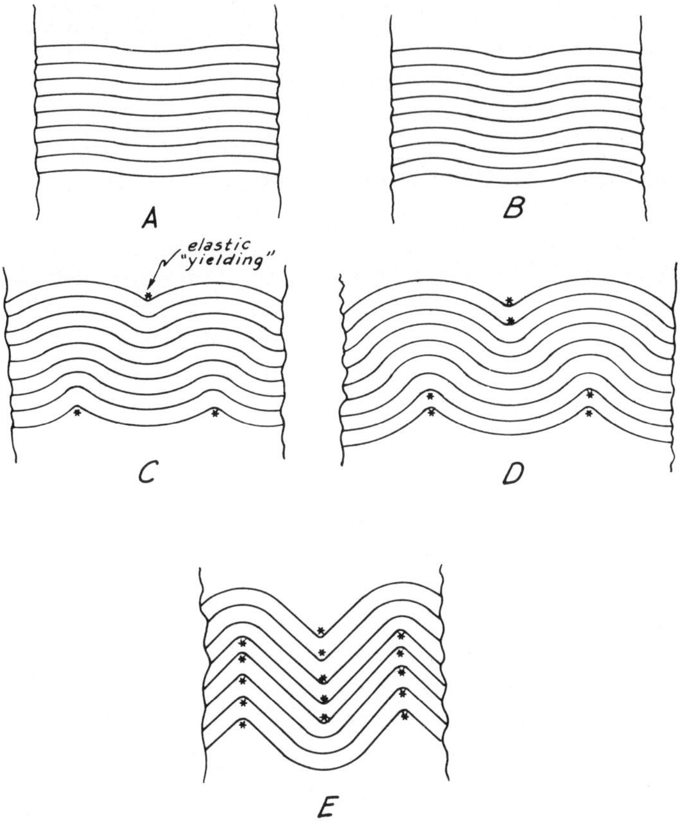

Fig. 14. Transition from sinusoidal to concentric-like to chevron folding. A. Sinusoidal pattern. B. Beginning of concentric-like pattern, characterized by different wavelengths at different places in multilayer. C. Elastic "yielding" in stiff layers at cores of folds initiates development of chevron pattern. D. Chevron pattern grows at expense of concentric pattern. E. "Yielding" of hinges throughout much of the multilayer produces a pattern of chevron folds, leaving only remnants of concentric-like folds.

in the stiff layer at the time of sinusoidal folding, the lower the amplitude of the fold when the chevron form develops (e.g., Figs. 5 and 8). Chevron folds in multilayers tend to form first in cores and propagate toward crests of concentric-like folds because the wavelengths are shortest in the cores (Figs. 10). However, according to our concept, chevron forms can develop even in single elastic layers in soft media if the media are sufficiently stiff to force wavelengths to be short. Thus, chevron folds can be a result of concentric-like folding, but they do not require concentric-like folding in order to develop.

We imagine the evolution of typical fold forms in a multilayered stack of layers in soft confining media to be as shown in Figs. 14A—E. Folding begins as low-amplitude, essentially sinusoidal wave forms (Fig. 14A), the wavelength of which requires the minimum axial compression. As the sinusoidal folds grow in amplitude, with further overall shortening, they induce stresses which cause them to become unstable. The wavelength of the stiff layer at midheight remains essentially constant, as measured along the layer, but the wavelengths of layers above in anticlines and below in synclines increase (Fig. 14B).

As overall shortening of the concentrically folded multilayer continues, a combination of the initial axial stress, of the second-order effect of the gradient of shear stress across stiff layers, and of the bending of individual stiff layers causes stiff layers to "yield" elastically at cores of folds. Chevron folding has thus begun (Fig. 14C). Further shortening causes layers deeper within the multilayer to "yield" elastically and produces a pattern of chevron folds below cores of concentric-like folds (Fig. 14D). Ultimately, the "yielding" and the resulting chevron pattern migrate through most of the multilayer, leaving only a vestige of the concentric pattern to suggest the history of the chevron pattern. The wavelength of the final chevron folds, measured along layers, should return to roughly the same value as the wavelength of the first-formed, sinusoidal folds (Fig. 14A).

In this way we can understand the transition from sinusoidal to concentric-like to chevron folds in elastic multilayers.

DISCUSSION

The understanding of the experiments with rubber and gelatin gained by means of the linearized theory encourages us to generalize the conclusions somewhat to include some folds in rocks, bearing in mind the uncertainties and problems mentioned in the Discussion of Part II. The experiments with multilayers suggest that the common assumption we structural geologists (e.g., DeSitter, I, 1956, p. 189, Ragan, 1973, p. 67) make that one can estimate the depth of folding by determining where radii of curvature of concentrically folded layers intersect, is of limited value. A combination of a concentric-like pattern and a chevron pattern (Figs. 10 and 14) can extend to indefinite depths. Thus, there appears to be no *geometrical* constraint on maximum depths of a combination of concentric-like and chevron folding. There are, of course, mechanical constraints. For example, the folding in a multilayered sequence may terminate at the contact with a layer that behaves unusually thick or stiff mechanically, especially if slippage is readily accommodated at the contact. Thus, surfaces of décollement probably are results of mechanical boundaries, not of geometrical constraints of fold patterns.

Another general conclusion is that chevron-like folds imply yielding at hinges and probably imply straightening of limbs that earlier were bent. The amount of straightening required depends on the properties of the stiff layers (e.g., Johnson, I, 1970, p. 301—311), but some type of yielding, even if reversible, as in rubber strips, apparently is required to concentrate bending in a hinge as in a chevron fold. Further, although our analyses and experiments suggest that sinusoidal folds may transform directly into chevron folds, the typical pattern in multilayers involves an evolutionary sequence from sinusoidal folds to concentric-like folds to chevron folds and that one can expect to find a remnant of a concentric-like fold above a chevron antiform and below a chevron synform.

Finally, the linearized theory provides indications of wavelengths of folds one can expect in elastic materials with various thickness and modulus ratios, but it also provides a warning that appropriate thickness and modulus ratios cannot necessarily be estimated readily by field observations of fold forms and bedding characteristics. As indicated by Currie et al. (I, 1962), the layers defined in theory are mechanical rather than sedimentary layers, because sedimentary layers can be largely color or textural banding. In addition, however, the linearized theory indicates that mechanical layering may not be necessary to fold color or textural banding in thick sequences of rock into concentric-like forms. Rather, the state of stress itself can allow folding of such homogeneous materials, effectively by inducing anisotropic properties which may be normal or otherwise inclined to sedimentary layers.

ACKNOWLEDGEMENT

The research reported in this part was financed largely by the National Science Foundation, grant no. GA-36917, but partly by the John Simon Guggenheim Memorial Foundation, N.Y. We thank these organizations for their generous support.

We wish to thank Atilla Aydin, graduate student at Stanford, who helped with the experiments and with other tasks that made the research proceed smoothly. We are especially grateful to Peter Gordon who designed and built the experimental equipment which worked perfectly. Dr. W.C. Luth of the Geology Department at Stanford kindly allowed us to use his PDP-8 minicomputer for performing the numerous calculations. We thank Marcella Flynn for improving the grammar and punctuation. Part of the research was conducted while Arvid Johnson was visiting the University of Uppsala, Sweden, where he was provided support facilities and an office, arranged by Professor Hans Ramberg. Discussions with Professor Hans Ramberg and Professor Ove Stephansson at Uppsala and with Dr. David Pollard, U.S. Geological Survey, helped to formulate and revise several of the concepts discussed here. In particular, we are indebted to Professor Ramberg for finally convincing us that ideal concentric folds cannot exist theoretically and to Dr. Pollard for criticizing and editing the manuscript.

REFERENCES NOT CITED IN CHAPTER 3

Bayly, M.B., 1964. A theory of similar folding of viscous materials. Am. J. Sci., 262: 753—766.

Bayly, M.B., 1970. Viscosity and anisotropy estimates from measurements on chevron folds. Tectonophysics, 9: 459—474.

Bayly, M.B., 1971. Similar folds, buckling, and great-circle patterns. J. Geol., 79: 110—118.
Biot, M.A., 1965. Theory of similar folding of the first and second kind. Geol. Soc. Am. Bull., 76: 251—258.
Gere, J.M. and Weaver, W., Jr., 1965. Matrix Algebra for Engineers. Van Nostrand, Princeton, New Jersey, 165 pp.
Hatschek, E., 1933. The elastic properties of glycerin-gelatin gels and of gelatin gels hardened with formaldehyde. Trans. Faraday Soc., 29: 1108—1131.
Johnson, A.M. and Ellen, S.D., 1974. A theory of concentric, kink and sinusoidal folding and of monoclinal flexuring of compressible, elastic multilayers. I, Introduction. Tectonophysics, 21: 301—339.
Johnson, A.M. and Honea, E., 1975. A theory of concentric, kink and sinusoidal folding and of monoclinal flexuring of compressible, elastic multilayers. II, Initial stress and nonlinear equations of equilibrium. Tectonophysics, 25: 261—280.
Ragan, D.M., 1973. Structural Geology. Wiley, New York, 208 pp.
Ramsay, J.G., 1974. Development of chevron folds. Geol. Soc. Am. Bull., 85: 1741—1754.
Ward, A.G. and Saunders, P.R., 1958. In: F. Eirich (Editor), The Rheology of Gelatin, 2. Academic Press, New York, pp. 313—357.

Chapter 6

SINUSOIDAL AND REVERSE CONJUGATE KINK FOLDS *

SUMMARY

This part concerns folding of elastic multilayers subjected to principal initial stresses parallel or normal to layering and to confinement by stiff or rigid boundaries. Both sinusoidal and reverse-kink folds can be produced in multilayers subjected to these conditions, depending primarily upon the conditions of contacts between layers. The initial fold pattern is always sinusoidal under these ideal conditions, but subsequent growth of the initial folds can change the pattern. For example, if contacts between layers cannot resist shear stress or if soft elastic interbeds provide uniform resistance to shear between stiff layers, sinusoidal folds of the *Biot wavelength* grow most rapidly with increased shortening. Further, the Biot waves become unstable as the folds grow and are transformed into concentric-like folds and finally into chevron folds. Comparison of results of the elementary and the linearized theories of elastic folding indicates that the elementary theory can accurately predict the *Biot wavelength* if the multilayers contain at least ten layers and if either the soft interbeds are at most about one-fifth as stiff as the stiff layers, or there is zero contact shear strength between layers.

Multilayers subjected to the same conditions of loading and confinement as discussed above, can develop kink folds also. The kink fold can be explained in terms of a theory based on three assumptions: each stiff layer folds into the same form; kinking is a buckling phenomenon, and shear stress is required to overcome contact shear strength between layers and to produce slippage locally. The theory indicates that kink forms will tend to develop in multilayers with low but finite contact shear strength relative to the average shear modulus of the multilayer. Also, the larger the initial slopes and number of layers with contact shear strength, the more is the tendency for kink folds rather than sinusoidal folds to develop. The theoretical displacement form of a layer in a kink band is the superposition of a full sine wave, with a wavelength equal to the width of the kink band, and

---

* Originally published as: Honea, E. and Johnson, A.M., 1976. Part IV, Development of sinusoidal and kink folds in multilayers confined by rigid boundaries. Tectonophysics, 30: 197—239.

of a linear displacement profile. The resultant form resembles a one-half sine curve but it is significantly different from this curve. The width of the kink band may be greater or less than the Biot wavelength of sinusoidal folding in the multilayer, depending upon the magnitude of the contact shear strength relative to the average shear modulus. For example, in multilayers of homogeneous layers with contact strength, the Biot wavelength is zero so that the width of the kink band in such materials is always greater than the Biot wavelength. In general, the higher the contact strength, the narrower the kink band; for simple frictional contacts, the widths of kink bands decrease with increasing confinement normal to layers. Widths of kink bands theoretically depend upon a host of parameters — initial amplitude of Biot waves, number of layers, shear strength of contacts between layers, and thickness and modulus ratios of stiff-to-soft layers — therefore, widths of kink bands probably cannot be used readily to estimate properties of rocks containing kink bands. All these theoretical predictions are consistent with observations of natural and experimental kink folds of the reverse variety.

Chevron folding and kink folding can be distinctly different phenomena according to the theory. Chevron folds typically form at cores of concentric-like folds; they rarely form at intersections of kink bands. In either case, they are similar folds that develop at a late stage in the folding process. Kink folds are more nearly akin to concentric-like folds than to chevron folds because kink folds form early, commonly before the sinusoidal folds are visible. Whereas concentric-like folds develop in response to higher-order effects near boundaries of a multilayer, kink folds typically initiate in response to higher-order shear, as at inflection points near mid-depth in low-amplitude, sinusoidal fold patterns. Chevron folding and kink folding are similar in elastic multilayers in that elastic "yielding" at hinges can produce rather sharp, angular forms.

INTRODUCTION

This fourth part of our series of papers * on folding of elastic multilayers primarily deals with folds in multilayers subjected to principal initial stresses oriented parallel or normal to layers and to confinement normal to layers by very stiff or by rigid boundaries. The spectrum of fold forms can be idealized as combinations of sinusoidal and kink end members. We have discussed sinusoidal folding at some length in Part III *. *Kink bands*, as defined by Dewey (1965, p. 460), are tabular zones of folding that transect foliation such as bedding, cleavage or schistosity. The foliation rotates within kink bands. Further, the kink bands we will investigate are of the *reverse* variety (Dewey, op. cit.).

Sinusoidal folding of multilayers confined by rigid boundaries has been discussed by Biot (I **, 1964; I, 1965a), by Johnson (I, 1969; I, 1970), and by Ramberg (I, 1970). Here we develop solutions based on the linearized theory introduced in Part II * and compare the solutions to those developed

---

* "Part I" will be used to abbreviate the reference: Johnson and Ellen, 1974. "Part II" is the reference: Johnson and Honea, 1975a; and "Part III" is the reference: Johnson and Honea, 1975b.

** References cited in Part I will be noted with an "I", followed by the date, such as I, 1955, and not be repeated in the References at the end of this paper.

Note: Part I = Chapter 3, Part II = Chapter 4, Part III = Chapter 5.

with elementary folding theory (Johnson, I, 1970).

There has been a large number of papers written about experimental and natural kink bands. The literature is conspicuous for its scarcity of theory relevant to kink folding, but many of the experimental observations are valuable and several conditions of kink folding have been deduced. Some investigations have been concerned with conditions of kinking as opposed to slippage or faulting between layers inclined to principal stress directions in foliated materials (e.g., Jaeger, 1960; Donath, 1961; Gay and Weiss, 1974). We shall postpone theoretical analysis of such conditions to Part VI of this series of papers.

Many investigators have studied the directions of kink bands relative to original foliation (e.g., Dewey, 1965; Paterson and Weiss, I, 1966; Weiss, I, 1968; Clifford, 1968; Anderson, I, 1968; Ellen, I, 1971). The theory that we shall develop in this part is not concerned with directions of kink bands, and we believe that the directions are controlled by higher-order processes which we are unable to describe with either elementary or linearized theories. In Part VI we shall discuss one process that predicts orientations of kink bands where dilitation within the bands is negligible or known.

Only a few investigators have concerned themselves with theories of kink folding. Biot (I, 1965a) indicated that his theory of folding of multilayers predicted kink folding in situations where layering is inclined at angles of about 45° to the principal axes of stress. Cobbold et al. (I, 1971) slightly modified the conclusions of Biot's theory, suggesting either that a direction of anisotropy is inclined to principal stress axes, or that layering is rotated to steep angles to the stress axes during folding, in order to explain kink folding where principal stresses initially are parallel or normal to layering. Ellen (I, 1971) also attempted to modify Biot's theory to explain some natural and experimental kink folds, but his results are unsatisfactory also. Neither Biot's original theory nor modifications by Cobbold et al. and by Ellen can explain kink folding in multilayers in numerous experiments where principal stresses are roughly parallel or normal to layering.

Nobody, therefore, has developed a theory of kink folding for conditions of maximum compression parallel to surfaces of layers or has explained why kink bands develop instead of smooth, repetitive folds, such as sinusoidal or concentric forms. Also, nobody has explained why chevron forms rather than kink forms develop in some multilayers, although Cobbold et al. suggested that chevron folds can transform from either kink folds or sinusoidal folds and Paterson and Weiss (I, 1966) developed chevron forms locally in experiments where kink bands intersected.

The purposes of this part are to introduce the rudiments of a theory that describes mechanisms by means of which reverse-kink, sinusoidal, and chevron folds can develop in elastic or elastic—plastic multilayers, and to describe experiments with elastic multilayers in which forms of folds can be changed by changing properties of multilayers while keeping loading conditions comparable. All the experiments were made with rubber strips as the competent layers because the rheological properties of rubber are relatively well known. Folds that might be described as sinusoidal, concentric and chevron forms have all been produced in the same material, as described in Parts I and III. In this way we eliminate the possibility that kink forms are uniquely a result of peculiar rheological properties of the layers themselves, such as plastic or brittle yielding, a possibility that prevents clear interpretations of results of experiments with materials with inelastic behaviors (e.g., experiments by Weiss, I, 1968; Donath, I, 1968; Ghosh, I, 1968; Johnson, I, 1970; Cobbold et al., I, 1971; and Ellen, I, 1971). A requirement of a suitable theory, of course, is that it provides explanations for kink folds that are not inconsistent with the results of experiments with these other materials. Some of the conclusions derived from the theory presented here and most of the conclusions derived in Part VI are independent of the rheological properties of the layers of a multilayer, so part of the theory satisfies this requirement.

We begin by describing a few of our experiments. Then we develop theories to explain mechanisms of initiation of kink bands and sinusoidal folds in the experimental materials. Finally, we combine the experimental and theoretical observations in order to predict conditions under which kink, sinusoidal and chevron folds should develop, and in order to explain the evolution of fold forms in multilayers confined by rigid boundaries.

One of the common criticisms of our research into theories of folding is that we appear to deal with materials whose rheological properties are contrary to properties currently attributed to rock. Thus, the argument commonly made is: rocks permanently deform, so how could a theory based on elasticity possibility be relevant to understanding of folds in rock? We explained as clearly as we are able in Part II that the theory of folding assumes merely that the small, incremental strains are linearly related to the small, incremental stresses associated with the buckling of the layers (Part II, fig. 5), *during the folding process.* These strains may later become permanent. Also, the initial strains need not be recoverable for the theory to be valid (e.g., Novozhilov, I, 1953). It is popular in folding theory to assume that rocks behave roughly as viscous materials. However, we believe that many rocks that fold at high levels of the earth's crust behave more as elastic or elastic—plastic materials than as viscous materials. It is for this reason that we develop a theory for folding of elastic materials. We discuss the problem of high ratios of initial stress to tangent modulus of elasticity in Part VII, where we deal with a specific field example in California.

EXPERIMENTS

We have performed several experiments with multilayers of rubber and gelatin, lubricated rubber, or dry rubber placed in the apparatus shown in Fig. 1. The multilayer is confined on one side by a stiff, "collapsible" metal bar which can shorten without increasing in width. The position of the bar is controlled by two metal screws. A platen is advanced into an open side of

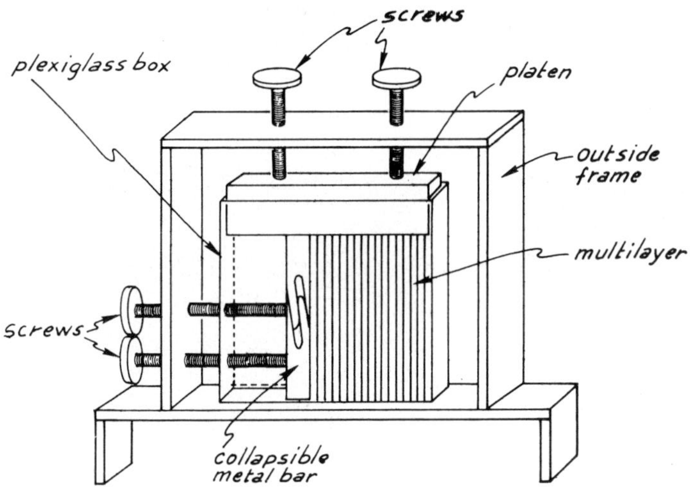

Fig. 1. Apparatus used to deform experimental multilayers.

the box by means of two screws, longitudinally shortening the multilayer. As the platen advances, the multilayer expands laterally due to Poisson's effect. In order to accommodate the expansion, the lateral confinement is released by backing out the screws controlling the position of the collapsible metal bar. Thus, displacements normal and parallel to layers can be adjusted during an experiment. The multilayer is confined front and back with plexiglass sheets about 1.3 cm thick, approximating plane-strain conditions, at least where stresses are compressive.

The apparatus we use has some advantages and some disadvantages in comparison with those used by others. Weiss' (I, 1968) apparatus, for example, allows some control of average stresses normal and parallel to layering. Ours does not, because displacement rather than load is controlled. On the other hand, our apparatus allows one readily to observe propagation of kink bands whereas Weiss' does not, because kink bands propagate stably under conditions of constant boundary displacement as in our experiments, but they propagate unstably under conditions of constant load as in Weiss' experiments.

Interbedded gelatin and rubber

All the types of fold patterns we have observed in multilayers of elastic materials confined by rigid boundaries can be shown with four experiments. Three of these were made with the same rubber strips used for experiments described in Part III; the primary differences among the three were the nature of interbeds between rubber strips. In one experiment (Fig. 2A) the interbeds were gelatin. In another (Fig. 2B) the interbeds were thin smears of silicone grease used to lubricate the layers. In the third (Fig. 2C) the interbeds were absent and the rubber strips were in frictional contact with one another. The fourth experiment (Fig. 2D) was with thinner rubber strips in frictional contact.

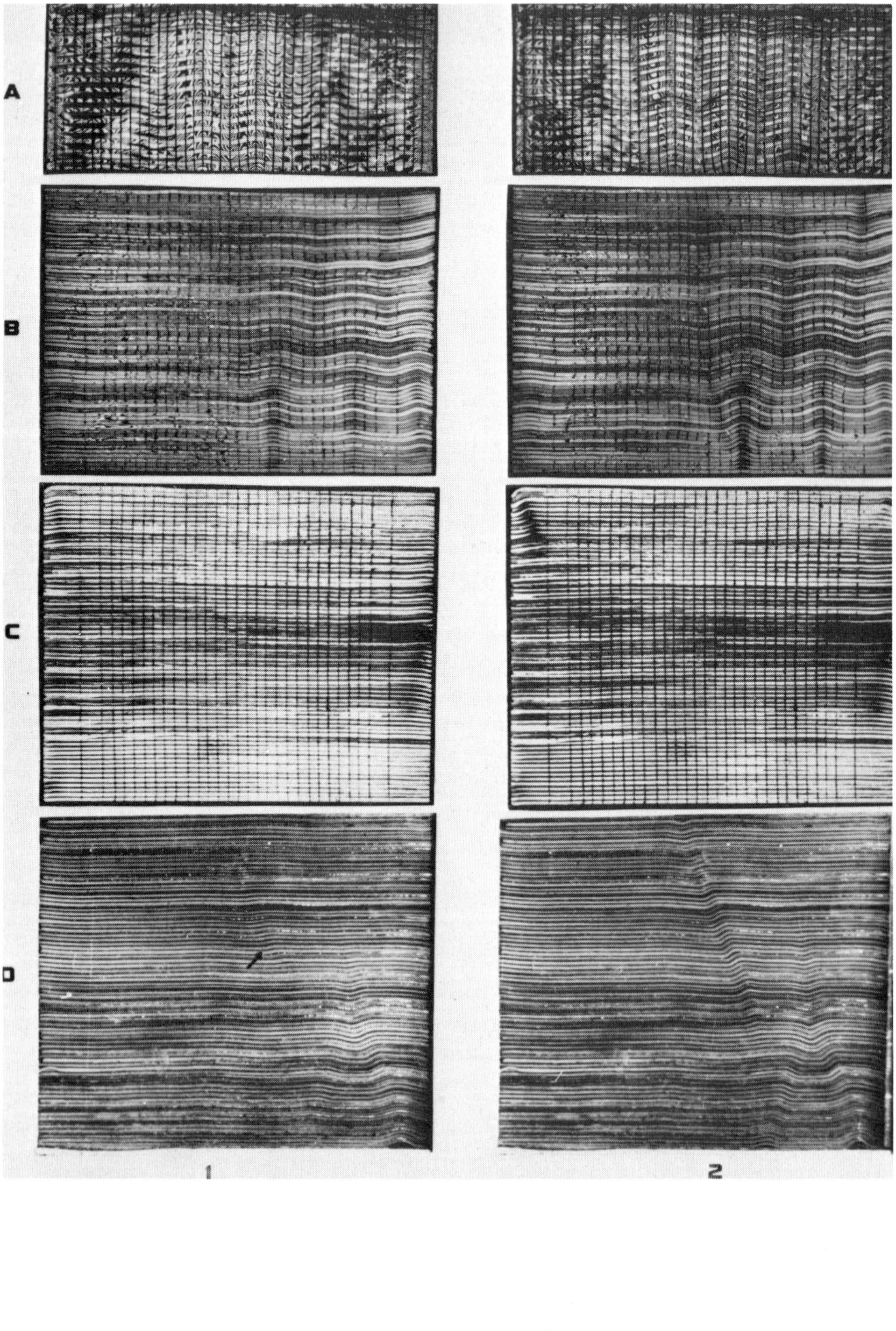

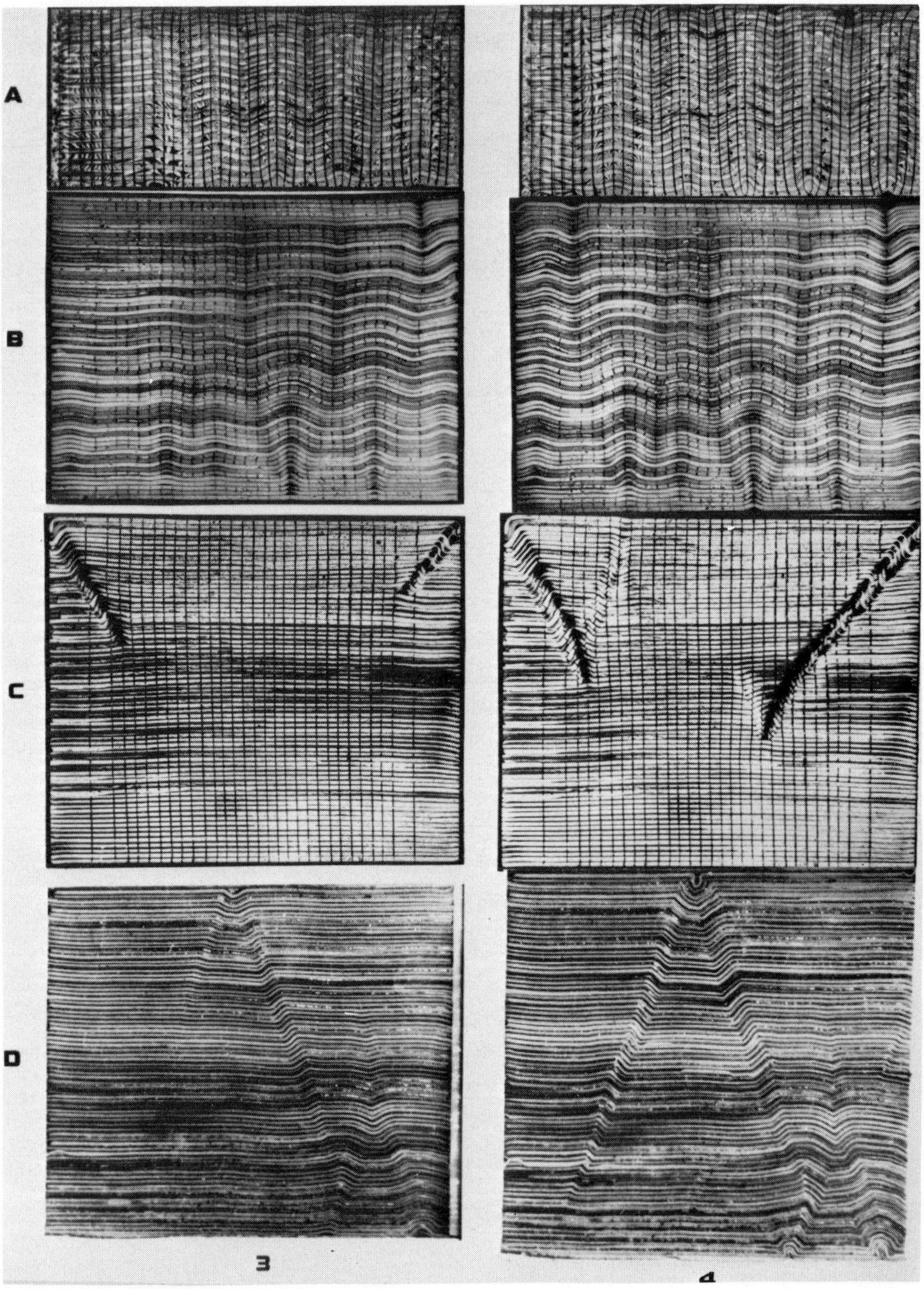

Fig. 2. Folds developed in four multilayers of elastic materials. A. Interbedded, stiff photoelastic rubber and soft gelatin. Modulus ratio about 45. Thickness ratio of unity. B. Photoelastic rubber strips with contacts lubricated with silicone grease. C. Dry, photoelastic rubber strips in frictional contact. D. Thin strips of pure gum rubber in frictional contact. Thickness of strips 0.8 mm.

In the experiment shown in Fig. 2A, a multilayer of relatively stiff rubber and soft gelatin was subjected to confinement normal to layers and to shortening parallel to layers. The modulus ratio of rubber to gelatin was about 45 (see Part III) and the thickness ratio was about unity. The first folds were rounded throughout the multilayer, resembling sinusoidal waves, with amplitudes decreasing from maxima at mid-depth to zero at the relatively rigid, planar upper and lower boundaries. The fold pattern closely resembles that produced in a multilayer confined by stiff gelatin, shown in fig. 10C2 of Part III. With further shortening the multilayer shown in Fig. 2A began to pull away from the planar boundaries, and chevron-like folds developed in the cores of the folds (Fig. 2A3). The wavelengths of folds were different near the planar boundaries than at mid-depth (Fig. 2A4). The fold pattern throughout much of the multilayer, however, continued to closely resemble sinusoidal waves.

Two similar experiments were reported in fig. 22 of Part I. In those experiments the multilayer stacks were thinner relative to the wavelengths of folds and the folds became concentric-like throughout the multilayer as amplitudes increased.

Grease between rubber layers

The multilayer comprised of rubber strips lubricated with silicone grease (Fig. 2B) was first subjected to moderate confinement and axial shortening. Then the axial shortening was increased until the series of folds shown in Fig. 2B1 developed. The folds near mid-depth were smooth, rounded, sinusoidal-like forms. Those closer to the upper and lower boundaries were a combination of a short, warped form and a long, straight form. The short, warped forms became more chevron-like as shortening progressed (Fig. 2B2). Then the chevron-like forms propagated toward mid-depth from the upper and lower boundaries and new folds initiated elsewhere within the multilayer (Figs. 2B3 and 2B4). A conscious effort was made to produce kink folds in the multilayers with grease between individual layers, but only the forms shown in Fig. 2B could be produced.

Lines were drawn across layers, normal to layering, before the experiment. *On the average* they remained normal to the original direction of layering although, in detail, the lines did rotate with individual layers. Displacements between ends of these lines indicate that the layers slipped past each other all along the limbs of the folds.

Layers with frictional contacts

Thick layers

In several experiments the multilayers were comprised solely of dry rubber strips in frictional contact with one another. Figure 2C shows one of the multilayers comprised of the same thick rubber strips used for the experiments shown in Figs. 2A and 2B. The multilayer was first subjected to confinement and longitudinal shortening. Then the confinement was gradually reduced. Ends of layers in the upper corners of the multilayer deflected (Figs. 2C1 and 2C2), apparently caused by drag between the ends of the multilayer and the loading frame. With further release of confinement a crimp occurred in one (Fig. 2C2) and then the other corner and began to propagate as a kink band into the multilayer (Fig. 2C3). Weiss (I, 1968) described the same phenomenon in experiments with paper cards. The direction of the kink band was first nearly normal to layers (Fig. 2C2), but as the band propagated, it rotated slightly and intersected bedding at a lower angle (Fig. 2C3). The two kink bands propagated through about one-third to one-half the thickness of the multilayer. Further downward propagation was by growth of a chevron-like fold at the tip of the kink band. Near the top of the chevron-like fold new kink bands initiated and propagated diagonally across the layers (Fig. 2C4). As confinement was reduced further, the lengths of limbs of the kink bands increased and the hinges of kinks transformed from relatively sharp to rounded (Fig. 3B).

The layers slipped relative to each other only within the kink bands, as shown by displacements between lines that originally were normal to the original direction of layering.

Thin layers

Several experiments were made with multilayers comprised of thinner rubber strips. The thin strips were of pure gum rubber, 0.8 mm thick. The edge of every other strip was darkened so that the layering would be clear. One experiment was performed under conditions similar to those for the experiment shown in Fig. 2C, where kink bands propagated during decreasing confinement normal to layering. The pattern and growth kink bands were essentially the same as those shown in Fig. 2C except that widths of kink bands were much smaller in the thinner strips.

Another experiment is shown in Fig. 2D. The lateral confinement initially was low and lateral displacements were not allowed during the experiment. Instead, longitudinal shortening was increased monotonically. Rounded folds developed in the lower right-hand quadrant of the multilayer as longitudinal shortening progressed (Fig. 2D1). In addition, a kink band had begun to develop near the center of the multilayer (near the arrow shown in Fig. 2D1). Fig. 2D2 shows the same kink band after it propagated to the upper boundary of the multilayer in response to further shortening. Then the kink band reflected at the top boundary and started to propagate across the multilayer again (Fig. 2D3). Further longitudinal shortening did not produce further growth of kink bands. Therefore, the experiment was continued by releasing confinement normal to layers (Fig. 2D4). Ultimately the complex pattern shown in Fig. 3A developed. The forms of the kink bands closely resemble those produced by Weiss (I, 1968) in decks of paper cards.

The gross forms of the kink bands shown in Figs. 2C and 2D are similar

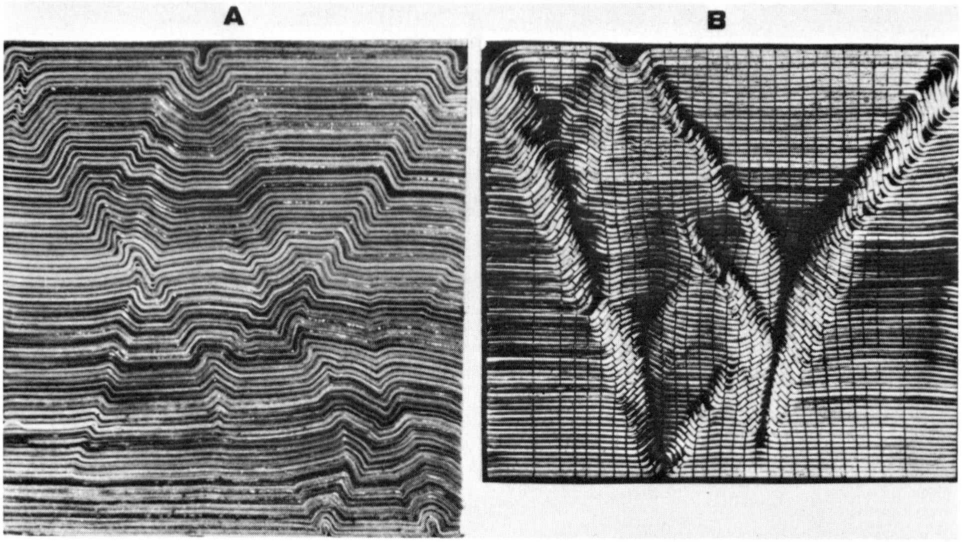

Fig. 3. Complex patterns of kink folds in rubber strips at late stages of experiments. B. Late stage of experiment with thick rubber strips shown in Fig. 2C. A. Late stage of experiment with thin rubber strips shown in Fig. 2D.

but details of propagation were somewhat different. Figures 4A and 4B show detailed views of propagating kink bands in the thin rubber strips subjected to different boundary conditions. In each experiment the multilayer was compressed by advancing both the loading platens. Then in one experiment (Fig. 4A), the kink folds developed as the platen supplying confinement was retracted. In the other (Fig. 4B), the kink folds developed as longitudinal shortening was increased. The kink band that developed under conditions of reduction of confinement initiated at one corner of the loading apparatus, and propagated by changing form of a short segment of each layer (Figs. 4A1 and 4A2). The layer near the tip of the band folded roughly symmetrically as a pucker or chevron fold. Layers within the band folded more asymmetrically, the degree of asymmetry increasing with distance from the tip until the fold in a layer consisted of a long, straight limb, a short, tilted limb, and another long, straight limb. There was slippage only on the short limb. Thus, there appeared to be a transition in form in each layer as a kink band propagated through it, from a small buckle to the short limb of a kink band. Further, the kink band reflected at a pucker within the multilayer and propagated back as a new kink band (Figs. 4A3 and 4A4). A pucker formed at a rigid boundary and another kink band initiated and began to propagate into the multilayer (Fig. 4A5). A characteristic feature of the kink band was that it widened as lateral confinement was reduced, as shown by comparing the first-formed band in Figs. 4A2 and 4A5.

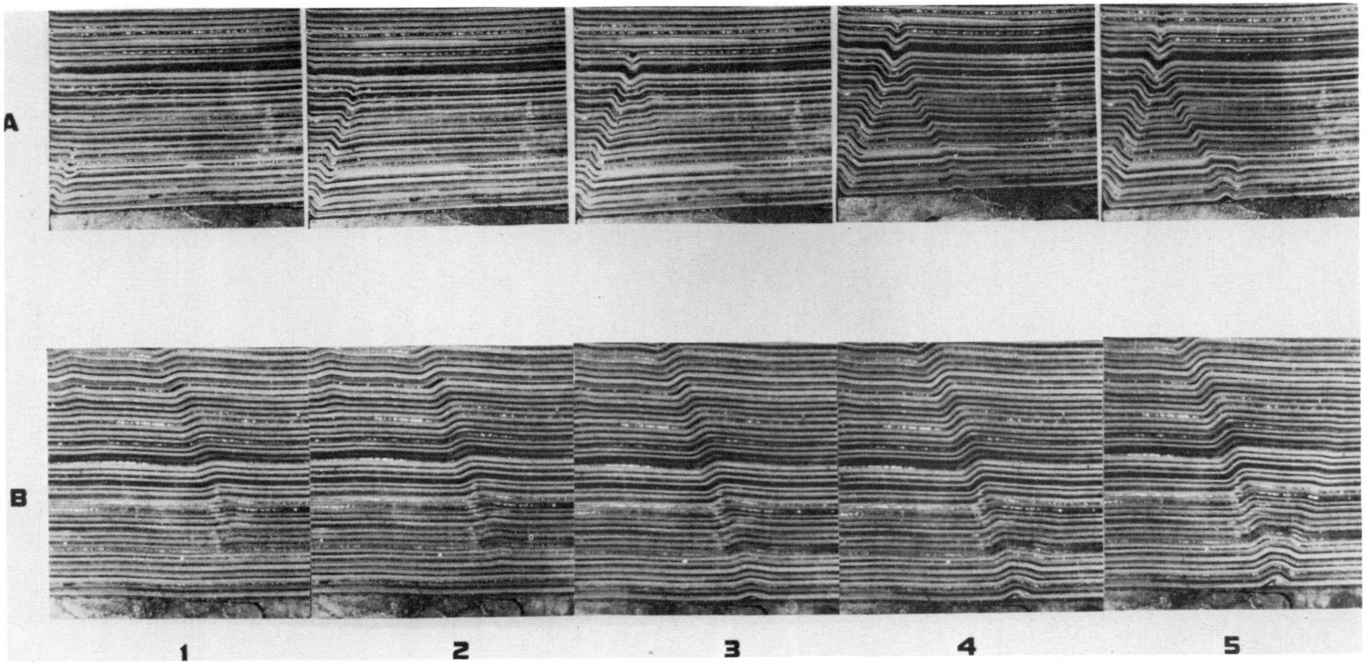

Fig. 4. Details of propagation of kink bands. A. Propagation of kink band initiated at corner of apparatus under conditions of *decrease* of confinement. Propagation characterized by pucker in layer near termination of kink band. B. Propagation of kink band initiated near midheight in multilayer under conditions of *increasing* shortening parallel to layering. Amplitude of kink fold gradually decreases toward termination.

In contrast is the propagation and reflection of a kink band in the same multilayer subjected to conditions of *increase* of longitudinal shortening (Fig. 4B). Figure 2D shows the gross pattern of folds developed in this experiment (Fig. 4B3 is part of Fig. 2D2). The pucker did not form at the ends of the kink band (Fig. 4B1). Instead, the amplitudes of the folds in the kink band gradually decreased and disappeared distally so that the tip of the kink band was an indistinct feature (Fig. 4B2) until the band reached a planar boundary (Fig. 4B3), where a pucker developed and the kink band reflected (Figs. 4B4 and 4B5). Furthermore, the kink band appeared to maintain the same width rather than to widen as the one shown in Fig. 4A.

THEORY

The four experimental multilayers shown in Figs. 2 were subjected to the same general boundary conditions, but they developed quite different fold forms. The multilayer comprised of interlayered soft gelatin and stiff rubber developed repetitious, rounded folds that resemble sinusoidal waves. The multilayers in which dry rubber strips were in direct frictional contact with one another developed discontinuous, rather angular kink forms. The multilayer in which friction between layers was reduced somewhat by silicone grease developed a combination of rounded and angular forms. There are also similarities among the experiments. Fold forms having straight segments parallel to planar, upper and lower boundaries developed in all four experiments. In the experiments with dry rubber strips the straight segments persisted through the entire multilayer as boundaries of kink-like folds. In the other two experiments, where there was grease or gelatin between layers, the straight segments were largely restricted to layers near the upper and lower rigid boundaries and appeared to be flattened crests and troughs of rounded folds.

Now we will develop theories to explain the mechanisms of folding as exemplified by the experiments. First we will develop the theory of sinusoidal folding of elastic multilayers confined by rigid boundaries. Then we will develop a theory to account for short, folded limbs and long, straight limbs of individual layers.

Sinusoidal folding of multilayer between rigid confining media

Finite thickness

The linearized theory of folding of multilayers confined by elastic media was presented in Part III, eqs. III, 19 through III, 25. Only one modification is required to transform those into the equations for buckling of a multilayer confined by rigid boundaries. One specifies that the vertical displacements, v_0, rather than normal boundary stresses, $\tau_{\tilde{y}y}$, are zero at the planar boundary. In order to satisfy the condition of no displacement normal to the rigid boundary, the coefficients in the second of eqs. III, 23b:

$$a_{21}C_{51} + a_{22}C_{61} + a_{23}C_{71} + a_{24}C_{81} = 0 \qquad \text{(III, 23b)}$$

become:

$$\begin{aligned} a_{21} &= \sinh(m_{11}lt_1/2) \\ a_{22} &= \cosh(m_{11}lt_1/2) \\ a_{23} &= \sinh(m_{21}lt_1/2) \\ a_{24} &= \cosh(m_{21}lt_1/2) \end{aligned} \qquad (1)$$

where m_{11} and m_{21} are constants defined in eqs. III, 22a, l is the wave number,

$$l = 2\pi/L$$

L is the wavelength and t is the thickness of the layer. Thus, as explained briefly in Part III, we solve eqs. III, 25 for wavelength as a function of axial loading of multilayers and determine the wavelength for which axial compression is minimized. For a multilayer confined by rigid boundaries, this is the *Biot wavelength* (Johnson, I, 1969), L_B, named after Maurice Biot, who has contributed greatly to folding theory.

The upper heavy curve in Fig. 5 is a generalization of calculations of wavelength/thickness ratio L_B/t_1, as a function of the number of layers in the multilayer for conditions where the thickness of stiff (1) and soft (2) layers is the same, $t_1 = t_2$, where the ratio of shear moduli is $G_1 = 50\ G_2$, and where Poisson's ratio is 0.49 for both stiff and soft layers. The upper dashed curve in Fig. 5 and the continuation of the upper solid curve for numbers of layers greater than 20 were derived with the elementary theory (Johnson, I,

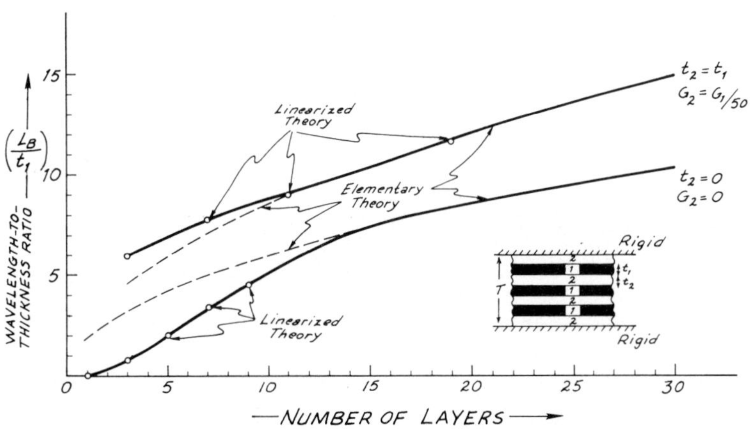

Fig. 5. Theoretical relations between Biot wavelength and number of layers in multilayer. One set of lines is for multilayers with thickness ratio of unity and modulus ratio of 50. The lower is for multilayer of homogeneous layers in frictionless contact. Open circles represent solutions according to the linearized theory. Solid lines are best estimates of correct relations between wavelength and number of layers.

1969; I, 1970, p. 243). To a close approximation this is:

$$L_B/t_1 = [\pi N_p]^{\frac{1}{2}} [4(B^* + t^*)/(3B^*)]^{\frac{1}{4}} \tag{2}$$

where:

$$t^* = t_2/t_1$$

$$B^* = B_2/B_1 = (G_2/G_1)(1 - \nu_1^2)(1 + \nu_2)/[(1 - \nu_2^2)(1 + \nu_1)] \tag{3}$$

and N_p is the number of *pairs* of stiff (1) and soft (2) layers. The numbers of layers plotted for the upper curves in Fig. 5 are twice N_p in eq. 2. Comparison of the upper dashed and solid curves in Fig. 5 indicates that the Biot wavelength derived from the elementary theory, eq. 2, should closely estimate the wavelength calculated with the linearized theory for multilayers containing more than ten layers.

Now we can compare theoretical wavelengths with those observed in the experimental multilayer shown in Fig. 2A. The apparent ratio of moduli for the rubber and gelatin was about 45 and the thickness ratio was about unity. There were 16 rubber strips and 17 gelatin strips so that $N_p = 16.5$ and the number of layers was 33. According to a slight extrapolation of the upper curve in Fig. 5, the theoretical Biot wavelength is approximately:

$$L_B/t_1 \simeq 15 \text{ or } 16$$

The ratio measured in Fig. 2A1 ranged from 14 to 15. Thus the theory roughly estimates the wavelength.

Essentially the same theory can be applied to lubricated multilayers where the friction between stiff layers is zero. The appropriate elementary theory is derived by letting $t_2 \to 0$ so that $t^* \to 0$ in eq. 2 in which case the Biot wavelength becomes (Biot, I, 1964; Johnson, I, 1970, p. 244):

$$L_B/t_1 \simeq 1.90\ (N)^{\frac{1}{2}} \tag{4}$$

where N is the number of stiff layers. The lower dashed curve and the part of the lower solid curve in Fig. 5 corresponding to numbers of layers greater than 14 are based on eq. 4. The remainder of the lower solid curve is based on the linearized theory. The linearized theory was derived by changing boundary conditions III, 24 as follows: vertical displacements, v_0, and vertical, boundary normal stresses, $\tau_{\tilde{y}y}$, are matched at interfaces as in eqs. III, 24b and 24d. Horizontal displacements, u_0, need not be matched because slippage is allowed, so eq. III, 24a is dropped. The two additional boundary conditions at each interface required to specify the arbitrary constants are derived by setting the boundary shear stresses, $\tau_{\tilde{y}x}$, equal to zero at the bottom of one layer and the top of the other. Thus, eqs. III, 24 become replaced by:

$$
\begin{aligned}
(\tau_{\tilde{y}x})_1 &= 0 = a_{31}C_{51} + \ldots + a_{34}C_{81} + 0 + \ldots + 0 \\
v_{01} - v_{02} &= 0 = a_{41}C_{51} + \ldots + a_{45}C_{52} + \ldots + a_{48}C_{82} \\
(\tau_{\tilde{y}x})_2 &= 0 = 0 + \ldots + 0 + a_{55}C_{52} + \ldots + a_{58}C_{82} \\
(\tau_{\tilde{y}y})_1 - (\tau_{\tilde{y}y})_2 &= 0 = a_{61}C_{51} + \ldots + a_{64}C_{81} + a_{65}C_{52} + \ldots + a_{68}C_{82}
\end{aligned} \tag{5}
$$

Eqs. III, 23b with coefficients defined in eq. 1 remain valid for this problem. In this way the necessary number of boundary conditions is specified and the matrix of coefficients is evaluated with a computer in order to determine wavelengths for which axial compression is minimized (see Part III).

The results of computations for several multilayers are generalized as the lower solid curve in Fig. 5 for numbers of layers less than ten. The results of the linearized theory indicate that the elementary theory significantly overestimates wavelengths only where there are fewer than about 13 layers in the multilayer. The reason for the deviation between wavelengths predicted by the two theories is that the elementary theory considers deformation due to bending alone, whereas, for short wavelengths, buckling is primarily a shearing phenomenon, as shown in Part III. If there is only one layer, for example, the linearized theory predicts a wavelength approaching zero whereas the elementary theory predicts a wavelength of about twice the thickness of the layer.

Now we will attempt to apply the theoretical results to predict wavelengths for the multilayer of rubber strips lubricated by silicone grease, shown in Fig. 2B. We assume tentatively that the grease reduces friction between layers to zero. There were 68 rubber strips used in the experiment, $N = 68$, and the elementary theory should adequately predict wavelengths. According to eq. 4, the theoretical wavelength/thickness ratio L_B/t_1, should be 15.7. The wavelength/thickness ratio for the relatively smooth waveforms near mid-height (Fig. 2B1) was about 21, which is much higher than the predicted value. Some waves that developed late in the experiment in the lower left-hand side of the multilayer (Fig. 2B3) had a wavelength/thickness ratio of about 13. Thus, some waves were longer and some shorter than predicted theoretically. We believe that imperfect lubrication between layers in the experiment produced the discrepancy between theory and experiment; whereas the theory assumed zero contact strength, the silicone grease at contacts probably had some strength. We shall consider effects of contact strength in the following pages.

Infinite thickness

Before proceeding to theories of kink folding, we will discuss briefly the sinusoidal folding of an infinite multilayer. According to the elementary theory, eqs. 2 and 4, the Biot wavelength approaches infinity as the number of layers in a multilayer becomes boundlessly large. The linearized theory provides the same answer if the thickness of soft interbeds is relatively small. Where thicknesses of soft interbeds are relatively large, however, there is a

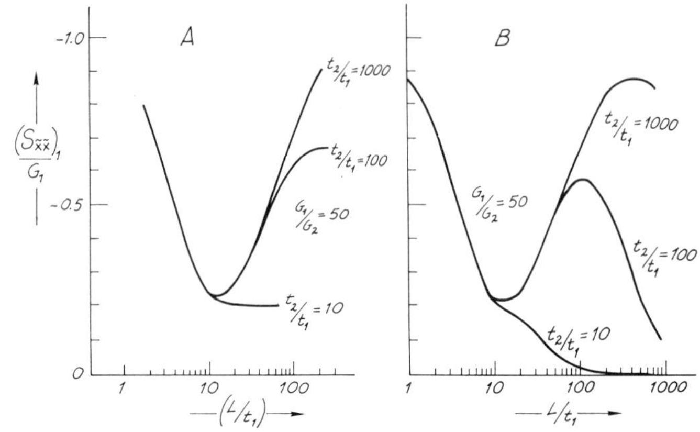

Fig. 6. Relations between initial axial compression and wavelength of sinusoidal folds for multilayers of infinite thickness. A. Layers buckled together, without slippage. B. Layers with frictionless contacts.

finite wavelength for which the axial compression is minimized. In this case the stiff layers are acting more or less as single, isolated layers in infinite, soft media so that they do not interact significantly. Figure 6A shows relations, according to the linearized theory, between initial axial compression, $S_{\bar{x}\bar{x}}$, and wavelength, L, for thickness ratios, t_2/t_1, of soft-to-stiff layers of 10, 100 and 1000. We have assumed bonded contacts and $\nu_1 = \nu_2 = 0.49$, $G_1/G_2 = 50$, and $S_{\bar{y}\bar{y}} = 0$. The results are similar to those derived by Ramberg (I, 1963a, figs. 5 and 6; 1961, fig. 5) for viscous layers. According to Fig. 6A, there is no finite wavelength for which axial compression is minimal unless the thickness ratio, t_2/t_1, exceeds a certain value. For the conditions assumed for the calculation, the thickness ratio, t_2/t_1, must be somewhat greater than 10 for there to be a minimum. According to Ramberg (op. cit.), the thickness of soft interbeds must be greater than the wavelength a stiff layer would assume if it were isolated in an infinite soft medium in order for there to be a finite wavelength for an infinitely thick multilayer. This conclusion is consistent with our results because, according to fig. 8 in Part III, the wavelength for an isolated layer, where $G_1 = 50\ G_2$, would be about 13. Thus, if $t_2/t_1 > 13$, approximately, there should be a wavelength for which there is minimum axial compression and this is consistent with the results in Fig. 6A. Finally, Fig. 6B shows relations between axial compression and wavelength for the same infinite multilayer except that the contacts between layers are frictionless. The figure shows that even a multilayer with free slippage between layers should develop infinite wavelengths unless the thickness ratio, t_2/t_1, is somewhat greater than ten. The figure shows, further, that if a multilayer with frictionless contacts is infinitely long, the wavelength requiring least axial stress is infinite. Only if the multilayer is finite in length will the wavelength corresponding to the minimum be developed.

Thus, the theory indicates that very thick, elastic multilayers containing relatively thin, soft interbeds will tend to buckle with very long wavelengths if the fold forms are sinusoidal. Many field observations suggest that wavelengths of folds can be relatively small in thick multilayers, and we would suggest that these folds are non-sinusoidal, involving processes not considered in traditional, sinusoidal folding theories. We shall now discuss one of these processes.

Kink form

The linearized and elementary theories of sinusoidal folding of multilayers derived thus far describe satisfactorily the wavelengths of experimental folds in one multilayer of rubber and gelatin, but they provide no explanation for the kink folds developed in experiments with dry rubber strips. One difference between the experiments is that there is frictional drag between dry rubber strips (Figs. 2C and 2D) whereas the soft gelatin layers allow easy slippage between stiff rubber layers in the first experiment (Fig. 2A). In order to generalize the theoretical problem of buckling of multilayers with frictional contacts, we will assume that slippage between layers requires that the shear stress between layers be:

$$|\tau| = \sigma'_n \tan\phi + C \tag{6}$$

where σ'_n is the effective compressive stress normal to the surfaces of layers, C is cohesive strength, and $\tan\phi$ is the coefficient of friction between layers. Equation 6 is the *Terzaghi-Coulomb* equation of strength.

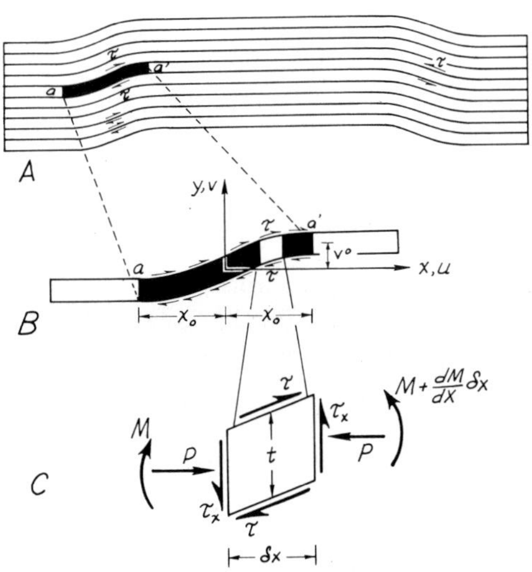

Fig. 7. Equilibrium of small element of multilayer folded into kink band. A. Multilayer with kink band inclined to original layering. B. Single layer containing kink is isolated for analysis. C. Element of single layer, showing forces, moments and stresses acting upon it.

We shall assume in the derivations in the following pages that each layer within a kink band deforms similarly so that a single layer can be analyzed as a representative example of any layer in the multilayer (Fig. 7A). We also assume that layers do not thicken or thin appreciably within kink bands, which requires that the kink band propagate diagonally across the multilayer (Fig. 7A), as shown by Paterson and Weiss (I, 1966). We ignore the terminations of kink bands in the analyses.

Elementary theory

First we will use elementary folding theory and then the linearized theory developed in Part III to analyze kink forms. Figure 7B shows the geometry and coordinates of part of a layer within a stack of layers folded into a kink band (Fig. 7A). The shear stress, τ, resisting slippage acts over a length $\pm x_0$ of the layer. The shear stress is applied to the segment of the layer by layers above and below. We ignore other interactions among layers for now. Figure 7C shows forces and moments acting on a small element of the layer, according to a slight modification of elementary plate theory (e.g., Johnson, I, 1970, p. 92). As in elementary plate theory, deformation is assumed solely to be caused by bending, and slopes of the deformed layer are assumed to be less than about 15°. The element is acted upon by an axial load, P, which is resisted by the induced bending moment, M, and the shear strength, τ. Another driving or resisting stress is the shear stress, τ_x, the origin of which will be considered in following pages. Equilibrium of bending moments requires (Fig. 7C):

$$\mathrm{d}M/\mathrm{d}x + P(\mathrm{d}v/\mathrm{d}x) - \tau_0 tb = 0 \tag{7a}$$

or:

$$M + Pv - \tau_0 tbx = C_0 \tag{7b}$$

where:

$$\tau_0 = \tau - \tau_x \tag{8}$$

b is width and t is thickness of the layer, v is displacement and C_0 is an arbitrary constant. The shear stresses have been assumed implicitly to be constant, independent of x. Replacing the bending moment with its expression in terms of curvature and rearranging, eq. 7b becomes:

$$\mathrm{d}^2v/\mathrm{d}x^2 + (P/BI)v - (\tau_0 tb/BI)x = C_0/BI \tag{9}$$

where

$$B = E/(1 - \nu^2) \quad \text{and} \quad I = bt^3/12$$

The origin of coordinates, $x = 0$, is taken to be the inflection point of the layer (Fig. 7B), where the curvature and displacement are zero, so $C_0 = 0$ according to eq. 9. Thus, the general solution of eq. 9 is:

$$v = C_1 \sin(x\sqrt{P/BI}) + C_2 \cos(x\sqrt{P/BI}) + (\tau_0 tb/P)x \tag{10}$$

where C_1 and C_2 are constants. The deflection is zero at $x = 0$ (Fig. 7B) so $C_2 = 0$. The slope of the layer is zero at each end of the kink band, at $x = \pm x_0$ (Fig. 7B), so:

$$C_1 = -(\tau_0 tb)/[P\sqrt{P/BI}\cos(x_0\sqrt{P/BI})] \tag{11}$$

Let us assume, further, that the curvature is zero at each edge of the kink band, at $x = \pm x_0$, beyond which the layer remains straight. Thus, from eq. 10:

$$\mathrm{d}^2v/\mathrm{d}x^2 = 0 = -C_1(P/BI)\sin(x_0\sqrt{P/BI})$$

which requires that:

$$\sqrt{P/BI} = \pi/x_0 \tag{12}$$

and that the wavelength, L, of the sinusoidal part of the deflection is $2x_0$, that is:

$$L = 2\pi\sqrt{BI/P} = 2x_0 \tag{13}$$

Combining eqs. 10, 11 and 12, the expression for the deflection is:

$$v = (\tau_0 tbx_0/P)[(1/\pi)\sin(\pi x/x_0) + (x/x_0)] \tag{14}$$

Finally, the maximum deflection, v^0, is at $x = x_0$ (Fig. 7B):

$$v^0 = (\tau_0 tb/P)x_0 \tag{15}$$

so that eq. 14 can be written as:

$$v = v^0[(1/\pi)\sin(\pi x/x_0) + (x/x_0)] \tag{16}$$

Equation 16 is the sum of two terms; the first is a sine wave which passes through zero at the origin and at $x = \pm x_0$, and the second is linear in x (Fig. 8). The sum of the two produces a *kink form* that closely resembles one-half a sine wave and closely resembles the kink folds in individual layers in kink bands. Unlike the one-half sine wave, the kink form is discontinuous. This is because the kink form is undefined for $-x_0 > x > +x_0$. It is significantly different from one-half a sine wave in other ways:

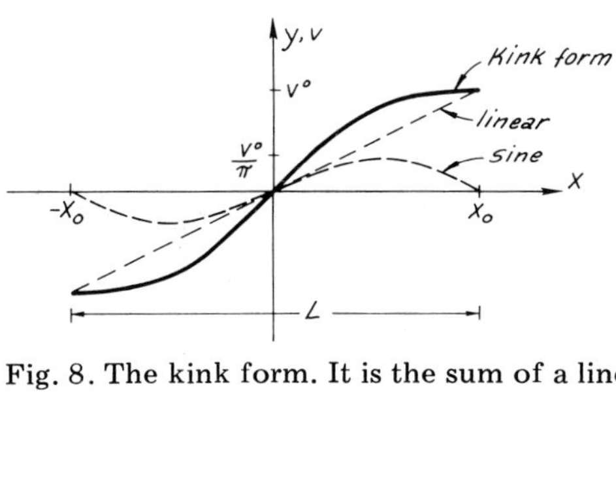

Fig. 8. The kink form. It is the sum of a linear displacement profile and a full sine wave.

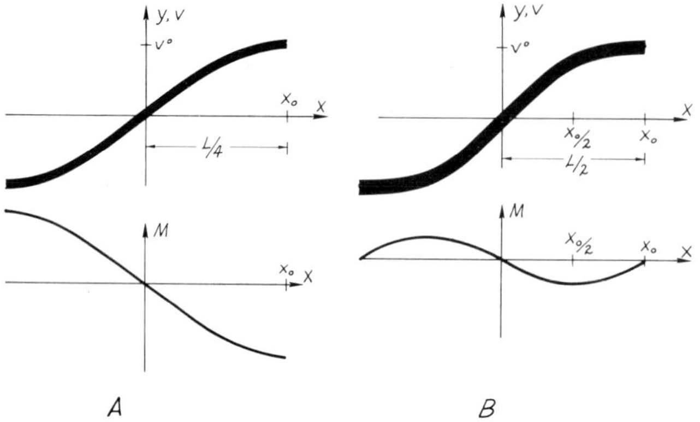

Fig. 9. Comparison of kink form and half-sine wave. A. Half-sine wave. Bending moment, M, maximal at edges of wave, $x = \pm x_0$. B. Kink form. Bending moment maximal at 1/4 and 3/4 of width, zero at edges.

(1) The critical load for the kink form is four times that for the one-half sine wave. If x_0 were one-quarter wavelength of a sine curve, the critical load would be (e.g., Johnson, I, 1970, p. 88):

$$P_{\text{sine}} = BI(\pi/2x_0)^2 \tag{17}$$

whereas the critical load for the kink form is (eq. 13):

$$P_k = BI(\pi/x_0)^2 \tag{18}$$

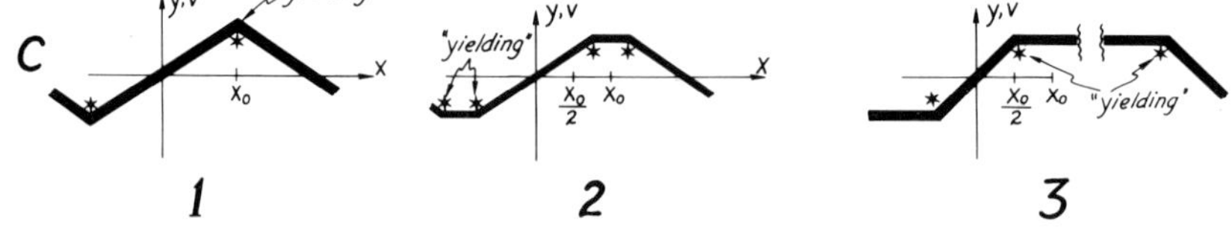

Fig. 10. Distribution of bending moment as a function of contact strength between layers. A. Waveforms. 1. Zero contact strength; waveform is one-half sine wave. 2. Low contact strength; waveform is intermediate between kink form and one-half sine form. 3. High contact strength; kink form. B. Distribution of bending moment. For 1 and 2 the bending moments are not zero at edges, $x = \pm x_0$, so continuous waveform normally is required for equilibrium. C. Forms of "yielded" elastic layers. 1. Chevron fold. 2. Fold with two hinges. 3. Kink fold with sharp hinges. Width of "yielded" kink band is one half width of kink form.

(2) The bending moment is maximal at the ends of one-half a sine wave (Fig. 9A), but it is zero at the end of the kink form (Fig. 9B).

(3) The one-half sine form would require special conditions to maintain the straight layers on each side of the kink band. The kink form, however, simply requires that the contacts between layers have strength, such as friction. This can be understood in terms of Figs. 10 and equations we have derived. Let us assume that the bending moment at the ends of the wave form is variable. The magnitude of the bending moment, M, can be calculated by combining eqs. 10 and 11:

$$v = -(\tau_0 tb/P)[\sin(x\sqrt{P/BI})/\{\sqrt{P/BI}\cos(x_0\sqrt{P/BI})\} - x] \tag{19}$$

and differentiating twice with respect to x:

$$M = BI(\mathrm{d}^2v/\mathrm{d}x^2) = \tau_0 tb \sin(x\sqrt{P/BI})/[\sqrt{P/BI}\cos(x_0\sqrt{P/BI})\} - x] \tag{20}$$

Now we can visualize changes in wave form and in bending moments as the shear strength, τ_0, ranges from zero to a certain critical value, τ_c, required to maintain the kink form shown in Fig. 9B. In order to examine the changes we will assume that the amplitude, v^0, and the width, x_0 of the wave form are constant, where v^0 is calculated from eq. 19 by replacing v with v^0 and x with x_0:

$$v^0 = -(\tau_0 tb/P)[\sqrt{BI/P}\tan(x_0\sqrt{P/BI}) + x_0] \tag{21}$$

Equation 21 shows that, if v^0 and x_0 are constant, a change in shear strength requires a change in the axial load. Further, eq. 20 shows that bending moment changes with both shear strength and axial load.

Let us suppose that shear strength is zero. In order for sinusoidal deflection to be possible, $\cos(x_0\sqrt{P/BI})$ in the denominator of eq. 19 must be zero, which requires that the axial load satisfy eq. 17. Further, the bending moment is proportional to the negative of the deflection. Figure 10A1 shows the deflection shape and Fig. 10B1 the distribution of bending moments as functions of x. The bending moment is maximal at each end of the one-half sine wave. Perhaps most important, however, the bending moment must be supplied by an external source or *the wave form must be repetitious*, as shown by dashed lines in Fig. 10A1.

Now let us suppose that the shear strength is at some critical value. The axial load is defined by eq. 18 and the combination of the axial load and the shear strength causes the bending moment, eq. 20, to vanish at $x = \pm x_0$. The result is the *kink form*. The deflection shape is shown in Fig. 10A3 and the distribution of bending moments is shown in Fig. 10B3. The most important observations are that the bending moment is zero at the ends of the kink form and that *the wave form may be discontinuous* (Fig. 10A3).

Finally we will consider an example where the shear strength is between zero and the critical value. Again, the axial load is calculated with eq. 21, assuming the amplitude and width of the fold are the same as those in the previous examples, and the deflection form and bending moment are computed with eqs. 19 and 20. The deflection form, Fig. 10A2, is intermediate between that of the half-sine wave and the kink form. The distribution of bending moments also is intermediate (Fig. 10B2). The most interesting feature is that the bending moment is maximal at some distance from the ends of the wave form. Most important, there is a bending moment at the ends of the wave form which must be balanced by external loading or by another wave form which is a mirror image of the first (Figs. 10A2 and 10B2). Thus, in general, the wave form *cannot be discontinuous.* Accordingly, the kink form is unique among the forms we have considered in that it can be discontinuous. Indeed, this is why we have called it a kink form.

(4) Growth of the one-half sine form requires that the axial load approach a critical value and that the layer either be disturbed or contain a slight initial deflection (e.g., Johnson, I, 1970, p. 118). Growth of the kink form, on the other hand, requires both that the axial load approach a critical value and that the slope of the layer have a certain, minimum, initial value. For example, consider eq. 7a. If the axial load, P, were less than the value:

$$P = \tau_0 tb/(\mathrm{d}v/\mathrm{d}x)_0 \tag{22}$$

there would be insufficient shear stress to make the layer slip relative to its neighbors even at the inflection point. If P exceeds this value, there is slippage and a gradient in the bending moment along the layer equilibrates the layer. The minimum slope at the inflection point, $x = 0$, can be obtained by substituting the critical axial load, eq. 18, into eq. 22 and rearranging:

$$(\mathrm{d}v/\mathrm{d}x)_0 > (\tau_0 bt/BI)(x_0/\pi)^2 \tag{23}$$

Equation 23 defines an upper bound for the width of the kink form as can be explained as follows. Suppose that the shear strength, τ_0, of the contacts between layers and the initial slope, $\mathrm{d}v/\mathrm{d}x$, of a waveform are pre-determined. Then, eq. 22 provides an expression of the maximum width $(2x_0)$ corresponding with that axial load. That is, according to eq. 23, the width is less than:

$$2x_0 < 2\pi\sqrt{(\mathrm{d}v/\mathrm{d}x)_0 BI/\tau_0 bt} \tag{24}$$

According to eq. 24, the upper limit of the width of the kink form is increased by increasing the slope $(\mathrm{d}v/\mathrm{d}x)_0$, and by decreasing the strength of contacts among layers. Also, because the strength increases with increasing effective stress normal to layers, the upper limit of the width of the kink form decreases with increasing stress normal to layers. Equation 24, incidentally, is valid whether the kink form has a positive or negative slope because the sign of τ changes with that of $(\mathrm{d}v/\mathrm{d}x)_0$ (Fig. 7A).

(5) The amplitude of the one-half sine wave is independent of wavelength. For the kink form, however, the amplitude and width are closely related, as can be shown by substituting eq. 18 for P into eq. 15 and rearranging. The width $2x_0$, of the kink form is:

$$2x_0 = \sqrt[3]{(Bt^2\pi^2 v^0/3\tau_0)} \tag{25}$$

According to eq. 25, the width changes slowly, with the cube root of the amplitude, v^0, of the kink form. As the shear strength becomes small, the width increases markedly. However, the width cannot change independently of axial load because eqs. 18 and 25 must be satisfied simultaneously.

These are the primary differences between the kink form and the half-sine form. It is well known that the effect of a soft confining medium on the buckling of an enclosed stiff layer is to reduce the wavelength of the stiff layer below that which it would assume if it were isolated (e.g., Johnson, I, 1970, p. 158, figs. 4.12A and B). Here we have shown that the effect of strength of contacts between layers in a kink band is similar.

The experimental multilayers containing kink folds were sharply bent at hinges of folds. The theory presented in Part III suggested that the sharp hinges of chevron folds in elastic materials form as a result of elastic "yielding" which is much like plastic yielding except that strains are recovered if the loading is removed. The "yielding" in rubber layers was explained in Part III as a result of high axial compression superimposed upon high incremental compression induced by bending. The hinges developed where bending moments were maximal, in crests and troughs of folds. We can, on the same basis, predict where hinges will develop in kinked layers by determining where bending moments are maximal. Figures 10B1 through 10B3 show distributions of bending moments in a layer with different contact strengths, and Figs. 10C1 through 10C3 show forms of layers that have "yielded" elastically at points corresponding to maximal bending. Figure 10C1 shows the chevron form discussed in Part III. Figure 10C3 shows a sharp-kink form, *the width of which* in the "yielded" form *is one-half* that in the nonyielded form (Fig. 10A3). Figure 10C2 shows the "yielded" form of the layer with intermediate contact strength; the peculiar feature of the form is that each wave has four hinges rather than two as in the chevron form. Layers with all three forms shown in Fig. 10C can be recognized in the experimental folds shown in Figs. 2.

Linearized theory

The linearized theory developed in Part III provides a similar solution to the problem of buckling of a layer with frictional contacts. This is not surprising in light of the similar predictions of the elementary and the linearized theories discussed in Part III. Nevertheless, we will briefly discuss the linearized theory here.

The solution for the displacements of an elastic plate subjected to initial stresses and to shear stresses on the upper and lower surfaces of the plate can be derived by superposition. The displacements for an isolated layer, free of surface tractions, are given by eqs. III,7b and III,9b, which simplify because $C_5 = C_7 = 0$, as required by symmetry. The displacements due to shear alone are:

$$v_0 = C_0 x \quad \text{and} \quad u_0 = -(C_0 - \tau/G)y$$

where C_0 is an arbitrary constant. Summing displacements:

$$v_0 = C_0 x + [C_6 \cosh(m_1 ly) + C_8 \cosh(m_2 ly)] \sin(lx) \quad (26a)$$

$$u_0 = -(C_0 - \tau/G)y + [C_6 M_1 \sinh(m_1 ly) + C_8 M_2 \sinh(m_2 ly)] \cos(lx) \quad (26b)$$

where m_1, m_2, M_1 and M_2 are constants defined in eqs. III,8 and III,10, and $l = 2\pi/L$, where L is wavelength. C_6 and C_8 are arbitrary constants.

The value of C_0 is determined by setting the slope of the layer equal to zero at $x = \pm x_0$, and $y = 0$:

$$(\partial v_0/\partial x)_{x=x_0} = C_0 + l[C_6 + C_8] \cos(lx_0)$$

However, $2x_0 = L$ (Fig. 8), so:

$$\cos(lx_0) = -1 \quad \text{and} \quad C_0 l = (C_6 + C_8) \quad (27)$$

The next step in the derivation is to set the boundary shear stress equal to the strength of the contact. Thus, at the boundary of a layer:

$$\tau_{\bar{y}x} = S_{\bar{y}\bar{y}}(\partial u_0/\partial y) + G(\partial u_0/\partial y + \partial v_0/\partial x) = \tau \quad \text{at} \quad y = \pm\, t/2 \quad (28)$$

Substituting eqs. 26 into eq. 28 we obtain:

$$C_6 = -C_8[(1 + m_2 M_2) \cosh(m_2 lt/2)/(1 + m_1 M_1) \cosh(m_1 lt/2)]$$

which is identical to that obtained by assuming zero shear at the boundary of the layer in Part III. Further, if we set the normal boundary stress, $\tau_{\bar{y}y}$, equal to the initial confining stress $S_{\bar{y}\bar{y}}$, as in Part III, we obtain the relation between initial axial stress and wavelength that we derived in Part III, as shown there in fig. III,2. Thus, according to the linearized theory, we replace L by $2x_0$ in fig. III,2 to determine relations between initial axial stress and width of kink forms. As shown in Part III, the elementary and linearized theories provide similar estimates of wavelengths if wavelength/thickness ratios are greater than about ten (fig. III,2). We will not proceed further with the linearized theory because its increased accuracy does not seem to warrant its increased labor.

Criteria for kink folding and sinusoidal folding

The solutions discussed in previous pages show that theoretical sinusoidal forms can be derived by assuming free slippage between stiff layers or by assuming soft interbeds between stiff layers, and that theoretical kink forms can be derived by considering contact strength between layers. The solutions do not indicate, however, which type of fold we might expect to observe in elastic multilayers. This is because the theory does not suggest unique conditions for kink and sinusoidal folding. For example, for multilayers of interbedded soft and stiff layers with contact strength, we can derive both a Biot wavelength for sinusoidal folding and a maximum band width for kink folding. The problem is that two distinct modes of folding are possible in such materials. The problem is not unique to the theory we are using; rather, in most analyses of folding, investigators have implicitly ignored the problem and studied conditions of sinusoidal folding as though no other fold form exists.

In order to derive conditions under which kink folds will be prominent we will make a further assumption. We assume that kink folding is normally preceded by sinusoidal folding in multilayers subjected to principal stresses parallel and normal to layering; the sinusoidal folding provides zones of sloping layers, near inflection points, where contact strength can be overcome and where kink folding can be initiated. Thus we assume that kink folds normally initiate as a result of slopes induced by growth of sinusoidal folds just as we assumed in Part III that concentric-like and chevron folds in elastic multilayers are results of higher-order, normal-stress effects.

First criterion (K_1)

A simple but useful way to compare conditions favorable to the development of prominent kink folds rather than prominent sinusoidal folds is to calculate the slope of layers at inflection points of folds required to initiate slippage, that is, to initiate kinking. We merely equate the shear stress between layers to the shear strength of contacts between layers. The shear stress between layers is the slope times the average shear modulus, $(\partial v_1/\partial x)G_a$, where v_1 is the additional deflection, that is, the deflection caused by buckling of initial irregularities (Johnson, I, 1970, p. 114). The slope of layering where slippage initiates is, approximately:

$$\partial v_1/\partial x \simeq \tau/G_a \tag{29a}$$

where:

$$G_a = (t_1 + t_2)G_1G_2/(t_1G_2 + t_2G_1) \tag{29b}$$

is the average shear modulus and where 1 refers to stiff and 2 to soft layers in eq. 29b. Here τ is the contact strength, defined in eq. 6. The derivation of the expression for the average shear modulus is explained elsewhere (Johnson, I, 1970, p. 231, 232 and 236), but we will discuss the modulus briefly here because it is a key parameter in most of the analyses in the following pages. The modulus is derived by assuming that shear stress is transferred from one layer to the other as a result of contact strength between layers and that the shear stress is the same in the stiff and soft layers. Therefore, if contact strength, τ, is zero, eq. 29b becomes:

$$G_a = 0; \quad \tau = 0 \tag{29c}$$

regardless of thicknesses or elastic properties of the layers. If the soft layers vanish in thickness, t_2, but there is contact strength:

$$G_a = G_1; \quad t_2 = 0; \quad \tau > 0 \tag{29d}$$

so that the average shear modulus is equal to the modulus of the stiff layers.

If the strength of contacts is zero, eq. 29c would describe the average shear modulus. Finally, if the modulus of the soft layers is very small:

$$G_a \simeq [1 + (t_1/t_2)]\,G_2; \quad G_2 \ll (t_2/t_1)G_1; \quad \tau > 0 \tag{29e}$$

in which case the average modulus depends upon the thickness ratio of stiff to soft layers and the modulus of the soft layers, but not upon the modulus of stiff layers.

Now, if the ratio τ/G_a is small, according to eq. 29a, the slopes of layers are small when slippage occurs, and the sinusoidal waveform may be indistinct at the time kink bands develop. If the ratio is zero, however, slippage can occur easily everywhere and we would not expect a kink form to develop. We note that, if contact strength is zero, the average shear modulus is zero so that the slope, eq. 29a, is indeterminant. Finally, if the ratio is large, slippage may not occur, and the sinusoidal waveforms may transform into concentric-like and then chevron forms rather than into kink forms. Thus, if:

$$G_a \gg \tau; \quad \tau \neq 0 \tag{30a}$$

kink forms should be prominent. And, if:

$$\tau = 0 = G_a$$

or, more generally, if:

$$G_a \lessgtr \tau \tag{30b}$$

sinusoidal folds should be prominent.

Another way to compare conditions favorable to kink rather than to sinusoidal folding is to compute the critical axial load required to initiate slippage between layers at inflection points of Biot waves. The load computed in this way is slightly less than the critical load required for kink folding, as discussed in connection with the kink form in earlier pages. The shear stress at inflection points due to growth of Biot waves is equal to the shear strength of contacts at mid-depth in the multilayer:

$$(\partial v_1/\partial x)G_a = \tau \tag{31}$$

$$x = 0$$

$$y = 0$$

The additional deflection, v_1, of the multilayer is (Johnson, I, 1970, p. 243):

$$v_1 = [\delta_0/(P_B/P - 1)]\ \sin(2\pi x/L_B)\ \cos(\pi y/T) \tag{32a}$$

where δ_0 is the amplitude of initial sinusoidal waves with the Biot wavelength, L_B, T is the total thickness of the multilayer and P_B is the critical load for Biot waves:

$$P_B = (2b\pi/T)\sqrt{B_v(t_1 + t_2)B_1 t_1^3/12} + b(t_1 + t_2)G_a \tag{32b}$$

in which G_a is defined in eqs. 29, L_B in eq. 2, and:

$$B_v = (t_1 + t_2)B_1B_2/(t_1B_2 + t_2B_1) \tag{32c}$$

is the average modulus for strains normal to layering.

Substituting eq. 32a into eq. 31 and rearranging:

$$(P/P_B) = 1 - (\partial v_0/\partial x)(G_a/\tau) = 1 - K_1 \tag{33a}$$

$$K_1 = (\partial v_0/\partial x)(G_a/\tau) \tag{33b}$$

where $(\partial v_0/\partial x)$ is the initial slope of Biot waves, before loading of the multilayer:

$$(\partial v_0/\partial x) = 2\pi\delta_0/L_B \tag{33c}$$

K_1 in eq. 33b will be called an index of kink folding of multilayers. If $K_1 \simeq 0$, we would expect folds to be sinusoidal, because the axial load, P, must approach the critical load, P_B, for growth of Biot sinusoidal folds before slippage can occur between layers. As indicated in eq. 32a, if $P_B/P \to 1$, $v_1 \to \infty$, so that the Biot waves grow without bound if $K_1 = 0$. On the other hand, if $K_1 > 0$, P can be significantly smaller than P_B at the time slippage occurs, and this condition favors growth of prominent kink folds. The significance of the variation of K_1 can be understood most clearly in terms of relative amplitudes of Biot waves. The amplitude is the sum of the initial amplitude, δ_0, and the growth of amplitude, δ_1, due to buckling. The latter is given by v_1 in eq. 32a, in which $x = L_B/4$ and $y = 0$. Thus the total amplitude, δ, is:

$$\delta = \delta_0 + \delta_1 = \delta_0/(1 - P/P_B) = \delta_0/K_1 \tag{34}$$

Following are some numerical values:

K_1	P/P_B	δ/δ_0
$\to 0$	$\to 1$	$\to \infty$
0.1	0.9	10
0.25	0.75	4
0.5	0.5	2
0.9	0.1	1.11

The higher the value of the index of kink folding, K_1, therefore, the lower is the amplitude of the sinusoidal waveform at the time kink folding occurs.

According to eq. 33b, the index of kink folding increases, that is the tendency for growth of prominent kink folds increases, as the initial slope of layers increases and as the ratio of the average shear modulus to the contact strength increases. The criterion based on critical axial loads, therefore, is a slight modification of the criterion stated earlier in eqs. 30a and 30b. We would note that, although eq. 33b indicates that $K_1 = 0$ if the initial slopes of layers are zero, this is merely a condition of indeterminancy. This condition does not favor growth of sinusoidal folds either because they do not grow unless there are initial waves, as shown in eq. 32a.

Second criterion (K_2)

The first criterion of kink and sinusoidal folding was based on the assumption that the axial load required to *initiate* slippage at inflection points of layers with contact strength approximates the critical load required to initiate kink folds. Now we will derive a second criterion based on the assumption that an estimate of the critical load for kink folding is the load required to overcome the contact strength all along the waveform in the layer at mid-depth in a multilayer. First we will compute the critical load for sinusoidal folding assuming no slippage at contacts, and then we will slightly modify the analysis to place limits on the critical load required to induce slippage.

The basic differential equations are (Johnson, I, 1970, p. 227 and 228):

$$\partial M/\partial x + (\sigma_{xy} - \sigma_{yx})(t_1 + t_2)b + P(\partial v/\partial x) = 0 \tag{35a}$$

$$\partial \sigma_{xy}/\partial x = -\partial \sigma_{yy}/\partial y \tag{35b}$$

where:

$$\sigma_{yx} = G_a(\partial v/\partial x) \tag{35c}$$

$$\sigma_{yy} = B_v(\partial v/\partial y) \tag{35d}$$

$$M = B_1 I_1(\partial^2 v/\partial x^2) \tag{35e}$$

$$I_1 = bt_1^3/12 \tag{35f}$$

and where G_a is defined in eq. 29 and B_v is defined in eq. 32c. The notation used here is the same as that used by Johnson (op. cit.). One of the assumptions Johnson used to derive eqs. 35 is that the value of $\partial u/\partial y$ is negligible (op. cit., p. 232). We have shown experimentally here and in Part III that this assumption is quite accurate for folding of multilayers, where lines drawn originally vertical, normal to layering, remain on the average vertical during folding (e.g., Figs. 2A and 2B).

As shown elsewhere (op. cit., p. 227), the sinusoidal solution of eqs. 35 for folding of the multilayer initially deflected into low, sinusoidal waves, and confined by rigid boundaries is:

$$v = [\delta_0/(1 - P/P_C)] \cos(\pi y/T) \sin(2\pi x/L_0) \tag{36a}$$

where:

$$P_C = B_1 I_1(2\pi/L_0)^2 + bB_v(t_1 + t_2)(L_0/2T)^2 + bG_a(t_2 + t_2) \tag{36b}$$

is the critical load and δ_0 is the amplitude of initial waves with a wavelength of L_0. According to eq. 36a, as the axial load increases, the amplitude of the fold increases and as the axial load approaches the critical value, P_C, the deflection grows without bound. According to eq. 36b, the critical load is determined by the bending rigidity (first term), resistance of the multilayer to thinning and thickening (second term) and resistance of the multilayer to shear (third term). Further, the wavelength that requires the minimum axial load for maximum amplification is the Biot wavelength, defined in eq. 2. The Biot wavelength was derived by differentiation of eq. 36b with respect to wavelength, setting the result equal to zero and solving for the wavelength. Finally, the critical load for the Biot waves is derived by substituting the Biot wavelength into eq. 36b. The result is eq. 32b.

Now we will derive the load required to fold the multilayer after the contact strength between layers at mid-depth has been overcome. We use all of eqs. 35 except eq. 35c which becomes replaced by the contact strength, eq. 6:

$$\sigma_{yx} = \tau$$

Thus eq. 35a becomes:

$$\partial M/\partial x + (\sigma_{xy} - \tau)(t_1 + t_2)b + P(\partial v/\partial x) = 0 \qquad (37)$$

Putting eqs 35b, 35d and 35e into eq. 37 and solving the resulting equation:

$$v = [C_1 \sin(\alpha x) + C_2 \sin(\beta x) + C_3 \cos(\alpha x) + C_4 \cos(\beta x)] \cos(\pi y/T) \qquad (38a)$$

where $\cos(\pi y/T) = 1$ at mid-depth and where:

$$\alpha, \beta = \left[(P/2B_1I_1) \pm \left\{ (P/2B_1I_1)^2 - \frac{B_v b(t_1 + t_2)}{B_1 I_1} (\pi/T)^2 \right\}^{1/2} \right]^{1/2} \qquad (38b)$$

in which the value of α corresponds with the plus root and β with the minus root. If we select the origin of coordinates at the inflection point and assume symmetry about that point, $C_3 = C_4 = 0$, in eq. 38a. Finally, we assume that the slope, $\partial v/\partial x$, of the layer is zero at $x = \pm x_0$ and that the change of curvature balances the shear strength at $x = \pm x_0$ where $\sigma_{xy} = 0$ and $\partial v/\partial x = 0$. These assumptions evaluate the constants and provide the solution:

$$v = [\tau(t_1 + t_2)b/B_1I_1(\alpha^2 - \beta^2)] [\{\sin(\beta x)/\beta \cos(\beta x_0)\} - \{\sin(\alpha x)/\alpha \cos(\alpha x_0)\}] \qquad (39)$$

The solution and the resulting waveform is the sum of two sine curves so that the waveform is not the same as the kink form, eq. 16, which is characterized by a sine term plus a linear term. Further, the waveform described by eq. 39 cannot satisfy both the conditions of zero slope and zero bending moment at the ends of the waveform. The kink form can satisfy these conditions simultaneously. The differences between the two solutions are results of effects of lateral resistance to deflection, which was assumed to be zero in the derivation of the kink form and assumed to be related to the modulus of vertical deformation, eq. 35d, for the waveform described by eq. 39. However, the solutions are equivalent for small values of β. In this case, $\sin(\beta x) \to \beta x$ and eq. 39 reduces to eq. 14, which is the equation for the kink form.

As with the solution for the kink form discussed in earlier pages, the solution derived here does not provide unique values for widths of wave forms and axial loads. For both sine components in eq. 39 to be positive, as is re-

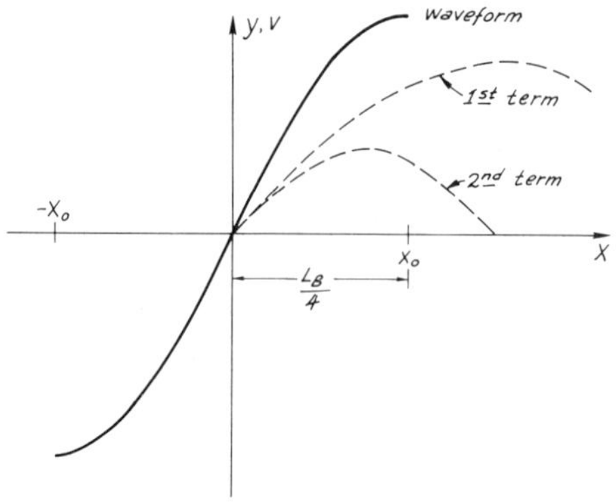

Fig. 11. Theoretical waveform for layer at midheight in multilayer with contact strength. The waveform is sum of two sine waves with different wavelengths.

quired to produce the waveforms shown in Fig. 11:

$$\cos(\alpha x_0) < 0 \qquad \text{or} \qquad \alpha > \pi/2x_0 \tag{40a}$$

Also:

$$\beta < \pi/2x_0 \tag{40b}$$

Combining eqs. 40 and 38b, we determine that the axial load must be greater than a certain value:

$$P > B_1 I_1 (\pi/2x_0)^2 + B_v(t_1 + t_2)b(2x_0/T)^2 \tag{41a}$$

and evaluation of inequality 41a indicates that the axial load is minimized for widths, $2x_0$, of the waveform that are equal to one-half the Biot wavelength. That is:

$$2x_0 = L_B/2 \tag{41b}$$

where L_B is defined in eq. 2, minimizes the axial load required to produce the waveform shown in Fig. 11.

One reason this conclusion is important is that it means that the waveform defined in eq. 39 is compatible with Biot waves throughout a multilayer. The Biot waves are sinusoidal and the waveform defined in eq. 39 is not, but the two types of waves will tend to have the same wavelength. Thus, the solution implies that the form of folds in a multilayer within which contact strength is overcome completely along the layer at mid-depth and partly overcome along layers elsewhere is gradational between the waveform defined in eq. 39 and the Biot sinusoidal waveform.

Now let us return to evaluation of the waveform. In order for the roots, α and β, in eq. 38b to be real and distinct, the quantity under the inner square-root sign must be positive:

$$(P/2B_1I_1)^2 > [B_v(t_1 + t_2)b/B_1I_1](\pi/T)^2 \qquad (42a)$$

which implies that:

$$P > (2b\pi/T)\sqrt{B_v(t_1 + t_2)bB_1t_1^3/12} = P_0 \qquad (42b)$$

Inequality 42b is equivalent to 41a, which can be demonstrated by substituting 41b and then 2 into 41a.

Now, P_0 defined in eq. 42b is equal to the first term on the right-hand side of eq. 32b, which defines the critical load for the Biot wavelength of sinusoidal folding. Therefore, whether the load required for the waveform shown in Fig. 11 to develop is greater or less than the critical load for Biot waves depends upon the magnitude of the average shear modulus, G_a (eq. 24b). The average shear modulus, again, plays a critical role.

In order to proceed further with the interpretation of the analysis of the waveform shown in Fig. 11 we must make a further assumption. We assume that conditions required for development of the waveform at mid-depth in multilayers with contact strength also represent the conditions required for initiation of kink folding throughout the multilayer. Thus, we imagine that the waveform shown in Fig. 11 first develops and that kink folds propagate from the waveform through the multilayer to form kink bands. Thus, if the shear modulus, G_a, is relatively large compared to resistances to lateral deflection and to bending, the axial load, P, will be low relative to the critical value, P_B, and the amplitudes of Biot waves will be small at the time inequality 42b is satisfied. This, presumably, describes conditions favoring kink folds. On the other hand, if the shear modulus is relatively low, the Biot waves will have high amplitudes at the time inequality 42b is satisfied, and the kink form might not be visible or not even develop.

Based on this analysis we propose a second index, K_2, of conditions favorable for kink folding of multilayers. The index is a measure of the relative values of the critical load, P_B, for development of sinusoidal waves and of the minimum load, P_0, eq. 42b, for slippage to occur all along the layer at mid-depth in the multilayer, which we interpret to be the minimum load for kink bands to develop. Thus, in order for kink folds rather than sinusoidal folds to be prominent within the multilayer, eq. 42a and the discussion above indicate that:

$$P_0 < P < P_B$$

or:

$$1 < (P/P_0) < (P_B/P_0) = K_2 \qquad (43)$$

which is the second index. Conditions favorable to kink folding are where:

$$K_2 >> 1 \qquad (44a)$$

and conditions favorable to development of prominent sinusoidal folds are where:

$$K_2 \simeq 1 \qquad (44b)$$

Substituting eqs. 32b and 42b into eq. 43, the index becomes:

$$K_2 = 1 + N_p[(1 + t^*)^2\sqrt{3}/\pi][G^*/(t^* + G^*)][\sqrt{(B^* + t^*)/B^*}]\,G' \qquad (45a)$$

where:
$$G^* = G_2/G_1$$
$$G' = G_1/B_1 = (1 - \nu_1^2)/[2(1 + \nu_1)]$$
$$B^* = B_2/B_1$$
$$t^* = t_2/t_1$$

and N_p is the number of pairs of stiff (1) and soft (2) layers in the multilayer.

If the materials are nearly *incompressible*, Poisson's ratio is nearly 0.5 and eq. 45a simplifies to:

$$K_2 \simeq 1 + N_p(1 + t^*)^2\sqrt{3G^*/(t^* + G^*)}/(4\pi) \qquad (45b)$$

Thus, the stiffer the interbeds, that is, the larger G^*, and the larger the number of pairs, N_p, of layers, the larger is the index and the more likely would a multilayer with contact strength develop kink folds.

Let us consider an extreme. Suppose that the soft layers are very thin, so that $t^* = t_2/t_1 = 0$. If the contacts were frictionless, G_a would not enter the expression for critical load, $K_2 = 1$, and we would predict sinusoidal folding of the multilayer. However, if the contacts have strength, eq. 45b becomes:

$$K_2 = 1 + N_p\sqrt{3}/(4\pi); \qquad t^* = 0 \qquad (45c)$$

Thus, for large numbers of layers K_2 is much greater than unity and conditions should be favorable to kink folding. For small numbers of layers, presumably, sinusoidal folding could be prominent because K_2 would be only slightly greater than unity.

Discussion of criteria

In many situations there may be a mixture of rounded, continuous, sinusoidal folds and angular, discontinuous, kink folds rather than the pure end members to which most of the discussion here has been devoted. There will be conditions where contact strength is overcome along only parts of the layers, rather than along all or none of the layers, as assumed in the theoretical analyses. Thus, there will be gradational forms intermediate between the end-member forms, and we have implied such gradational conditions by means of inequality signs and of unspecified gaps between the various criteria. We suggest that the folds developed in the experimental multilayer of greased rubber strips (Fig. 2B) are of this mixed type, and that it is because of imperfect lubrication between layers that some lengths of the waveforms are greater and some are less than the Biot wavelength, computed by assuming zero contact strength.

As far as we know, nobody has previously derived these or similar criteria for different types of folding. Johnson (I, 1970, p. 327) and Cobbold et al. (I, 1971) derived different sets of conditions for kink folding based on Biot's (I, 1965a) theory which we now believe to be unsound, as mentioned in Part I and as explained in some detail by Ellen (I, 1971). Indeed, the criterion developed by Cobbold et al. is in direct contradiction to the criteria presented above and with our experimental results. They consider a measure of anisotropy of a multilayer which is roughly equivalent to the average shear modulus, G_a, of our analysis divided into the average modulus of longitudinal compression. We will define the anisotropy, A_0, as that ratio:

$$A_0 = 2(1 + t^*G^*)(G^* + t^*)/[(1 + t^*)^2 G^*] \qquad (46)$$

where t^* is t_2/t_1 and $G^* = G_2/G_1$. If the shear modulus of the interbeds is very small:

$$A_0 \simeq 2t^*/[(1 + t^*)G^*]$$

so that the softer the interbeds, the higher the anisotropy. That is, *the smaller* G_a, eq. 24b, *the higher the anisotropy*. Cobbold et al. conclude that high anisotropy favors kink folding but our conclusion, as stated in eqs. 30 and 45 above, is opposite. Also, if the thickness of interbeds vanishes, the anisotropy becomes:

$$A = 2; \qquad t^* = 0$$

Cobbold et al. indicate that such conditions favor forms other than kinks, but our criterion indicates that such conditions favor kinking if contact strength between layers is much less than the shear modulus of the stiff layers. Further, an anisotropy of two describes closely that of the multilayers of rubber strips in frictional contact, which developed nice kink forms, as our criterion suggests (Figs. 2C and 2D).

The veracity of the criteria of conditions of development of kink forms in multilayers is supported by results of our experiments. The multilayers comprised of dry rubber strips (Figs. 2C and 2D) had very high shear moduli, approaching the shear moduli of the strips themselves, yet the strips could slip where friction was overcome. We performed an experiment with a multilayer comprised of rubber strips and gelatin interbeds, the thickness of which was on the order of $\frac{1}{10}$ the thickness of the rubber strips. The gelatin interbeds failed locally as the strips buckled and kink-like forms developed. Also, in one experiment reported in Part III (fig. III, 10A2), a kink-like form developed early in the experiment, controlled by the accidental alignment of air bubbles in the gelatin interbeds. The air bubbles presumably reduced the shear modulus of the gelatin interbeds locally, much as though there had been yielding and slippage between the layers. As the multilayer was shortened further, the kink form was overprinted with sinusoidal and concentric-like forms (Part III, fig. 10A4).

Widths of kink bands

We were able to estimate the maximum width of kink bands in terms of the slopes of layers, eq. 24, and in terms of axial load, eq. 18. Now we will discuss the widths of kink bands as compared to the Biot wavelength of sinusoidal folds in multilayers and derive an expression for widths of kink bands in multilayers where the properties and initial form of the multilayer are known. We will show that the widths of kink bands can be greater or less than the Biot wavelength, depending upon the rheological and dimensional properties of the multilayer.

Figure 12 shows the theoretical relationship between wavelength, L, and critical axial load, P_C, largely as determined by eq. 36b. The dotted curve in Fig. 12 represents the middle term in eq. 36b. That term increases with the square of the wavelength. The point labelled P_s represents the last term in eq. 36b, which is independent of wavelength but dependent upon the average shear modulus. The dashed curve in Fig. 12 represents the first term in eq. 36b, which arises from bending resistance of individual stiff layers. It varies according to the inverse square of the wavelength, but, according to the linearized theory, it has maximum value equal to $P_y = t_1 b G_1$, which is shown in Fig. 12. The dashed curve also represents the critical load, P_k, for kinking, as indicated in eq. 18, where $2x_0 = L$. Thus, the dashed curve is labeled "kink" in Fig. 12. The sum of the three resistances (dashed and dotted lines plus the constant P_s in Fig. 12) is the total resistance to sinusoidal folding, and is represented by the solid curve in Fig. 12. The total resistance has a minimum which has coordinates of the critical load, P_B, and the Biot wavelength, L_B.

Let us consider the wavelengths of kink folds and sinusoidal folds for various axial loads. As the axial load is increased, sinusoidal waves with the Biot wavelength, L_B, amplify most rapidly. For low loads the amplification is

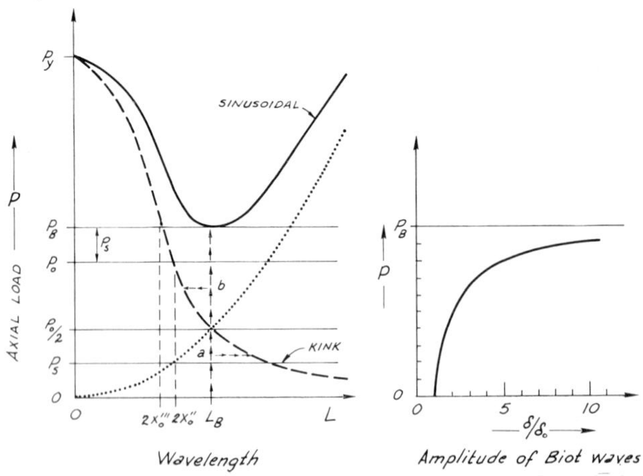

Fig. 12. Relations between axial load and wavelength for sinusoidal folds and kink folds. Also, relation between axial load and amplitude of Biot sinusoidal waves.

small. For example, if the load is one-half the critical value P_B, the amplitudes of small initial waves have merely doubled. If slippage occurs locally and a kink fold initiates before the axial load reaches one-half the value P_0 defined by eq. 42b and represented by the intersection of the dashed and dotted curves in Fig. 12, the width, $2x_0$, of the kink form will be greater than the Biot wavelength (e.g., point *a*, Fig. 12). If the kink fold initiates when the axial load exceeds $P_0/2$, however, its width will be less than the Biot wavelength (e.g., point *b*, Fig. 12).

We can make three estimates of the width, $2x_0$, of the kink band by considering the two criteria of kink folding developed in previous pages. If we assume that the kink band develops when the Biot waves have grown sufficiently so that the shear stress at inflection points is equal to the shear strength of contacts between layers, we can compute the width of the kink band by equating P in eq. 32a to P_k in eq. 18 and substituting the result into eq. 31 and rearranging:

$$2x'_0 = \pi t_1 \sqrt{(K_1 + 1)/(P_0^* + G_0^*)} \tag{47a}$$

where:

$$K_1 = (\partial v_0/\partial x)(G_a/\tau) \tag{47b}$$

$$P_0^* = (\pi/N_p)\sqrt{3B^*/(t^* + B^*)} \tag{47c}$$

$$G_0^* = 3G^*G'(1 + t^*)^2/(t^* + G^*) \tag{47d}$$

The dimensionless variables B^*, G^*, G' and t^* are defined in four equations following eq. 45a. Equation 47a indicates that the width, $2x'_0$, of the kink band is determined by many factors: the initial slope at inflection points of Biot waves, the number, N_p, of pairs of soft and stiff layers, the shear strength, τ, of contacts between layers, as well as the thickness ratio, t_2/t_1, modulus ratio, B_2/B_1, and shear-modulus ratio, G_2/G_1, of soft to stiff layers. Thus, although we conceptually could predict widths of kink bands, prediction would be most difficult in practice. Neither the strength nor the initial amplitude of the Biot waves are easy to measure.

The theoretical deduction that the width of a kink band would be sensitive to material properties is consistent with experimental observations. Donath (1970) and Weiss (I, 1968) report that the width of kink bands decreases with increasing confining pressure, as is predicted by eq. 47a. However, Donath notes that "the slightest difference in cohesion in the slate [material used in his experiments] could destroy the relationship". This is consistent with predictions of eq. 47a, which implies that the width of a kink band would be sensitive to many factors, including material properties such as cohesion.

We can make two further estimates of the width of kink bands by means of inequalities 43. According to those inequalities, the axial load at the time of kink folding should be greater than P_0 (Fig. 12) but less than the critical load for the Biot waves, P_B. Thus:

$$2x_0'' \equiv \pi t_1 \sqrt{1/(P_0^* + G_0^*)} < 2x_0 < \pi t_1 \sqrt{1/P_0^*} \equiv 2x_0''' \qquad (48a)$$

The two limits, $2x_0''$ and $2x_0'''$, of widths are shown in Fig. 12, and according to these limits, the widths of kink bands should be smaller than the Biot wavelength. Comparison of inequality 48a and eq. 47a indicates that the width, $2x_0$, of kink bands should satisfy the following inequality:

$$2x_0' \gtrless 2x_0 < 2x_0''' \qquad (48b)$$

which indicates that the widths of kink bands may be greater or less than the Biot wavelength. Thus, the widths of kink bands are not readily predicted according to our theory of kink folding.

The prediction becomes even more uncertain if we relax the requirement that the kink bands initiate at inflection points of Biot waves, an assumption which appears to be inappropriate for one type of experiment with multilayers of dry rubber strips. In this type of experiment the multilayer was first subjected to high longitudinal shortening and lateral confinement and then the lateral confinement was released to produce kink bands (Figs. 2C and 4A). Drag between the ends of the layers and the loading frame induced an initial slope (Fig. 2C1) which is unrelated to growth of Biot waves and the drag presumably induced shear between layers. We can obtain some insight into the processes in terms of eq. 22:

$$P = \tau_0 tb/(\mathrm{d}v/\mathrm{d}x)_0 \qquad (22)$$

where P is the critical axial load required to induce slippage, $(\mathrm{d}v/\mathrm{d}x)_0$ is the slope of layers locally and:

$$\tau_0 = \tau - \tau_x$$

in which τ is shear strength (eq. 6) and τ_x is a shear stress normal to layers (Fig. 7C). Thus, drag between the ends of the layers and the loading frame probably induced shear stress, τ_x, normal to layers and induced a slope $(\mathrm{d}v/\mathrm{d}x)_0$ locally. Both of these effects would lower the critical load, as shown in eq. 22, reproduced above, and the combination of the decrease in effective shear strength, τ_0, and the increase in slope presumably caused kink bands to initiate near the upper corners of the multilayer shown in Fig. 2C when the axial load reached the critical load for the kink band to form.

Other mechanisms of initiation of kink bands can be imagined, such as reduction of frictional resistance by increase in pore-water pressure (eq. 6), or inhomogeneous contact strength over much of the area of a potential kink band, but the mechanisms discussed here presumably were the most important in our experiments.

AN EXPLANATION FOR SINUSOIDAL AND KINK FOLDING OF MULTILAYERS CONFINED BY RIGID BOUNDARIES

The experiments with multilayers of gelatin and rubber described here and in Part I show the initiation and growth of folds that closely resemble ideal sinusoidal or kink folds. The linearized and elementary theories of folding of elastic multilayers with contact strength provide some understanding of mechanisms of folding. Now we will review the theoretical and experimental observations and explain why different fold forms developed under different conditions in the experiments.

We begin with a multilayered sequence of alternating stiff and soft layers, bounded by relatively stiff or rigid media and subjected to confinement normal to surfaces of layers and to axial shortening. The contacts between the confining media and the multilayer are nearly frictionless. The stiff and soft layers are subjected to axial compression of different amounts, depending upon the thicknesses and rheological properties of the layers. The stiffer layers support higher axial compressions than the softer layers. This is the state of initial stress of the multilayer confined by stiff media. The layers have a spectrum of initial irregularities in slope, the form of which might be described in terms of a Fourier sine series (e.g., Johnson, I, 1970. p. 119—123). The irregularities have very low amplitudes so that the initial stress state is essentially homogeneous within each layer.

As axial shortening is increased, the initial sinusoidal waveforms amplify, some more than others, approximately as indicated in eq. 36a. The waveform that amplifies most rapidly is the one with the Biot wavelength, which is defined approximately by eq. 2 or 4. The sinusoidal folds may not be visible, depending upon the magnitude of the axial compression relative to the critical load, defined approximately by eq. 32b. If the axial compression approaches the critical load, the Biot wavelength amplifies markedly. However, before the axial compression can approach the critical load, another fold form might develop, so now we will consider two end members of fold patterns: sinusoidal form and kink form.

Analysis of the theory of kink folding of multilayers provides the following criteria for distinguishing between conditions that favor growth of prominent kink folds or sinusoidal folds in multilayers of elastic materials.

(1) *Zero contact strength between layers.* If the contact strength between layers is zero, that is, if the layers are perfectly free to slip relative to each other, only repetitious, sinusoidal folds are possible at initial stages of folding. Subsequently concentric-like and chevron folds develop as the sinusoidal folds amplify, as described in Part III. For this special case $\tau = 0$ and the average shear modulus, G_a, must be zero. The criterion K_1 is determinant only if:

$$\lim_{\tau \to 0} (G_a/\tau) = 0$$

(2) *Finite contact strength between layers.* If shear stress is required to cause the layers to slip relative to each other, folds in multilayers may be either kink forms or sinusoidal forms, depending upon the following parameters:

(a) *Shear modulus.* A high average shear modulus relative to contact strength between layers favors kink folding. A low average shear modulus relative to contact strength favors sinusoidal folding at early stages of folding and subsequently concentric-like and chevron forms.

(b) *Initial slope of layers.* High initial slope of layering favors kink folding; low initial slope favors sinusoidal folding. Factors a and b are contained in the first index, K_1, of kink folding:

$$K_1 = (\partial v_0/\partial x)(G_a/\tau) \simeq 0; \text{ sinusoidal folding}$$

$$K_1 > 0; \text{ kink folding}$$

Here $(\partial v_0/\partial x)$ is the initial slope of layers, G_a is average shear modulus, defined in eq. 29, and τ is the Terzaghi-Coulomb shear strength, defined in eq. 6.

(c) *Number of layers.* The number of layers is an important factor determining the type of fold only if the kink folds initiate at inflection points of Biot, sinusoidal folds. In that case, the larger the number of layers with contact strength, the greater the tendency for kink forms rather than sinusoidal forms to be the prominent folds in the multilayer (eqs. 45).

Sinusoidal form

Let us suppose: either, that the layers are bonded tightly together so that the contact shear strength is so high relative to the shear modulus of the multilayer that the slope of layers must be large before slippage can occur; or, that the contacts between layers are frictionless so that the layers can slip readily everywhere. These conditions favor growth of a sinusoidal end member of the spectrum of fold forms. Now, as the axial load approaches the critical load, eq. 32b, initial waves of the Biot wavelength markedly increase in amplitude. Initial wave components with other wavelengths do not amplify as markedly so they are largely overprinted by sinusoidal waves with the Biot wavelength.

The wavelengths of stiff layers in the subsequent fold pattern are largely determined by the early, Biot wavelength but they become modified by higher-order processes such as those described in Part III. Briefly, different normal stresses acting at right angles to surfaces of stiff layers throughout the multilayer cause wavelengths of individual stiff layers to change. Wavelengths of stiff layers decrease in cores of folds, where there is incremental tension so that confinement is reduced, and increase in crests of folds, where there is increase in confinement. The magnitudes of these incremental stresses increase with the amplitude of the waveforms. Thus, the sinusoidal pattern becomes unstable as it amplifies, and transforms into a concentric-like pattern. The incremental normal stresses are largest near the boundaries of the multilayer and are zero at mid-depth, so that if the multilayer is thick relative to the wavelengths, the wavelengths change markedly only near the boundaries and most of the fold pattern resembles a sinusoidal waveform (Fig. 2A4). If the thickness of the multilayer is small relative to the wave-

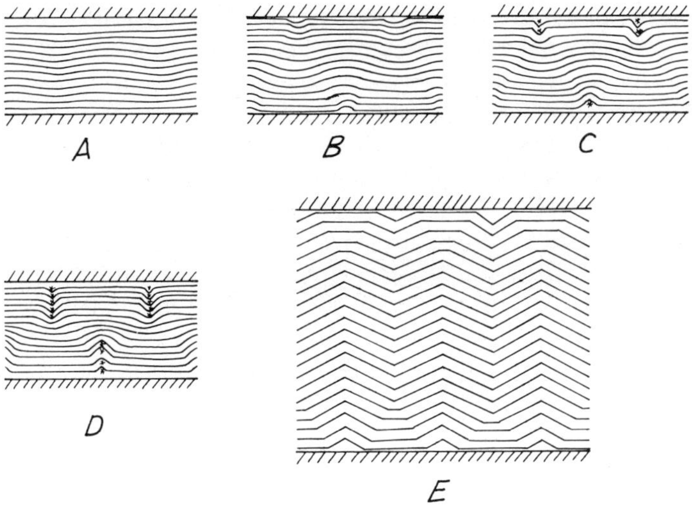

Fig. 13. Development of concentric-like and chevron folds. A. Irregularities in slope of layering with Biot wavelength are selectively amplified. B. Higher-order effects cause sinusoidal waves to be unstable so that wavelengths of folds are different at different places within multilayer, producing a flattened, concentric-like pattern. C. Elastic "yielding" at hinges of short waves produces chevron folds in layers near boundaries. D. Elastic "yielding" extends into layers above. E. If multilayer is relatively thick, much of the multilayer ultimately develops a chevron pattern.

length, the concentric-like pattern develops throughout most of the multilayer (fig. 22, Part I).

Finally, if the axial compression in a stiff layer becomes sufficiently large, the layer may "yield" elastically, forming a relatively sharp hinge, and transform layers in the core of a concentric-like fold or of a sinusoidal fold into a chevron pattern (Figs. 2A and 2B). The final pattern of ideal folds is a flattened variant of the combination of concentric-like and chevron forms shown in fig. 14, Part III.

Figure 13 shows an idealization of the transition of fold forms from sinusoidal (Fig. 13A) to concentric-like (Fig. 13B) to chevron forms in thin (Fig. 13D) or thick (Fig. 13E) multilayers. The mechanisms of the transitions from one fold form to another were discussed more fully in Part III.

Kink form

Let us return to the scenario begun in earlier paragraphs, where we were loading a multilayer, and where sinusoidal forms with the Biot wavelength were beginning to amplify. Now, however, rather than continuing to amplify with increasing axial compression, the layers slip at inflection points of the waves and a pattern of kink bands develops spontaneously, propagating to both edges of the multilayer, and the sinusoidal waves disappear.

Our theory of kink folding is based on two fundamental assumptions. One is that *kink folding in elastic materials is a process of buckling.* Another is that the *layers can slip relative to each other in some places but not in others.* One characteristic of the kink forms we have observed in elastic multilayers is that slippage between layers is concentrated in limbs of the kink bands themselves. Kink forms were developed in multilayers of rubber strips in frictional contact with one another (Fig. 2C and 2D) apparently because the multilayers could slip where the shear strength between layers was overcome locally, but not everywhere within the multilayer. However, the assumption of slippage locally may be too restrictive. A kink-like fold developed early in the buckling of one multilayer of interbedded gelatin and rubber confined by soft elastic media as reported in fig. 10A, Part III, apparently because the interbeds of gelatin in one part of the multilayer contained many more air bubbles and, therefore, sheared more easily than in other parts of the multilayer. In general, however, the development of kink folds probably does imply that the contact strength between layers was overcome. We have defined the shear strength as the Terzaghi-Coulomb strength of materials, which depends upon effective normal stress across layers as well as cohesive strength between layers (eq. 6).

Our experimental observations are consistent with the criteria stated in earlier paragraphs. The average shear modulus of multilayers of rubber and gelatin is small, roughly equal to that of the gelatin itself, which is about fifty times softer than the rubber strips. The contact strength, or bond strength, between rubber and gelatin must be quite high because rubber strips and gelatin shear relative to each other only in cores of chevron folds. We would expect sinusoidal folds to develop in such materials, which, of course, we observe. On the other hand, the effective shear modulus of the multilayer of dry rubber strips in direct contact is quite high, roughly equal to the shear modulus of rubber. The contact strength is variable, depending upon the confining stress normal to layers, but for low confinement the contact strength is small. Thus, the slopes of layers are small at the time slippage occurs locally. Such conditions favor development of kink folds according to one of our criteria. We do observe kink bands in the multilayers, as shown in Figs. 2 and 4.

Folds with short wavelengths in thick sequences of thin elastic layers probably indicate discontinuous folding, according to the theoretical analyses. Very thick multilayers should develop very long wavelengths of sinusoidal folds according to Fig. 6 and according to Ramberg's (I, 1963a; 1961) theory of folding of viscous multilayers. The second index, K_2, of kink folding derived here (eqs. 45) indicates that the larger the number of layers in a multilayer with contact strength between layers, the more likely are kink folds to develop. The widths of the kink folds may be much smaller than the wavelength of Biot sinusoidal waves (Fig. 12).

The two fundamental assumptions allow one to make certain predictions about widths of kink bands. An upper limit of the widths of kink bands in homogeneous multilayers without interbeds can be computed if we assume that the bending moment in each layer is zero at each side of the kink band, and that the only interaction among layers is contact friction. The analyses suggest that the only condition under which this is possible is where the kink band propagates diagonally across a multilayer, so that each layer in the band develops the same kink form. Paterson and Weiss (I, 1966) have shown that there is no thickening or thinning of layers within a kink band if the slope of the band is 90° minus half the slope angle of the layers. According to our theory, the upper limit of the width of the kink form is proportional to the square root of the product of the local slope of layers, the bending rigidity of the stiff layers, and one over the contact strength, eq. 24. The equation predicts that, the higher the contact strength, the narrower the kink band. This prediction is consistent with results of our experiments with multilayers of thin and thick rubber strips and with Weiss' (I, 1968) experiments with decks of paper cards. The experiments indicate that the higher the stresses normal to layering, which we interpret as the higher the frictional resistance to slippage at contacts between layers, the narrower the kink band.

Widths of kink bands can be related to the Biot wavelength of sinusoidal folding for multilayers of homogeneous layers of interbedded stiff and soft material with contact strength. If the contact strength is overcome at inflection points of Biot waves before the axial load reaches a value $P_0/2$, defined in eq. 42b, the width of the kink form will be greater than the Biot wavelength (Fig. 12). Otherwise, the width of the kink form will be less than the Biot wavelength. The width of the kink band is a function of many factors as indicated in eq. 47a. Indeed the factors are so abundant that we doubt that one can deduce much about the rheological properties of elastic multilayers containing kink bands by measuring widths of kink bands. Wavelengths of Biot sinusoidal folds in elastic multilayers depend upon the total thickness of the multilayer, and the thickness and modulus ratios of stiff to soft layers. Thus, one can estimate modulus ratios for multilayers in terms of the geometric properties of the multilayers and the wavelengths of folds developed within them. Widths of kink bands, unfortunately, depend upon all these factors plus the initial slopes of Biot waves and the contact strength, parameters which would be difficult to estimate in experiments and most difficult to estimate in rock.

An idealization of the development of kink folds in elastic multilayers is shown in Fig. 14. The multilayer contains a variety of initial waves. As the multilayer is shortened, initial waves with the Biot wavelength become selectively amplified (Fig. 14A). When slopes at inflection points of the amplified waves reach a critical value, depending on the average shear modulus and the contact strength, the layer at mid-depth slips, first at inflection points and then slippage spreads toward the crests and troughs of folds. With increased shortening the slopes and axial loads increase until the kink form becomes a stable configuration for individual layers (eq. 22) and the kink form appears

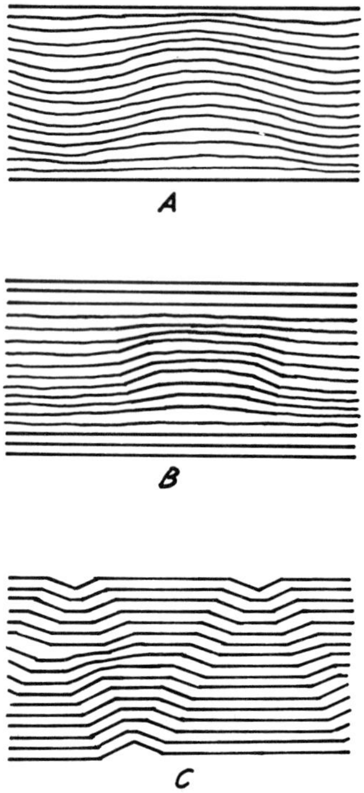

Fig. 14. Development of kink folds. A. Shortening selectively amplifies waves with the Biot wavelength. B. Shear stress between layers induced by growth of sinusoidal waves overcomes contact strength between layers at inflection points and kink bands begin to form there. C. With further shortening the two kink bands propagate and reflect at stiff boundaries to produce a relatively complex pattern of kink bands.

and propagates spontaneously through the multilayer (Figs. 14B and 14C). The subsequent development of the pattern of kink bands is complex. If sharp hinges develop as a result of elastic "yielding", the widths of the sharp kinks will be one-half that of the kink form (Fig. 10C3). The kink bands can form a complex pattern by initiating at different points and reflecting from stiff boundaries. Also, the kink bands may lock and widen to accommodate overall shortening of the multilayer. Our analyses apply only to the initial stages of development of kink bands in elastic multilayers.

ACKNOWLEDGEMENT

The research reported here was supported by the National Science Foundation, Grant no. GA-36917.

We are grateful to Peter Gordon, machinist in Earth Sciences at Stanford, who invented the collapsible bar which was the critical element of the experimental apparatus and who built the apparatus which worked perfectly. We thank Ze'ev Reches, graduate student at Stanford, for running some of the experiments and for criticizing and reviewing the manuscript. We thank Marcella Flynn for typing the manuscript and for improving the grammar and punctuation. Dr. David Pollard, U.S. Geological Survey, Menlo Park, again has provided invaluable service by editing and criticizing the organization and content of the manuscript. Dr. Peter Cobbold reviewed the paper and offered many useful suggestions for improving the accuracy and quality of reviews of previous papers.

REFERENCES NOT CITED IN CHAPTER 3

Clifford, P.M., 1968. Kink-band development in the Lake St. Joseph area, Northwestern Ontario. In: A.J. Baer and D.K. Norris (Editors), Proceedings Conference Research in Tectonics. Geol. Surv. Can. Pap., 68-52: 229—242.

Dewey, J.F., 1965. Nature and origin of kink-bands. Tectonophysics, 1: 459—494.

Donath, F.A., 1961. Experimental study of shear failure in anisotropic rocks. Bull. Geol. Soc. Am., 72: 985—990.

Donath, F.A., 1970. Some information squeezed out of rock. Am. Sci., 58: 54—72.

Gay, N.C. and Weiss, L.E., 1974. The relationship between principal stress directions and the geometry of kinks in foliated rocks. Tectonophysics, 21: 287—300.

Jaeger, J.C., 1960. Shear failure of anisotropic rocks. Geol. Mag., 97: 65—72.

Johnson, A.M. and Ellen, S.D., 1974. A theory of concentric, kink, and sinusoidal folding and of monoclinal flexuring in compressible, elastic multilayers. I. Introduction. Tectonophysics, 21: 301—339.

Johnson, A.M. and Honea, E., 1975a. A theory of concentric, kink, and sinusoidal folding and of monoclinal flexuring in compressible, elastic multilayers. II. Initial stress and nonlinear equations of equilibrium. Tectonophysics, 25: 261—280.

Johnson, A.M. and Honea, E., 1975b. A theory of concentric, kink, and sinusoidal folding and of monoclinal flexuring in compressible, elastic multilayers. III. Transition from sinusoidal to concentric-like to chevron folds. Tectonophysics, 27: 1—38.

Ramberg, H., 1961. Contact strain and folding instability of a multilayered body under compression. Geol. Rundsch., 51: 405—439.

Chapter 7

ASYMMETRIC FOLDING IN INTERBEDDED CHERT AND SHALE OF THE FRANCISCAN COMPLEX *

SUMMARY

The origin of tight, asymmetric, kink-like or chevron-like folds in interbedded shales and radiolarian cherts of the Franciscan Complex in the San Francisco Bay area has been somewhat of a mystery for many years. Stephenson Ellen provided many clues as to the origin and indicated that the folds became asymmetric as a result of layer-parallel shear. He believed that the original folds were conjugate kink folds.

As a result of reexamination of most of the folds studied by Ellen, of experimentation with elastic multilayers, and of the theories developed in Parts III and IV of this series of papers, we believe that the original folds were mostly chevron rather than kink folds. Thus, we suggest that the folds formed by a combination of layer-parallel shortening and layer-parallel shear when the rocks were soft and pore pressures were high.

Several lines of evidence suggest that typical folds in the Franciscan are asymmetric chevron folds. A combination of theory of finite simple shear and of experimentation with elastic multilayers indicates that the tight folds of the Franciscan could have been produced by smaller angles of simple shear if the original folds were typical chevron folds rather than typical kink folds. Several field observations, including thickening of shales but not of cherts in hinges of folds and lack of deformation of radiolaria in the cherts, indicate that the cherts were soft and the shales very soft at the time of folding. The pore-water pressures in the shales probably were high. Such conditions theoretically favor concentric-like and chevron folding, not kink folding. Finally, most of the asymmetric folds in a quarry exposure can be reconstructed geometrically as typical chevron folds but not as typical kink folds subjected to simple shear.

INTRODUCTION

Tight, asymmetric, kink or chevron folds (Fig. 1A) are common within interbedded shales and radiolarian cherts of the Franciscan Complex in the

---

* Originally published as: Ramberg, I.B. and Johnson, A.M., 1976. Part V, Asymmetric folding in interbedded chert and shale of the Franciscan Complex, San Francisco Bay area, California. Tectonophysics, 32: 295—320.

San Francisco Bay area of California. Asymmetric folds similar to those in the Bay area occur throughout the Franciscan Complex in California and southern Oregon (Talieferro, 1943; Bailey et al., 1964) and also have been reported in metamorphic, volcanic, and sedimentary terranes in other parts of the world (e.g., Dewey, 1965; Clifford, 1968; Ramsay, I *, 1967, 1974).

It is unclear whether the folds in the Franciscan began as kink folds or as chevron folds. Kink bands are tabular zones of folding that transect foliation (Dewey, 1965), whereas chevron folds are symmetric, with axial planes roughly normal to original bedding, and the limbs of equal lengths (Fig. 2). Many kink bands are arranged in sets inclined to each other and with opposite senses of displacement. Each pair of kink bands is termed box fold or conjugate fold (e.g., Ramsay, 1962).

Typical folds in the Franciscan do not fit easily into these categories of folds because they are asymmetric (Figs. 1A and 2C); one limb is longer than the other and apical angles are less than 90 degrees. Where one limb is much shorter than the other, the folds resemble overturned kink bands. The small apical angles, however, might suggest asymmetric chevron folds because the apical angle of most experimental and natural kink bands is on the order of 120 degrees, whereas the typical apical angle of chevron folds is about 60 degrees. Further, in some places, the lengths of the two limbs of the folds are nearly the same, again suggesting chevron folds.

Ellen (I *, 1971) made a detailed study of the folds in the Franciscan in Marin County, north of the Golden Gate, and found a few (e.g., Fig. 6) that resemble sheared conjugate kink folds, with one open and one tight hinge. Based on these observations and a series of experiments with multilayers, Ellen suggested that most of the folds originated by a two-phase deformation involving, first, lateral compression during which conjugate folds developed, and then subsequent layer-parallel shearing during which the folds became asymmetric (see Johnson, I, 1970, Chapter 8). Ellen was unable to produce the asymmetric forms experimentally, however, and alternative explanations certainly remain. In particular, the folds might have been derived from chevron folds.

The purposes of this paper are to reopen the question of the origin of the tight folds in the Franciscan, to provide a respite in the series of papers which has been largely theoretical thus far, and to provide an opportunity to summarize many of the results of Parts III ** and IV ** within the context of

* References cited in Part I will be noted with an "I", followed by the date, such as I, 1955, and not be repeated in References at the end of this paper. "Part I" is the reference, Johnson and Ellen, 1974.

** "Part III" will be used to abbreviate the reference: Johnson and Honea, 1975. "Part IV" is the reference, Honea and Johnson, 1975. "Part VI" is the reference, Reches and Johnson, in preparation.

Note: Part I = Chapter 3, Part II = Chapter 4, Part III = Chapter 5, Part IV = Chapter 6.

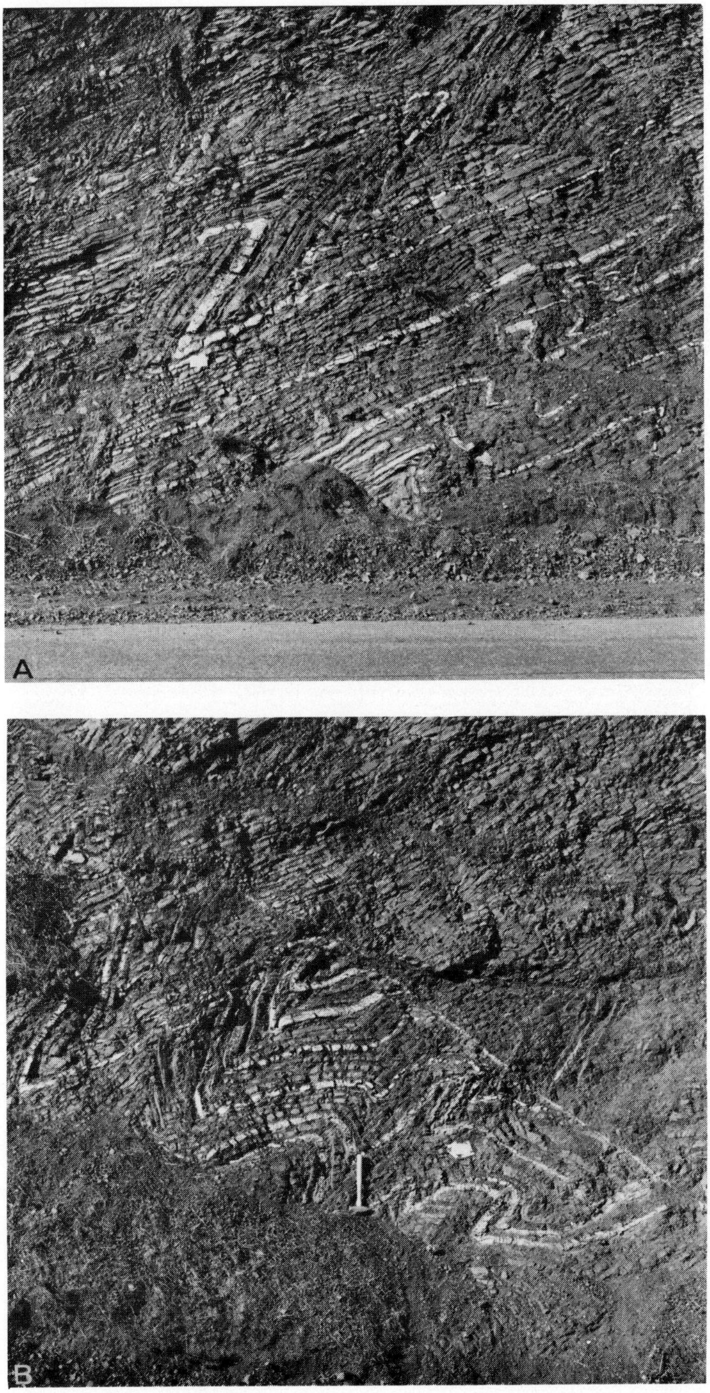

Fig. 1. Typical asymmetric folds in interbedded shale and radiolarian chert of the Franciscan Assemblage at the north side of the Golden Gate. Some chert beds have been marked with chalk. A. Tight angular folds. White card about 15 cm long. B. More open, rounded folds nearby.

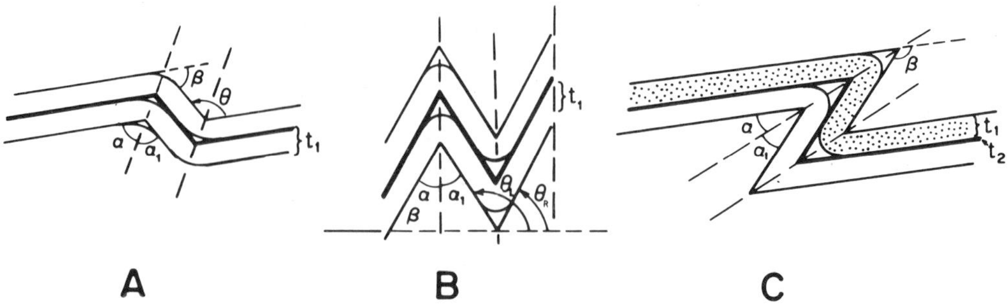

Fig. 2. Idealized folds. A. Ideal angular kink fold, showing slope angle, α, of axial trace, apical angle, $\alpha + \alpha_1$, slope angle, β, of layering and angle, θ, between nonfolded layering and layering in kink band. B. Ideal angular chevron folds, showing angle between non-folded layering and left limb, θ_L, and right limb, θ_R. C. Idealized tight, asymmetric fold of the type observed in Franciscan rocks. Apical angle, $\alpha + \alpha_1$, typically 40 degrees.

field observations, back to which, of course, the various theories ultimately lead. A final purpose is to set the theoretical problem to be discussed in Part VI, the effect of initial shear on the forms of folds.

We restudied many of the outcrops of Franciscan rocks examined by Ellen (I, 1971) and studied Franciscan outcrops elsewhere in the Bay area. Also, we made a series of experiments with simple elastic materials in order to study effects of superimposed shearing upon fold forms in multilayers with markedly different contact strengths between layers.

We begin by describing forms of folds in interbedded chert and shale of the Franciscan Complex. Then we briefly review the relevant theoretical predictions in Parts III and IV and derive equations to describe the ideal simple shear of preexisting fold forms. Then we describe some experiments and determine to what extent the theories are applicable for simple elastic materials. Finally, we bring together the various lines of evidence from the field, the theory and the experiments and present our best estimate of the evolution of fold forms in the interbedded chert and shale.

FORMS OF FOLDS IN THE FRANCISCAN CHERTS AND SHALES

Excellent exposures of the folded, interbedded chert and shale of the Franciscan Complex are at the north side of the Golden Gate, Marin County, at Twin Peaks, San Francisco County, and at Coyote Hills, Alameda County (Fig. 3). There, the typical brownish-red, thin-bedded shales and radiolarian cherts occur as irregular bodies or lenses interbedded on the scale of tens or hundreds of meters with metavolcanics and graywacke sandstones in the complex Franciscan Complex (Bailey et al., 1964; Ellen, I, 1971; Schlocker, 1974). Individual chert beds in the folded areas commonly are 1—10 cm thick. They are sharply bounded by shale interbeds, the thicknesses of which range from a few mm to about 3 cm. Pinching, swelling, and lensing of the chert beds produce gentle waviness of bedding, even where there are no

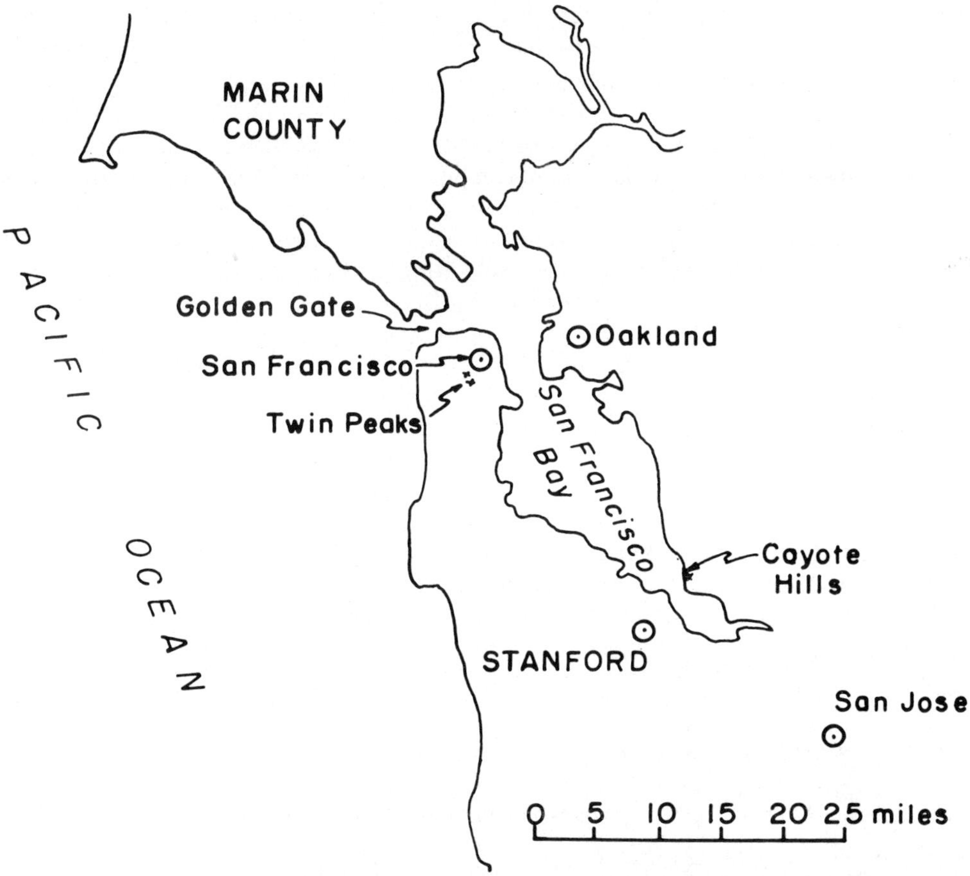

Fig. 3. Locations of areas where folds in Franciscan cherts and shales were studied.

folds. The contrast of the present physical properties of the chert, which is aphanitic, hard and brittle, and the smooth curvature of layers in tight hinges (Figs. 1A, 4A and 5A) suggests that the folding in the Franciscan took place at an early stage, when the rocks were relatively soft. This impression is supported by Ellen's (I, 1971) observation that delicate tests of radiolaria appear to be undeformed, even in tight hinges where strains must have been large.

The sense of asymmetry or overturning of the folds is essentially consistent, towards the northeast, throughout exposures at the north side of the Golden Gate. The sense of asymmetry is the sense one might expect if rocks overlying the interbedded cherts and shales were pushed northeastward, relative to rocks below. The sense of asymmetry apparently is not related to larger folds, with wavelengths on the order of hundreds of meters. The asymmetric folds occur on each limb of these larger anticlines and synclines, according to unpublished mapping of roadcuts by S. Ellen and A.M. Johnson, yet the sense of asymmetry is independent of position within the larger folds. Ellen (I, 1971) reported that the sense of asymmetry is quite consistent within a small part of the area, about 1 km^2, that he mapped in detail.

Figure 4A shows a close view of typical, tight folds. The limbs are straight or slightly undulating. Folding is confined to a narrow hinge zone, and individual chert beds can be followed continuously around the hinge zone. The chert beds maintain roughly constant thicknesses whereas the shale beds thicken in hinges (Fig. 5A). Open or cemented fractures are perpendicular to the bedding planes, even around the fold hinges (Figs. 4A and 5A). They apparently are joints formed after folding because the sides of the fractures are parallel.

A characteristic feature of the folds is the alternation of long and short fold limbs (Figs. 1A and 2C), leading to a conspicuously asymmetric or kink-band-like appearance. The ratio of short to long limbs commonly is about 1/3, but the range is wide. The short limb generally is 0.5—2 m long, so that the length/thickness ratio of chert layers in short limbs generally is 10—40. The angular folds in some places grade laterally into more open, low-amplitude folds, involving only a few chert layers (Fig. 1B), which typically have limb length/thickness ratios less than 10.

The interlimb angles of the high-amplitude angular folds are characteristically in the range 25—55 degrees, but may be as small as 10 degrees. Average

Fig. 4A.

Fig. 4. Close views of folds in Franciscan cherts and shales. A. Sharp folds with apical angles of about 43 degrees. Limbs are straight or slightly undulating. Shales thicken in hinges but cherts do not. B. Apical angles of about 35 degrees. In places the hinges have collapsed to produce very small apical angles.

values for folds shown in Figs. 4A and 4B are 43 degrees and 35 degrees, respectively. Smaller interlimb angles are accompanied by greater dilation in the hinge zone and by thickening parallel to axial planes of the folds. Figure 5A illustrates how the dilation was accommodated by flow of shale from the limbs into the hinge zone. A marked cleavage in the shale (Fig. 5A) has a curved configuration, convex towards the axial plane. Apparently where the amount of space-filling material was insufficient, the hinges collapsed (Fig. 4B) as reported by Ramsay (1974) for folds he interpreted to be of the chevron variety . This feature is common in the Franciscan rocks and can be recognized for instance in Fig. 1A, and especially above and to the right of the hammer shown in Fig. 4B. An extra-thick bed, under the head of the hammer, presumably caused a larger than usual hinge space, and the thinner chert beds outside collapsed to form interlimb angles of about 10 degrees. Folds with small interlimb angles typically occur near midlength of a folded band and more open folds occur at either end (Fig. 5B). Thus, the zone of folding widens from the narrow hinge zone to a broader arch.

Thus far we have described the typical folds in the interbedded chert and shale of the Franciscan Complex in the Bay area. In some places, however,

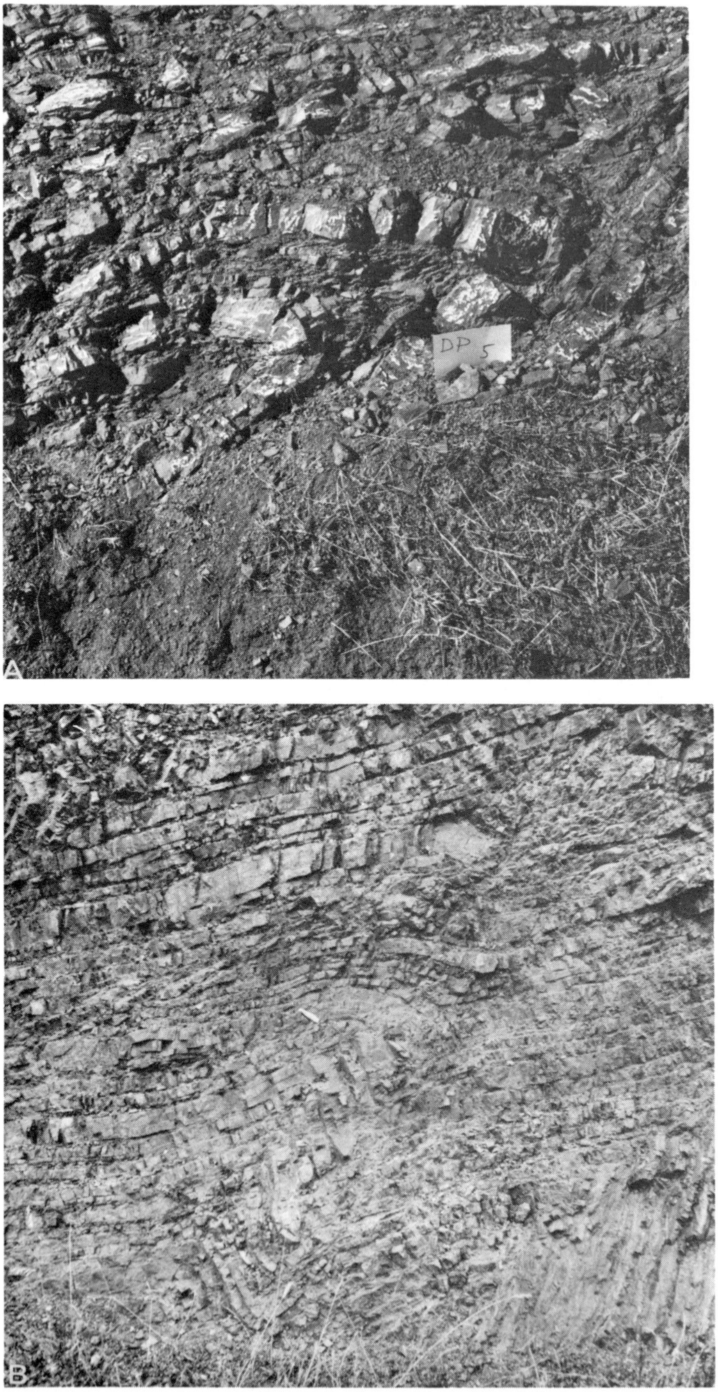

Fig. 5. Other features of folds in Franciscan cherts and shales. A. Thickening of shale in hinges of folds. Cleavage in the shale is curved, convex toward axial plane. B. Change of fold form along an axial trace, from tight fold near midlength, at lower left in photo, to more open fold at upper end of trace.

Fig. 6. Example of asymmetric conjugate fold with one tight hinge zone and one open hinge zone.

folds such as those shown in Fig. 6 can be recognized. The fold appears to be an asymmetric box fold, with a shorter, steeper limb on the left and a longer, more gently dipping limb on the right. This is the kind of fold Ellen (I, 1971) recognized in a few places and to which he attached great significance. He interpreted such folds to represent relicts of the first stage of development of most of the tight folds in the Franciscan cherts. Thus, he interpreted the fold shown in Fig. 6 to be the result of layer-parallel shearing of a symmetric-conjugate set of kink bands.

THEORY

The asymmetry and tighness of the folds in the interbedded chert and shale cannot be expained in terms of existing theories of folding. However theories do provide some understanding of conditions that favor the development of various forms of symmetric folds. Then, following Ellen (I, 1971), we can transform the symmetric forms into asymmetric forms by superimposing uniform shear, assuming that the layers behave passively relative to the shear. This is the approach we will follow here and it seems to be adequate to explain most of the experiments. We will present a theory of asymmetric folding of elastic multilayers in Part VI.

First we will briefly review the theories of chevron and kink folding presented in Parts III and IV for elastic and elastic—plastic materials. The chert and shale certainly did not behave elastically, but they may have behaved much as elastic—plastic materials. The problem of the rheological behaviors of the chert and shale of the Franciscan at the time of folding cannot be solved here.

The theories developed in Parts III and IV suggest that kink and chevron folding can be distinctly different phenomena. According to this view, chevron folds are the end product of a sequence of fold forms (Fig. 7) starting with sinusoidal, transforming into concentric-like, and finally into chevron forms (Fig. 7; and fig. 14, Part III; fig. 13, Part IV). If the stiff layers are widely separated, the stage of concentric-like folding may be indistinct (fig. 10B, Part III). Kink folds, however, normally develop directly from irregularities in slopes of layering (fig. 2C, Part IV) which might be induced by low sinusoidal waves (fig. 14, Part IV). Two criteria have been derived in Part IV to distinguish between conditions that favor kink folding on the one hand, and conditions that favor concentric-like and chevron folding on the other. One criterion compares the shear strength of contacts between layers and the bending rigidity and resistance to lateral deflection of layers. This criterion, in short, indicates that the larger the number of layers in a multilayer stack, the more likely are kink folds to develop.

The other criterion compares the shear strength, τ, of contacts between layers and the average shear modulus, G_a, of the multilayer:

$$K_1 = (\partial v_0/\partial x)(G_a/\tau) \quad (1)$$

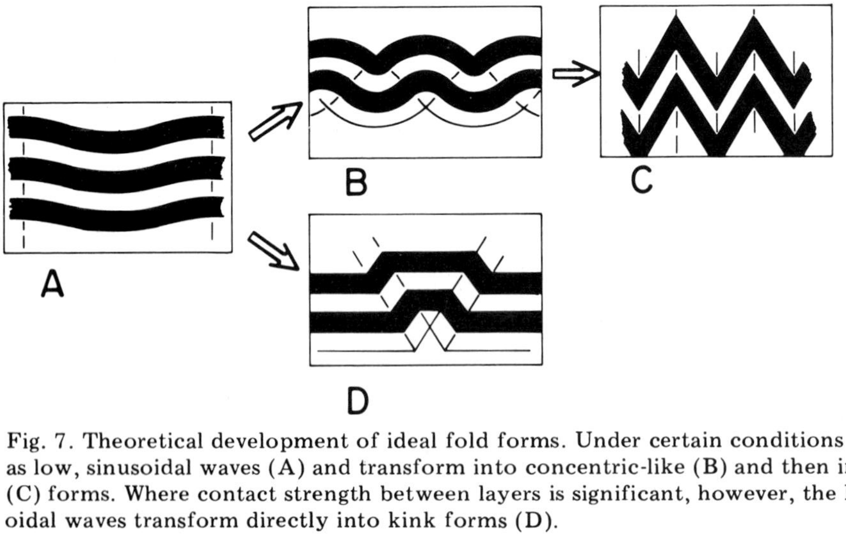

Fig. 7. Theoretical development of ideal fold forms. Under certain conditions folds begin as low, sinusoidal waves (A) and transform into concentric-like (B) and then into chevron (C) forms. Where contact strength between layers is significant, however, the low sinusoidal waves transform directly into kink forms (D).

where $(\partial v_o/\partial x)$ is the slope of initial irregularities in the layers. According to the analyses in Part IV, if:

$$\tau = 0$$

or, more generally, if:

$$K_1 \simeq 0 \tag{2}$$

sinusoidal, concentric-like and chevron folding is favored. If, however:

$$K_1 > 0 \tag{3}$$

kink folding is favored.

The parameters G_a and τ are critical so they will be discussed briefly. Parameter τ is the Terzaghi-Coulomb strength of contacts between layers:

$$|\tau| = C + \sigma_n' \tan \phi \tag{4}$$

where ϕ is angle of internal friction, C is cohesion and σ_n' is *effective* normal stress across contacts. Thus, the strength of contacts depends upon pore-water pressure, which probably is a critical parameter in folding of rocks. The parameter G_a, for elastic materials, is:

$$G_a = (t_1 + t_2)G_1G_2/(t_1G_2 + t_2G_1) \tag{5}$$

where t is thickness, G is shear modulus and 1 refers to stiff and 2 to soft layers. Let us consider two special cases. Suppose that the multilayer is comprised of stiff layers tightly bonded to interbedded, very soft layers. Then:

$$G_a \simeq (t_1 + t_2)G_2/t_2; \qquad (t_1/t_2)G_2 << G_1$$

that is, the average modulus is essentially equal to the modulus of the soft layers. In this case, $K_1 \simeq 0$, so chevron forms would be expected to develop.

If the soft layers are nearly as firm as the stiff layers, or if the thickness of soft layers is very small:

$$G_a \simeq G_1$$

that is, the average shear modulus is high so that, for moderate strengths of contacts, $K_1 > 0$, and kink forms would be expected to develop. These predictions are in accord with experimental observations with multilayers or rubber and gelatin, as shown in Parts I, III and IV, as well as with experimental observations of Weiss (I, 1968) who used decks of paper cards.

We cannot directly apply these criteria to folds in the Franciscan because we do not know how the rocks behaved at the time of folding. However, the criteria are qualitatively in accord with experimental results of Ghosh (I, 1968), who experimented with plasticene multilayers with different but unknown degrees of interlayer lubrication. The behavior of plasticene might be described as plasticoviscous. The correlations between Ghosh's experimental folds and the criteria suggest that the theory, derived for elastic and elastic—plastic materials, might be applicable to viscous materials as well, as are most of the other aspects of folding theory developed thus far. In any case, field evidence indicates that the shales were much softer than the cherts, because the shales thicken markedly in hinges of folds whereas the cherts maintain constant thicknesses. If the shales were indeed much softer than the chert, we would expect chevron folds to have developed. Further, if pore pressures were high in the shales at the time of folding, reducing contact strengths, we should expect chevron folds to have developed. On the other hand, where shale interbeds were very thin, we might expect kink folds to have developed. Honea (1975) found that the forms of folds in interbedded carbonates and shales near Darwin, California, correlate with relative thicknesses of carbonate and shale layers. Rounded forms developed where thickness ratios were about unity whereas angular, kink forms developed where shale layers were very thin or absent.

The notion that kink and chevron folding are distinctly different phenomena, as indicated by the theory and the criteria described above, seems to be consistent with our experiments, reported in Parts III and IV. However, chevron folds can develop in combination with kink folds. They can form locally where kink folds reflect, as at a stiff boundary as shown in Part IV and by Paterson and Weiss (I, 1966). Under unusual conditions they also can form by the intersection of many *regular* kink bands (e.g., Paterson and Weiss, I, 1966). These chevron folds probably can be differentiated by field relations from those related to concentric folding, however. For example, the folds in the Franciscan rocks commonly are unidirectional, with folded bands widely separated from one another, so they cannot be chevron folds which resulted from intersection of many kink bands. Further, the folds are not local, they extend through many meters of interbedded shale and chert.

Now let us assume that a pattern of conjugate kink bands or a pattern of chevron folds preexists in a multilayer, and investigate the changes of orientations of layers and axial planes as a result of large, layer-parallel shear, as suggested by Ellen (I, 1971). We will assume, for this part of the analysis, that the shear is simple shear and that *layers and axial planes deform passively*, much as lines drawn on a homogeneous material. Figure 8A shows an element of the multilayer and the trace of an axial plane, *ac*, for example, before simple shear. Figure 8B shows the element and the trace of the axial plane, *ac*′, after simple shear. The slope angle, θ', of the trace after shear is:

$$\theta' = \tan^{-1} [1/(\tan \psi + \operatorname{ctg} \theta)] \qquad (6)$$

where ψ is the angle of simple shear and θ is the original slope angle of the trace (Fig. 8A). Equation 6 is plotted in Fig. 9 for lines with a wide range of initial orientations, θ, showing the final slope angle, θ', of the lines and the angle of shear strain, ψ.

Let us consider the amounts of simple shear required to reduce the apical

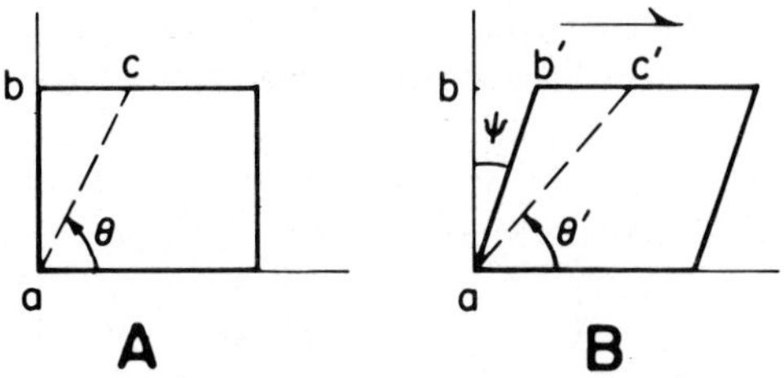

Fig. 8. Simple shear of a passive line. θ is orientation of line before simple shear, and θ' after simple shear. ψ is angle of simple shear. Simple trigonometry indicates that $\tan(\theta') = 1/[\tan(\psi) + \mathrm{ctg}(\theta)]$.

angle of kink and chevron folds to 40 degrees, the angle typically observed in the Franciscan folds. Kink bands typically have apical angles ($\alpha + \alpha_1$ in Fig. 2A) of about 120 degrees (e.g., Weiss, I, 1968). For such a fold, $\theta = 120$ degrees (Fig. 2A), and the amount of shear required to reduce the apical angle from 120 degrees to 40 degrees is about 62 degrees according to Fig. 9. A common apical angle for chevron folds is 60 degrees (e.g., Ramsay, I, 1967; Weiss, I, 1968). For such a fold, $\theta_R = 60$ degrees and $\theta_L = 120$ degrees (Fig. 2B), so that the amount of shear required to reduce the apical angle from 60 to 40 degrees is $\psi \simeq 38$ degrees. Thus, the amount of layer-parallel,

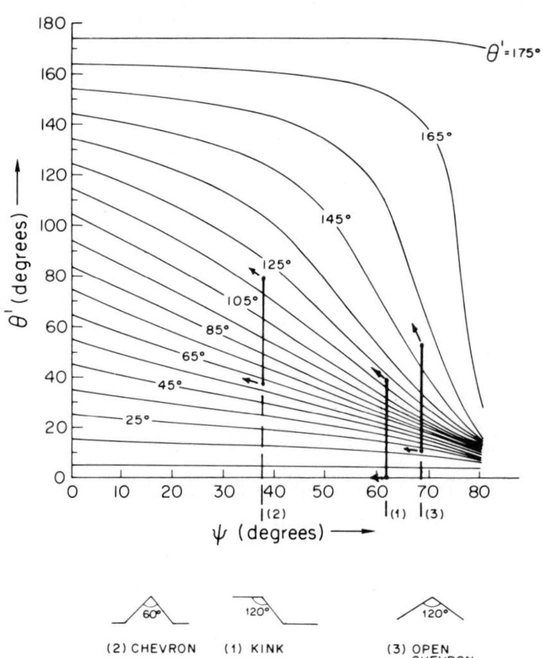

Fig. 9. Nomogram for computing final orientation, θ', of line, given angle of simple shear ψ, and original orientation of that line (θ' corresponding to $\psi = 0$).

simple shear required to reduce the apical angle of the typical symmetrical chevron fold to about 40 degrees is substantially less than that required to deform the typical kink fold into a similarly tight fold. Finally, for an open chevron fold with an apical angle of 120 degrees (θ_L = 150 degrees and θ_R = 30 degrees), ψ must be about 68 degrees (Fig. 9), which is slightly larger than that for the typical kink fold.

Therefore, the slope of the layers in the fold and the type of fold are both critical in determining the amount of layer-parallel shear required to produce an apical angle of a certain amount. We will use this simple model to analyze experiments and field observations described in the following pages.

EXPERIMENTS WITH ELASTIC MULTILAYERS

Multilayers of thick or thin elastic rubber strips were subjected to layer-parallel shortening and lateral compression and either subsequent or simultaneous layer-parallel shear. The thick strips were of photoelastic rubber about 2 mm thick, the properties of which were described in Part III. The thin rubber strips were of pure gum rubber, about 0.3 mm thick. In some of the experiments the rubber strips were lubricated with thin films of silicone grease or talcum powder. In other experiments the rubber strips were in direct, frictional contact.

The multilayers were deformed in an apparatus designed by Ze'ev Reches, graduate student at Stanford University. It will be described in detail in Part VI. Here it is sufficient to note that the apparatus is similar to that described in Part IV, fig. 1, except that the apparatus used for the experiments discussed here had four collapsible bars, which allow control of confinement normal to layers, axial shortening and shear. In most experiments the multilayer was shortened parallel to layers until a pattern of kink bands or chevron folds developed. Then the multilayer was subjected to conditions approximating simple shear.

The experimental results can be illustrated by seven examples. Three of these show the result of layer-parallel shear superimposed upon a pattern of kink folds (Fig. 10), whereas another three show the same effect superimposed upon chevron-like folds (Fig. 12). Unidirectional kink bands produced by simultaneously shortening and shearing layers parallel to layering are shown in Fig. 11.

Relatively high interlayer contact strength

A series of experiments was performed with multilayers characterized by relatively high contact strength. In accord with predictions, they all resulted in kink folds; chevron folds did not form. The relatively high contact strength was a result of friction between the rubber strips (Figs. 10B and 10C) and high normal stress, σ_n, due to a fixed distance between the layer-parallel collapsible bars (Figs. 10A through 10C). The thin rubber strips were powdered with talc to reduce friction, but the thick rubber strips were placed in direct contact.

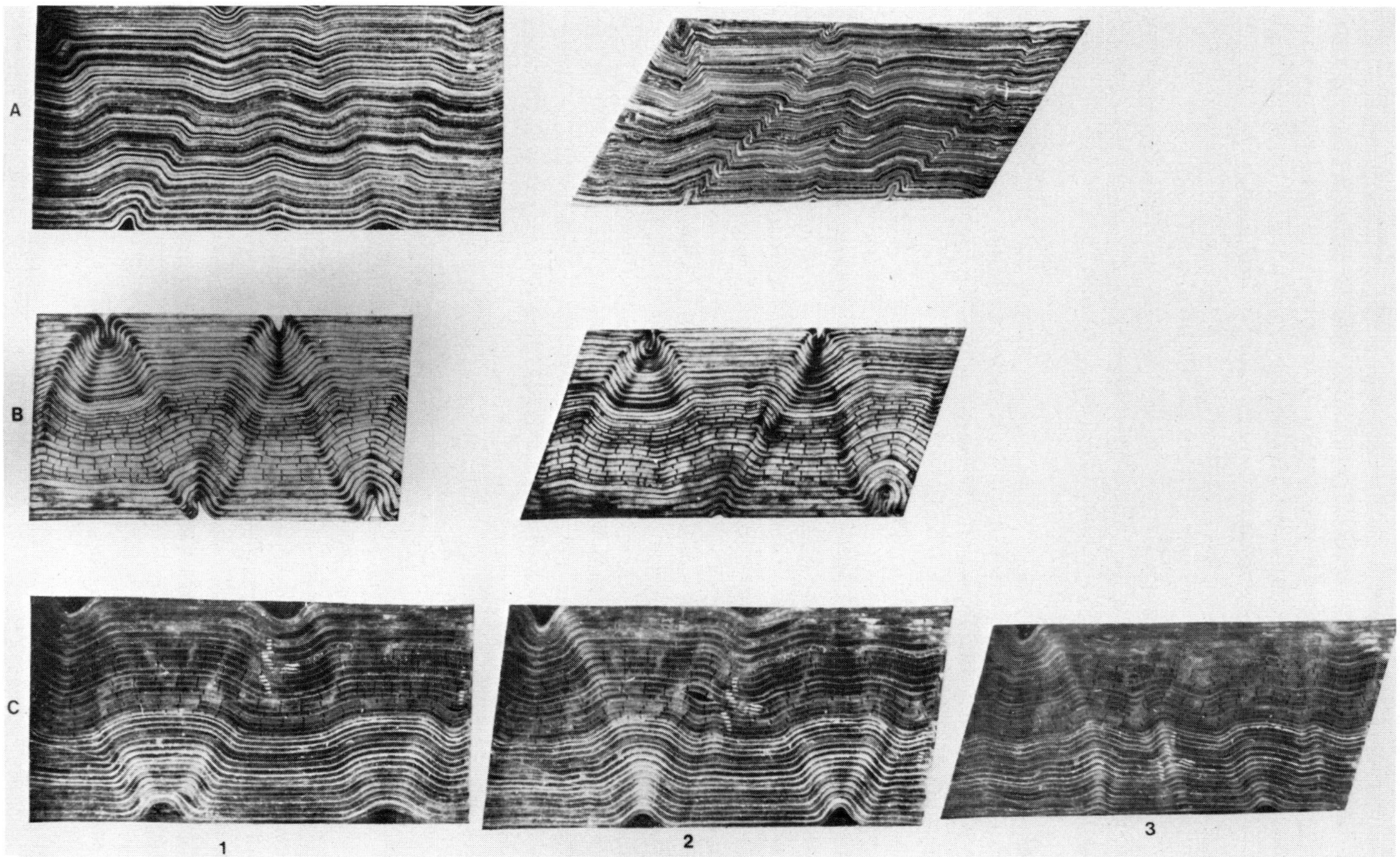

Fig. 10. Kink folds in multilayers of rubber with relatively high contact strengths. Multilayers first subjected to layer-parallel shortening and then to layer-parallel shear.

The multilayers were first shortened parallel to layers. Confinement increased during the shortening. Axial shortening was stopped when the kink folds developed, and then shear was begun. The conjugate axial planes and fold limbs systematically rotated relative to the layer-parallel bars. The right-hand axial plane, AP_R, made an increasingly smaller angle with the bars, the limb of the kink band shortened, and the interlimb angle became smaller, approaching 90 degrees. The limb of kink bands on the left side, AP_L, lengthened, the interlimb angle increased, and the slope of the layers within the band approached that of the nonfolded layers, nearly parallel to the bars. The sharpness of the hinges and the straightness of the fold limbs were unchanged throughout the deformation.

Figure 10B illustrates the same type of experiment but with thick rubber strips. Although the hinges were somewhat more rounded, the same type of conjugate kink folds developed and were deformed by layer-parallel shear into asymmetrical folds (Fig. 10B2).

Figure 10C shows the same type of experiment as in Fig. 10B, with thick rubber strips, except that confinement was lower. The kink bands (Fig. 10C1) had lower amplitudes and more gentle limb slopes when layer-parallel shear was started. As shear began, the kink folds became asymmetric (Fig. 10C2) as did those of the two previous experiments. Continued shear, however, led to interlayer slip, rearrangement of the folds, and disappearance of the right (AP_R) kink band.

This pattern of unidirectional kink folds is similar to that obtained with contemporaneous axial shortening and shear, as shown in Fig. 11. The effect of the simultaneous shortening and shear is similar to that of loading oblique to layering. The multilayers developed unidirectional kink bands, Fig. 11B. The same patterns have been reported by others (e.g., Dewey, 1965; Weiss, I, 1968; Johnson, I, 1970; Gay and Weiss, 1974). These patterns will be one of the subjects of Part VI, so they will not be discussed in detail here. We would merely indicate that the sense of unidirectional kinking shown in Fig. 11 is *opposite* to that shown in Figs. 10A and 10B, even though the sense of shear is the *same* in the three experiments.

Thus, combinations of layer-parallel shear and shortening can lead to quite different kinds of asymmetric kink patterns, apparently depending upon initial fold amplitudes, slope angles, strength of contacts between layers, and sequence of deformation. Nevertheless, a sequence of layer-parallel shortening, to produce a symmetric pattern, followed by layer-parallel shear, to produce steepening of layers in one set of kink bands is the most likely sequence of deformations to produce overturned kink bands, as suggested by Ellen (I, 1971).

Low interlayer contact strength

Multilayers with low contact strength developed chevron folds in response to layer-parallel shortening (Fig. 12), as predicted in Part IV. In all three

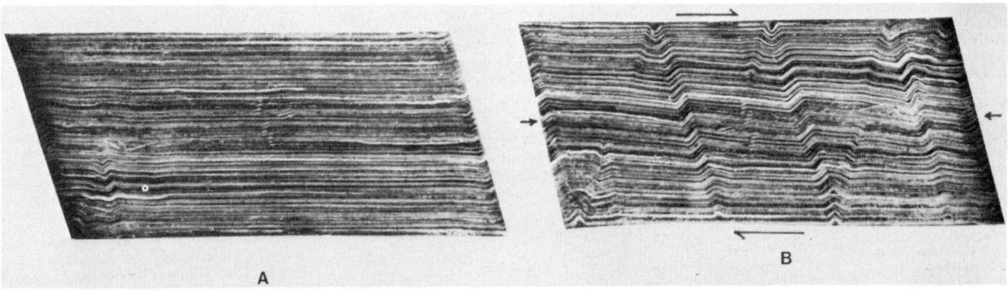

Fig. 11. Unidirectional kink bands produced by simultaneous layer-parallel shortening and shear of multilayer of rubber with relatively high contact strengths. Sense of shear indicated with arrows. Sense of asymmetry *opposite* to that for folds shown in Figs. 10A and 10B.

examples shown in Fig. 12, the shortening was accompanied by reduction of confinement. In the first two examples, Figs. 12A and 12B, thin rubber strips were lubricated with talcum powder. In the last example, Fig. 12C, thick rubber strips were used with silicone grease as a thin interbed film. The low confinement and the low friction between layers presumably made the contact strength nearly zero.

The same sequence of gross deformations was followed for the chevron folds shown in Fig. 12 as for the kink folds shown in Fig. 10. The open chevron folds shown in Fig. 12A1 were sheared into asymmetric folds with alternate long and short limbs. The short limbs developed steep slopes, and the interlimb angle decreased from approximately 135 to 120 degrees, as measured in the central part of the folded multilayers shown in Fig. 12A2. In another experiment, a thin lining of foam rubber between the rigid bars and the multilayers allowed growth of higher-amplitude folds, with substantially higher slope angles of the limbs (Fig. 12B1). In this case layer-parallel shear led to asymmetric, but much more highly appressed and even overturned, folds (Fig. 12B2). The tight folding was accompanied by the formation of open spaces in hinges and by hinge collapse.

The third example shows thick, greased rubber strips buckled into open chevron folds with somewhat rounded hinge zones. The layer-parallel shear first induced slight asymmetry (Fig. 12C1), as in previous examples. Further shear, however, resulted in a complete rearrangement, into a more nearly symmetric but irregular fold pattern (Fig. 12C2).

Thus, a sequence of layer-parallel shortening and shear can produce asymmetric chevron folds in multilayers of lubricated rubber strips if fold amplitudes are sufficiently large. The resulting folds resemble kink bands even though their origin is quite different from that of kink bands in multilayers or rubber strips in frictional contact. Apparently, the amplitudes of chevron folds must attain a certain magnitude in order for the layers to "lock" rather than slip. Thus, we can imagine that asymmetric chevron folds can form as a result of simultaneous layer-parallel shear and shortening, as well as first shortening and then shear as in the experiments.

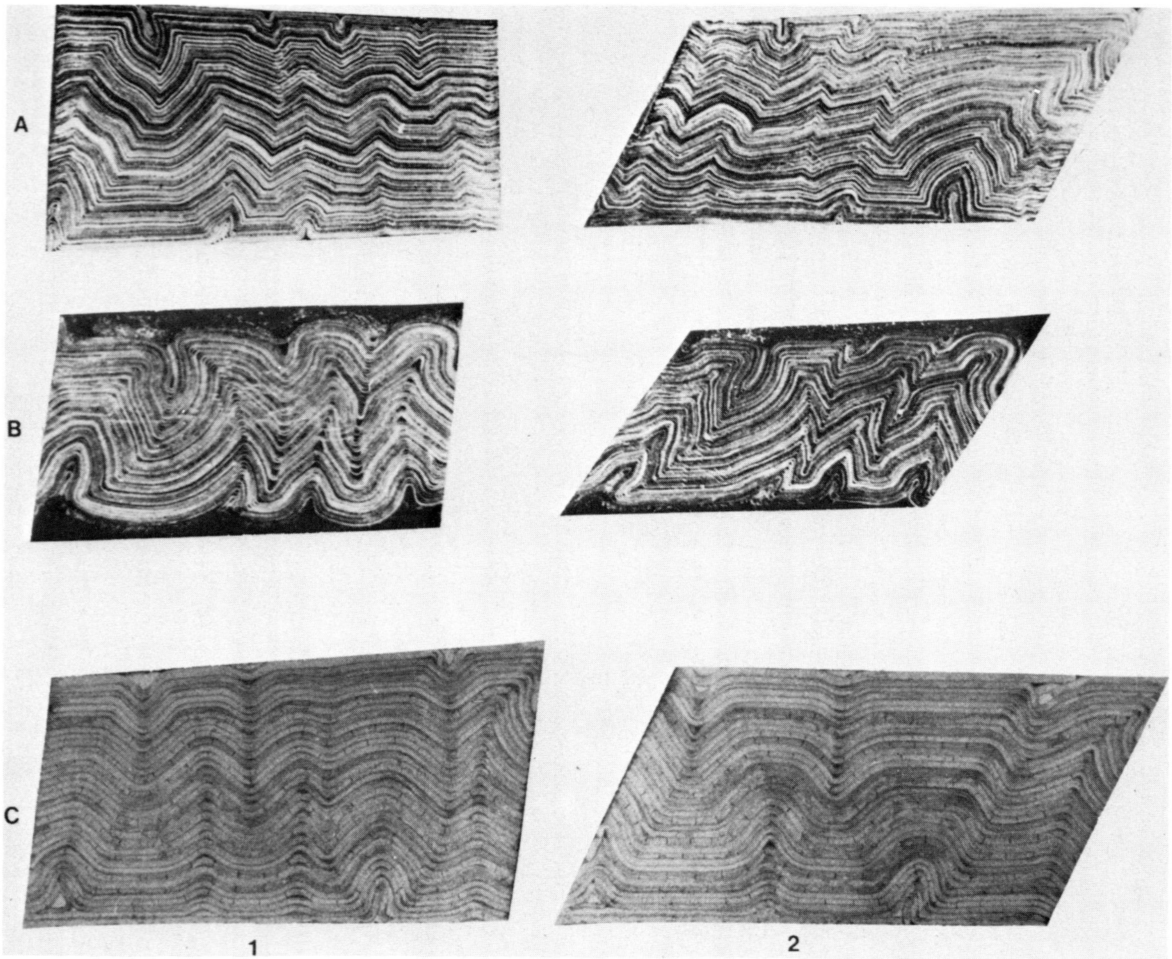

Fig. 12. Asymmetric chevron folds in multilayers of rubber with very low contact strengths. Multilayers first subjected to layer-parallel shortening and then to layer-parallel shear.

Analysis of superimposed shear

The total deformation in the multilayers shown in Figs. 10 and 12 is complex and cannot be classified simply. Nevertheless, in most experiments the deformation subsequent to initial, layer-parallel shortening closely approximated simple shear relative to nonfolded layering. Therefore, we will use the analysis of simple shear presented in earlier pages, eq. 6 and Fig. 9, to investigate the extent to which the asymmetric patterns developed in the experiments can be explained in terms of simple shear of passive lines. Hence, as in Fig. 9, the various slope angles, θ', of the axial planes and limbs of folds have been plotted relative to the angle of simple shear, ψ, for the different experiments in Figs. 13 and 14. Only folds in the center of the multilayers were used for the measurements. For most experiments, a series of four to seven photographs was used for the measurements. Fig. 13 shows the results obtained from the experimental kink folds (Fig. 10) and Fig. 14 shows results for the chevron folds (Fig. 12). The sense and amount of rotation of layers and axial planes in the experiments shown in Figs. 10A and 10B closely follow the predicted trends (Figs. 13A and 13B). The deviations for the experiment with thick rubber strips (Fig. 13B, at $\psi > 20$ deg.) indicates that the layers were not behaving passively to further shear. This is shown even more clearly in the experiment where unidirectional kink bands developed (Figs. 10C and 13C). The initial, passive rotation of the various limbs and axial planes was replaced by readjustment of layers at a shear angle of about 7 degrees. The axial planes both remained in fairly constant positions, independent of further shear, but the right limb, L_R, rapidly approached 180 degrees and vanished.

Figures 14A and 14B show that both the low- and the high-amplitude chevron folds in thin layers (Figs. 12A and 12B) closely follow the predicted paths of the superimposed simple shear. Both multilayers were lubricated, and were subjected to low confinement, so the results emphasize the importance of the slope angles rather than interlayer strength in producing asymmetric forms. In fact, all our experiments show that, at sufficiently steep initial limb dips, asymmetric folds will develop even if contacts have relatively low strengths. This is further emphasized by the plots for the experiment shown in Fig. 12C. Here initially low-amplitude chevron folds with gentle dips readjusted by interbed slip. This occurred mainly in two steps, at $\psi \simeq$ 12 degrees and then at 26 degrees (fig. 14C). Otherwise, the plots tend to follow the predicted course for simple shear of passive markers.

Figures 13 and 14 indicate clearly that experimental asymmetric folds with interlimb angles of about 40 degrees, as is typical of the Franciscan folds, would require very large angles of layer-parallel shear ($\psi \simeq 60$–70 deg.) of kink or open chevron folds in the multilayers of rubber. For initially tight chevron folds, as in Fig. 12B, however, the required angle of layer-parallel shear is about 35 degrees for these materials.

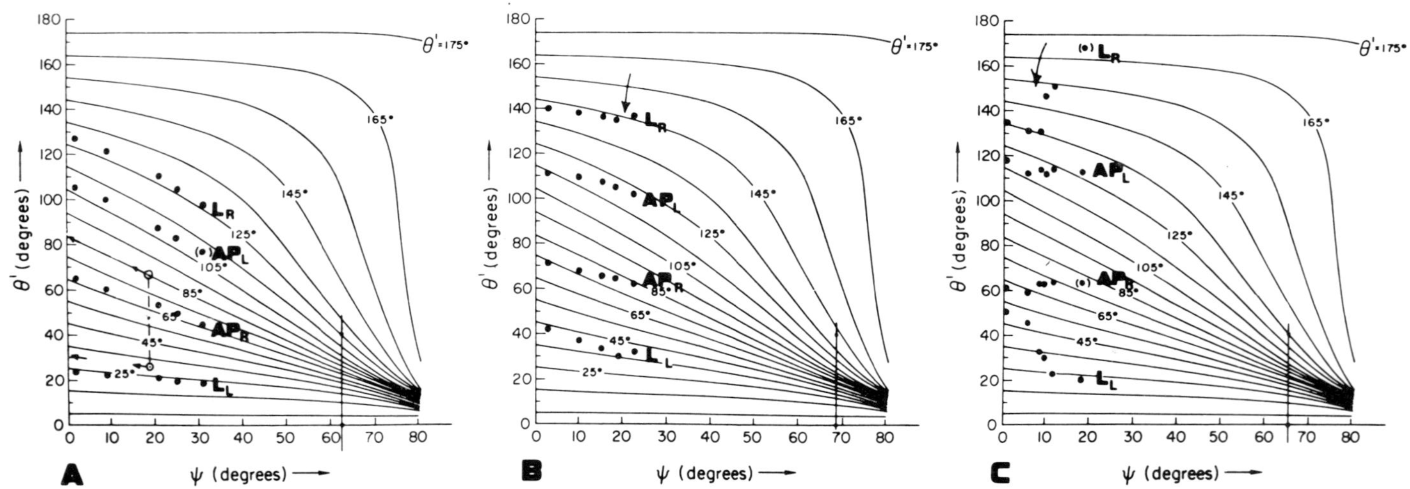

Fig. 13. Deformation paths of axial planes and folded layers in kink bands subjected to layer-parallel shear. Solid lines represent relations between angle of simple shear, ψ, and final angle, θ', of lines deformed passively in body subjected to simple shear. L is a limb facing left (L_L) or right (L_R), and AP is a trace of an axial plane of a kink band facing left or right. Points in Figs. 13A, B, C correspond with measurements of kink bands shown in Figs. 10A, B, C, respectively. If traces of axial planes and layers in kink bands were deforming passively in simple shear, points should parallel one of the solid lines.

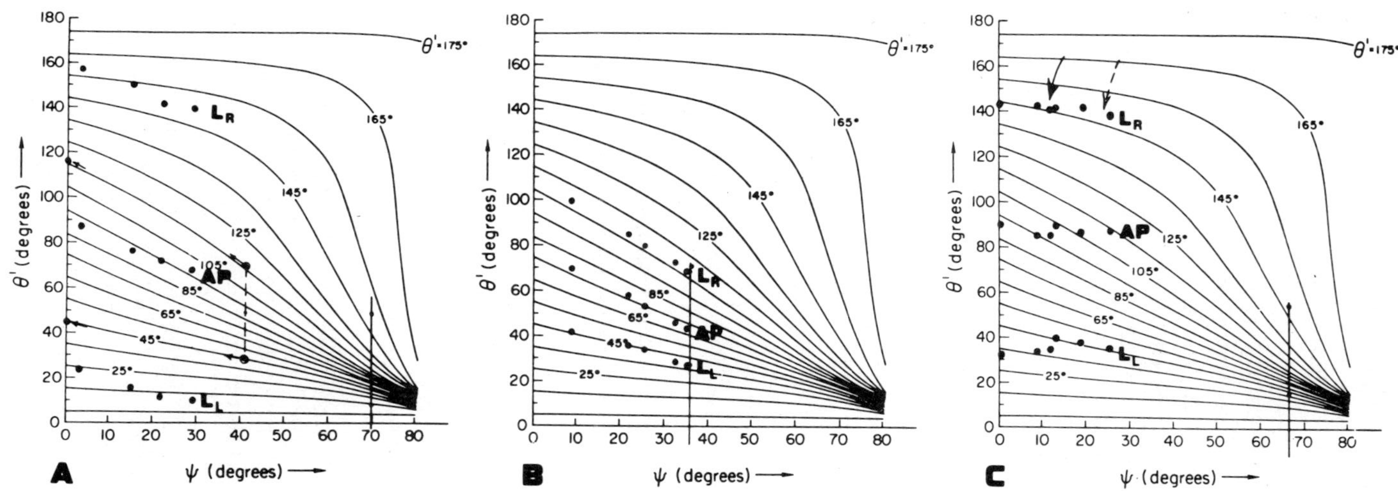

Fig. 14. Deformation paths of axial planes and limbs of chevron folds subjected to layer-parallel shear. Notation similar to that in Fig. 13. Points in Figs. 14A, B, C, correspond with measurements of hinges and limbs of folds shown in Figs. 12A, B, C, respectively.

Several lines of evidence suggest that many of the tight folds in the interbedded chert and shale of the Franciscan are sheared chevron folds, not, as suggested by Ellen (I, 1971), symmetric kink bands that became overturned by subsequent shearing. The lines of evidence can be derived from combinations of field, laboratory and theoretical observations.

The experimental evidence is suggestive, but not conclusive. It shows that asymmetric folds can be developed from both chevron folds and conjugate kink folds by the superposition of shear upon symmetric fold patterns. Chevron folds with initially steep limbs were deformed into tight, overturned folds with apical angles approaching those of about 40 degrees in the Franciscan. Further, the theory of finite deformation of passive lines in a body subjected to simple shear closely predicts the development of asymmetry in many of the experimental folds. This correlation between theory and experiment allows one to conclude that about twice the angle of simple shear would be required to produce apical angles of 40 degrees in sheared conjugate kink folds than in sheared chevron folds in the experiments. The same conclusion was drawn theoretically from an analysis of the typical kink fold, with an apical angle of 120 degrees, and the typical chevron fold, with an apical angle of 60 degrees. Thus, the combination of the theory of simple shear and the experiments indicates that tight folds such as those in the Franciscan could more easily form if the original folds were chevron rather than kink folds.

The experiments strongly suggest that the folds in the Franciscan did *not* form in a material with high contact strength subjected simultaneously to layer-parallel shortening and shear which is equivalent to loading where principal stresses are oblique to layering. The experimental folds formed under such conditions are unidirectional, but with a sense of asymmetry *opposite* to that predicted by the theory of passive deformation of lines in a body subjected to layer-parallel shear.

Another line of evidence suggesting that the original folds in the Franciscan were largely of the chevron variety is based on a combination of field observation and theoretical analysis. One of the criteria of kink folding indicates that very low contact strength between layers favors folds of the concentric-like and chevron types. Relatively low strength of the shales interbedded with the cherts is suggested by the marked thickening of the shales in fold hinges, whereas the chert beds maintain essentially constant thicknesses in hinges and limbs. Further, the chert layers themselves probably were soft at the time of folding because thin-section studies of samples of cherts from tight hinge zones by Ellen (I, 1971) indicate that delicate tests of radiolaria are unbroken by the intense deformation there. Thus, the cherts were probably soft and the shales very soft, suggesting that pore-water pressures were high in the shales at the time of folding. High pore-water pressures would imply low contact strength, a condition favoring concentric-like and subsequently chevron folding rather than kink folding. Finally, elastic or elastic—plastic multilayers with high contrasts in moduli tend to fold into concentric-like and chevron folds rather than into kink folds. Thus, this combination of field and theoretical evidence also suggests that many of the tight folds in the Franciscan began as chevron rather than as kink folds.

Perhaps the strongest evidence, however, derives from a combination of detailed field mapping by Ellen (I, 1971) and the theory of finite strain of passive lines in a body subjected to simple shear. Ellen (I, 1971) constructed a structural profile of folds in a quarry in Marin County by a combination of plane-table mapping and mapping on oblique photographs. Figure 15 is a slight modification of the structural profile (op. cit., fig. 3.20). It shows the basic outcrop pattern as well as projections of the upper and lower contacts of the folded chert and shale, based on Ellen's geologic map (op. cit., fig. 3.15). The layered chert and shale occur sandwiched between nearly massive chert, above, and thick greenstone, below. The whole sequence dips generally about 30 degrees to the soutwest. Superimposed upon this general dip are open folds with wavelengths of about 100—150 m. Thus, the mesoscopic folding of the chert and shale multilayer took place between relatively stiff, layer-parallel boundaries of greenstone and massive chert (Ellen, I, 1971).

Apical angles and slope angles of most of the well-defined, sharp folds were measured. The angle between a normal drawn from the projected boundary of the massive chert and the trace of the axial plane of each prominent fold is generally 35 to 40 degrees (e.g., see Trace 3, Fig. 15). The average is 41 degrees and the range is 30 to 60 degrees. Excluded from the average are two kink-like folds (nos. 1 and 2, Fig. 15) near the contact with the massive chert and two folds (nos. 9 and 10) in the complex, central part of the section.

The apical angles for the folds average 43 degrees, but the range is rather large, from 20 to 65 degrees. Most apical angles are scattered between the extremes and the average, rather than concentrated close to the average.

Now, if we assume that the original folds were chevron folds with axial planes normal to the contact with the overlying massive chert, the average amount of layer-parallel shear, ψ, was about 41 degrees, and an average apical angle of 43 degrees implies an original average apical angle of 72 degrees (Fig. 14A). This is somewhat larger than the value of 60 degrees for the typical chevron form.

In contrast, if we assume that the original folds were typical kink folds with axial planes with an initial slope of 60 degrees (30 deg. from the normal to the massive chert), the average shear would have been about 18 degrees. Further, with a final apical angle of 43 degrees, the initial apical angle would have been very small, about 57 degrees (Fig. 13A). Even with the maximum observed apical angle of 65 degrees, the initial apical angle would have to have been much smaller (about half) than the typical value of 120 degrees for kink folds. If the original folds were conjugate kink folds with apical

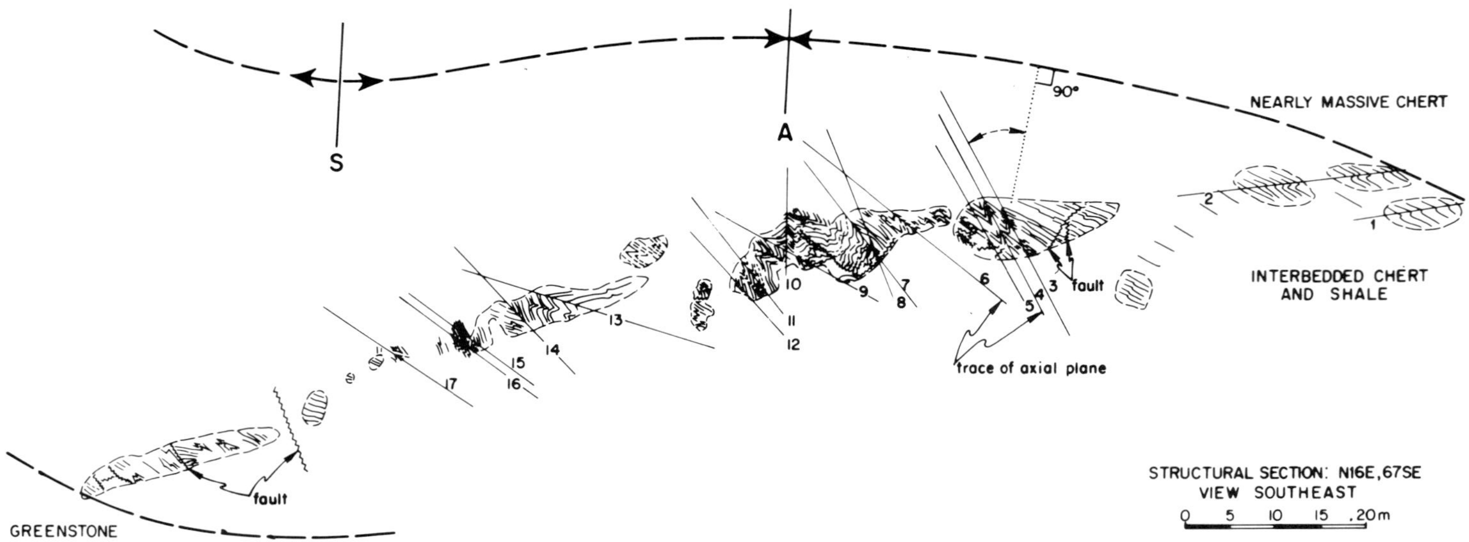

Fig. 15. Structural profile of folds in interbedded shale and radiolarian chert exposed in a quarry at the north side of the Golden Gate (Fig. 3). Profile slightly modified from Ellen (I, 1971, figs. 3.15 and 3.20). Folds 1, 2, 3 are interpreted to be unidirectional kink bands, similar to those shown in Fig. 11. Most of the others are interpreted to be asymmetric chevron folds. Interpreted in this way, the entire fold pattern is consistent with a single direction of layer-parallel shear in which layers above were sheared to the left relative to layers below.

angles of about 60 degrees, both sets should be clearly visible. They are not.

Thus, although kink folds certainly developed in some parts of the interbedded chert and shale, especially where the shale layers are thin (e.g., Fig. 6 and folds 1 and 2, Fig. 15), the various lines of evidence suggest that the typical, tightly appressed folds developed by layer-parallel shear of chevron folds.

We have implicitly accepted Ellen's assumption that the deformation involved first layer-parallel shortening and then subsequent shear. None of our observations is inconsistent with a deformation where layer-parallel shortening and shear were simultaneous. If contact strengths between layers were small, the layer-parallel shear would have no effect on folding until amplitudes became large, as we will show in Part VI. Thus the simplest explanation of the folds in the Franciscan would be that they are chevron folds that developed under conditions of simultaneous layer-parallel shortening and shear. This is what we suggest.

ACKNOWLEDGEMENT

The research reported in this part was financed by the National Science Foundation, Grant no. GA-36917, and by the University of Oslo, Norway. We are grateful to Ze'ev Reches for designing the experimental apparatus and to Peter Gordon for building it. Many of the ideas presented here were formulated during discussions with Ze'ev Reches and Elmont Honea, graduate students at Stanford University. The referees, L.E. Weiss and Peter Cobbold, suggested several improvements in the presentation. We accept full responsibility for errors in the text and do not mean to imply that any of the people mentioned above agrees with the conclusions we have reached.

REFERENCES NOT CITED IN CHAPTER 3

Bailey, E.H., Irwin, W.P. and Jones, D.L., 1964. Franciscan and related rocks and their significance in the geology of western California. Calif. Div. Mines Geol. Bull., 183, 177 pp.

Clifford, P.M., 1968. Kink band development in the Lake St. Joseph area, Northern Ontario. In: A.J. Bear and D.K. Norris (editors), Proceedings Conference Research in Tectonics. Geol. Surv. Can. Pap., 68-52: 229—242.

Dewey, J.F., 1965. Nature and origin of kink-bands. Tectonophysics, 1: 459—494.

Gay, N.C. and Weiss, L.E., 1974. The relationship between principal stress directions and the geometry of kinks in foliated rocks. Tectonophysics, 21: 287—300.

Honea, E., 1975. The Effect of Contact Strength on the Shape of Buckle Folds, Including Field Examples in South Darwin Canyon, California. Ph.D. Dissertation, Department of Geology, Stanford University, Stanford, Calif. 94 p.

Honea, E. and Johnson, A.M., 1975. A theory of concentric, kink and sinusoidal folding and of monoclinal flexuring of compressible, elastic multilayers. IV. Development of sinusoidal and kink folds in multilayers confined by rigid boundaries. Tectonophysics, 30: 197—239.

Johnson, A.M. and Ellen, S.D., 1974. A theory of concentric, kink and sinusoidal folding

and of monoclinal flexuring of compressible, elastic multilayers. I. Introduction. Tectonophysics, 21: 301—339.
Johnson, A.M. and Honea, E., 1975. A theory of concentric, kink and sinusoidal folding and of monoclinal flexuring of compressible, elastic multilayers. III. Transition from sinusoidal to concentric-like to chevron folds. Tectonophysics, 27: 1—38.
Ramsay, J.G., 1962. The geometry and mechanism of formation of 'similar' type folds. J. Geol., 70: 309—327.
Ramsay, J.G., 1974. Development of chevron folds. Geol. Soc. Am. Bull., 85: 1741—1754.
Reches, Z. and Johnson, A.M., in preparation. A theory of concentric, kink, and sinusoidal folding and of monoclinal flexuring of compressible elastic multilayers. VI. Asymmetric folding and unidirectional kinking.
Schlocker, J., 1974. Geology of the San Francisco North Quadrangle, California. U.S. Geol. Surv. Prof. Pap., 782, 109 p.
Talieferro, N.L., 1943. Franciscan-Knoxville problem. Bull. Am. Assoc. Pet. Geol., 27: 109—219.

Chapter 8

ASYMMETRIC FOLDING AND MONOCLINAL KINKING *

SUMMARY

One of the rules of thumb of structural geology is that drag folds, or minor asymmetric folds, reflect the sense of layer-parallel shear during folding of an area. According to this rule, right-lateral, layer-parallel shear is accompanied by clockwise rotation of marker surfaces and left-lateral by counterclockwise rotation. By using this rule of thumb, one is supposed to be able to examine small asymmetric folds in an outcrop and to infer the direction of axes of major folds relative to the position of the outcrop. Such inferences, however, can be misleading. Theoretical and experimental analyses of elastic multilayers show that symmetric sinusoidal folds first develop in the multilayers, if the rheological and dimensional properties favor the development of sinusoidal folds rather than kink folds, and that the folded layers will then behave much as passive markers during layer-parallel shear and thus will follow the rule of thumb of drag folding. The analyses indicate, however, that multilayers whose properties favor the development of kink folds can produce monoclinal kink folds with a sense of asymmetry *opposite* to that predicted by the rule of thumb. Therefore, the asymmetry of folds can be an ambiguous indicator of the sense of shear.

The reason for the ambiguity is that asymmetry is a result of two processes that can produce diametrically opposed results. The deformation of foliation surfaces and axial planes in a passive manner is the pure or end-member form of one process. The result of the passive deformation of fold forms is the drag fold in which the steepness of limbs and the tilt of axial planes relative to nonfolded layering are in accord with the rule of thumb.

The end-member form of a second process, however, produces the opposite geometric relationships. This process involves yielding and buckling instabilities of layers with contact strength and can result in monoclinal kink bands. Right-lateral, layer-parallel shear stress produces left-lateral monoclinal kink bands and left-lateral shear stress produces right-lateral monoclinal kink bands. Actual folds do not behave as either of these ideal end members, and it is for this reason that the interpretation of the sense of layer-parallel shear stress relative to the asymmetry of folds can be ambiguous.

Kink folding of a multilayer with contact strength theoretically is a result of both buckling and yielding instabilities. The theory indicates that inclination of the direction of maximum compression to layering favors either left-lateral or right-lateral kinking, and that one can predict conditions under which monoclinal kink bands will develop in

* Originally published as: Reches, Z. and Johnson, A.M., 1976. Part VI, Asymmetric folding and monoclinal kinking. Tectonophysics, 35: 295—334.

elastic or elastic—plastic layers. Further, the first criterion of kink and sinusoidal folding developed in Part IV remains valid if we replace the contact shear strength with the difference between the shear strength and the initial layer-parallel shear stress.

Kink folds theoretically can initiate only in layers inclined at angles less than $\pm(45^\circ - \phi/2)$ to the direction of maximum compression. Here ϕ is the angle of internal friction of contacts. For higher angles of layering, slippage is stable so that the result is layer-parallel slippage rather than kink folding.

The theory also provides estimates of locking angles of kink bands relative to the direction of maximum compression. The *maximum* locking angle between layering in a non-dilating kink band and the direction of maximum compression is $\pm(90^\circ - \phi/2)$. The theory indicates that the inclination of the boundaries of kink bands is determined by many factors, including the contact strength between layers, the ratio of principal stresses, the thickening or thinning of layers, that is, the dilitation, within the kink band, and the orientation of the principal stresses relative to layering. If there is no dilitation within the kink band, the *minimum* inclination of the boundaries of the band is $\pm(45^\circ + \phi/4)$ to the direction of maximum compression, or $\pm(45^\circ + \phi/4 - \alpha/2)$ to the direction of nonfolded layers. Here α is the angle between the direction of maximum compression and the nonfolded layers. It is positive if clockwise.

Analysis of processes in terminal regions of propagating kink bands in multilayers with frictional contact strength indicates that an essential process is dilitation, which decreases the normal stress, thereby allowing slippage and buckling even though slopes of layers are low there.

INTRODUCTION

Folds in multilayered rock commonly are asymmetric; one limb of a fold is shorter than the other, and the shorter limb normally has a steeper slope, relative to nonfolded layering, than the longer limb. Some asymmetric folds are drag features that occur within larger folds (Fig. 1A), and the sense of asymmetry of drag folds commonly is consistent with interpretations of shear between layers accompanying the growth of the larger folds (e.g., Billings, 1972, p. 54). Indeed, one of the rules of thumb of structural geology is that drag folds indicate where axial planes of larger synclines and anticlines lie with respect to the outcrop containing the drag folds (e.g., Ramberg, 1963; Dennis, 1972, p. 227; Billings, op. cit.). Unfortunately, this rule can lead to erroneous interpretations of asymmetric folds. One type of asymmetric fold which violates the rule is the *monoclinal* kink band, in which the kinked layers within an outcrop consistently face in one direction (Fig. 1B).

The sense of shear associated with the asymmetry of monoclinal kink bands has been misinterpreted, but some geologists have shown conclusively from field observations (Dewey, 1965) or from experiments (Weiss, I *, 1968; Donath, I, 1968, 1970; Ellen, I, 1971; Cobbold et al., I, 1971; Gay

* References cited in Part I, Johnson and Ellen, 1974, will be noted with an I, such as I, 1955, and not be repeated in references at the end of this paper.
Note: Part I = Chapter 3.

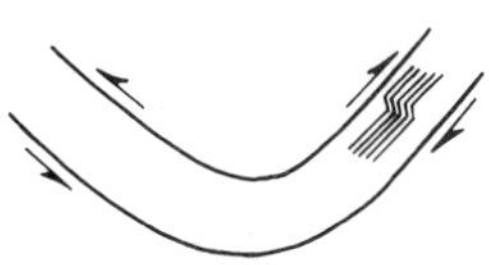

C

D

Fig. 4. Small kink bands on limb of syncline in Monterey Formation, southern end of Huasna Syncline, near Santa Maria, California. A. The syncline. Darkened beds are chert. Stippled bed is sandstone. Other layers primarily siliceous shale. Box shows location of B. B. Close view of kink bands beneath sandstone bed. Scale is about 18 cm long. Some of partings within siliceous shale are shown to indicate shapes of kink bands. Solid lines steeply inclined to bedding are small faults. C. Schematic view of syncline, showing sense of kinking and inferred sense of layer-parallel shear expected on limbs of a syncline. D. Schematic view of small anticline, showing sense of kinking and inferred sense of layer-parallel shear expected on limb of an anticline.

The problem posed by this field example is whether the sense of layer-parallel shear shown in Fig. 3 is consistent with the theory of asymmetric folding of different types of materials or whether one must postulate different senses of layer-parallel shear for the two types of asymmetric folds, as does Ellen (I, 1971), who studied the folds previously.

There are small asymmetric folds or kink bands, with widths of about 1 cm, on a limb of a syncline (Fig. 4A) in interbedded chert, sandstone, and siliceous shale, near the base of the Monterey Formation, at the south end of the Huasna syncline in California. Part VII deals with the larger folds in this area. The syncline shown in Fig. 4A is at location *A* in fig. 7 of Part VII. The asymmetric folds are within siliceous shale with paper-thin partings, some of which are shown in Fig. 4B. Associated with the asymmetric folds are small faults, some of which pass near the hinges of the folds (Aydin, 1973). A thin sandstone bed above the siliceous shale apparently deformed by faulting, even where there is no fault within the adjacent siliceous shale (Fig. 4B). Above and below the siliceous shale are chert beds, 5—10 cm thick.

The layer-parallel drag associated with development of the right side of the syncline apparently was right-lateral as shown in Figs. 1A and 4C. The sense of asymmetry of the small asymmetric folds would suggest left-lateral drag if we followed the rule of thumb. Even if we consider only the small anticline within which the tiny folds are well developed (Fig. 4B), the senses of drag and asymmetry appear to oppose one another (Fig. 4D). However, if the asymmetric folds are interpreted to be monoclinal kink bands, the sense of asymmetry and the inferred sense of layer-parallel shear are consistent according to the experimental research cited in previous paragraphs.

Thus, this last field example perhaps most clearly emphasizes the importance of understanding conditions under which monoclinal kink bands develop and of understanding conditions that favor asymmetric folding in general and monoclinal kinking in particular.

We will begin our analysis of asymmetric folding by reporting a few observations made during experiments with multilayers of rubber strips between which contact strength was varied, and of interbedded stiff and soft elastic strips, all subjected to layer-parallel shear and shortening and to confinement normal to layering. Then we will investigate possible processes of asymmetric folding of ideal multilayers. Finally, we will return briefly to the field examples.

EXPERIMENTS

We have experimented with several types of multilayers in order to obtain some insights into effects of shear on folding of multilayers. The multilayers include thick or thin rubber strips with lubricated contacts (fig. 12, Part V), thick or thin rubber strips with relatively high contact strengths (fig. 10, Part V, and Fig. 7), and interbedded thick rubber strips and soft gelatin (Fig. 9). The properties of the rubber and gelatin were reported in Part III.

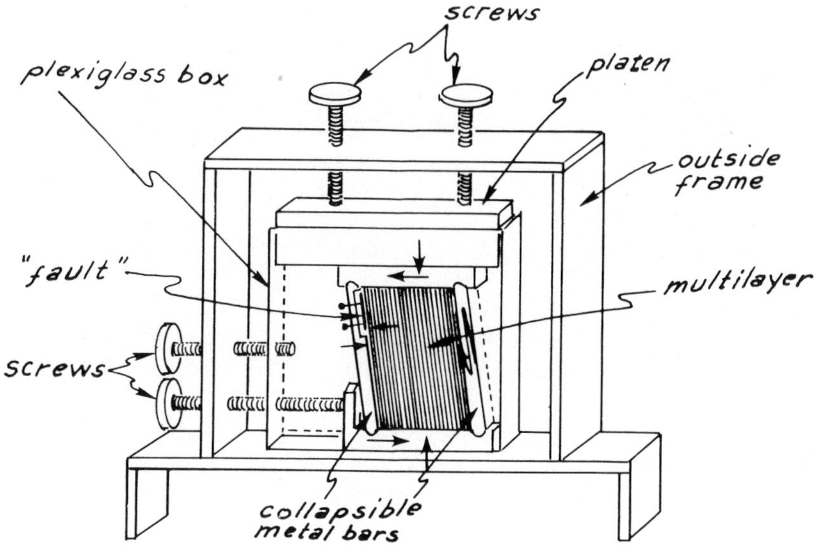

Fig. 5. Apparatus used to deform the multilayers. The multilayers are confined front and back with thick sheets of plexiglass. The other four sides are confined by collapsible metal bars, allowing confinement normal to layers and axial shortening as well as shear deformations to be controlled independently. Small piece in one of collapsible bars allows one edge of frame to be ''faulted''.

The multilayers were subjected to a variety of deformation histories which will be described in connection with each experiment.

All the multilayers were deformed in the apparatus shown in Fig. 5, which is capable of subjecting a multilayer to a combination of simple shear and of shortening normal and parallel to layers. The apparatus controls displacement boundary conditions so that stresses and strains within the multilayers are determined by an interaction between the boundaries and the properties of the multilayers. In particular, shear deformation of the apparatus does not insure uniform shear deformation of the individual layers. Indeed, if the surfaces of layers are perfectly lubricated and planar, the shear stress and strain within layers are zero. If the multilayer consists of interlayered stiff and soft layers, tightly bonded together, the shear strain is much higher in the soft layers than the stiff layers, although the shear stresses are nearly the same in the two materials.

In the typical experiment, the multilayer is first subjected to simple shear and shortening (Fig. 6B). These induce compression normal and parallel to layers and induce shear stress between layers if the layers have contact strength. If the layers are dry rubber strips, most of the shear stress is released by slippage. Then the multilayer is subjected to simple shear in the opposite sense (Fig. 6C), which increases shortening parallel to layers and decreases

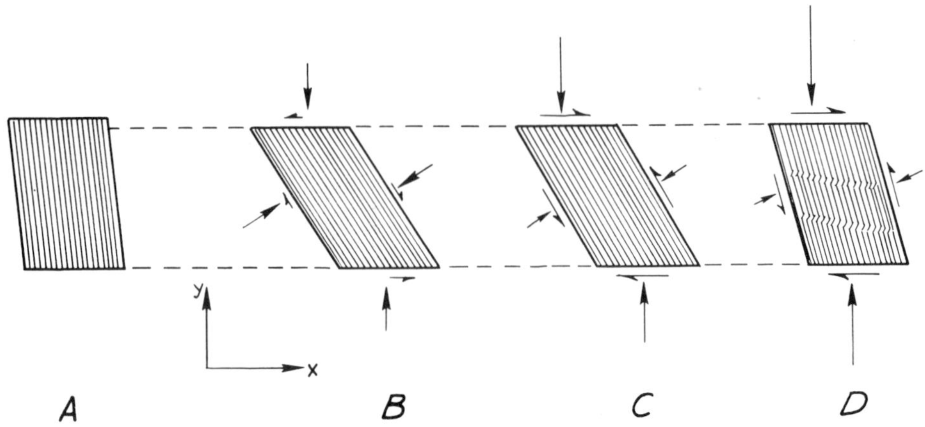

Fig. 6. Sequence of deformations during typical experiment. A. Multilayer before loading. B. Position of multilayer at beginning of a folding experiment. Small amounts of initial shear induced in multilayers with high contact strength. High normal stresses normal to layering and parallel to layering. C. First step of folding experiment. Shear stress reversed, confinement reduced and layer-parallel normal stress increased. Multilayer subjected essentially to simple shear. D. Beginning of folding.

confinement normal to layers, maintaining constant volume, as well as reverses the shear stress between layers if the layers are imperfectly lubricated.

In one series of three experiments, multilayers of unlubricated photoelastic rubber 2.5 mm thick were subjected to the sequence of deformations described in Figs. 6; however, the initial conditions and the fold patterns in the three experiments were different. In the first (Fig. 7A), the multilayer was subjected to relatively high initial compression normal to layers. Initial compression was lower in the second (Fig. 7B) and absent in the third (Fig. 7C). The first experiment, with high confinement, developed monoclinal kink bands, roughly equally spaced along the multilayer (Fig. 7A3). The sense of kinking shown in Fig. 7A3 will be termed *right lateral* because the relative displacements on opposite sides of the kink band are the same as those for right-lateral faults. The second and third experiments

---

Fig. 7. Conjugate and monoclinal kink bands produced in multilayers comprised of rubber strips in frictional contact subjected to loading sequence described in Fig. 6. A. Multilayer of rubber strips about 2.5 mm thick subjected to relatively high confinement normal to layers. Fold pattern is monoclinal. B. Same multilayer subjected to moderate confinement normal to layers. First-formed kinks are monoclinal, but kink bands reflect to produce part of a conjugate pattern of asymmetric folds. C. Same multilayer subjected to low initial confinement. First-formed kink bands are monoclinal, but subsequent bands are conjugate. D. Multilayer of soft rubber strips about 0.3 mm thick subjected to relatively high initial confinement normal to layers. Folds are monoclinal kink bands.

A
B
C
D
1
2
3

developed box folds, or conjugate kink bands, comprised of both right-lateral and left-lateral kink bands. Folding in the second experiment, with moderate confinement normal to layers, began with a monoclinal kink band (Fig. 7B2) and bands parallel to the first-formed band were more highly developed than the others.

Fig. 7D shows monoclinal kink bands in softer rubber strips about 0.3 mm thick under conditions of relatively high initial confinement normal to layers. The pattern is the same as that for the thicker layers (Fig. 7A), but the width and spacing of kink bands are smaller for the thinner strips.

The strong tendency for monoclinal kink bands to develop in multilayers with high contact strength is illustrated in a second series of experiments, two of which are reported in Figs. 8A and B. One side of the apparatus can be "faulted" (Fig. 5) by displacing one piece of it. The multilayers were loaded in the same way as that in the first experiment (Fig. 7A), and when incipient folding was suspected, the side bar was "faulted", by retracting one side of the small piece of the loading frame. In one experiment (Fig. 8A), the shear induced by the "faulting" was of the same sense as the shear applied by the loading frame, and a left-lateral, monoclinal kink band developed over the "fault". Subsequently, other parallel monoclinal kink bands developed as shown in Fig. 8A. In another experiment (Fig. 8B), the sense of shear induced by "faulting" was opposite to that induced by the loading frame. Two distinct flexures developed simultaneously, a short, right-lateral band over the "fault" and a longer, left-lateral band nearby, which extended from boundary to boundary of the apparatus.

Neither monoclinal nor conjugate kink bands developed in experiments where shortening parallel to layers was relatively small. Kink bands could not be produced in these multilayers subjected merely to "faulting" or to shear. The layer-parallel shortening was clearly required for folding.

The effect of initial shear on the folding of multilayers of interbedded, relatively soft gelatin and stiff rubber with a modulus ratio of about 1 : 40 was investigated in a third series of experiments. The gelatin and rubber strips were bonded tightly together in the process of fabricating the multilayer so that there was no slippage between layers until amplitudes of folds became quite large and the gelatin and rubber pulled apart in cores of folds.

Figure 9 shows two experiments in which the thickness ratio of stiff to soft layers was unity. The experiment shown in Fig. 9A was with a multilayer which was constructed essentially in the parallelogram shape shown in Fig. 9A1. The multilayer was subjected to simple shear so that layer-parallel shortening increased and confinement decreased during the experiment. A fold train developed and the folds initially were nearly symmetric (Fig. 9A2), so that the shear had no apparent effect on the shapes of the folds until the amplitudes became relatively large (Figs. 9A3 and A4). The sense of asymmetry is *opposite* to that for the monoclinal kink bands. The short limbs of the asymmetric, rounded folds face right, that is, they are left-lateral, whereas the monoclinal kink bands are right-lateral, under the same sense of applied

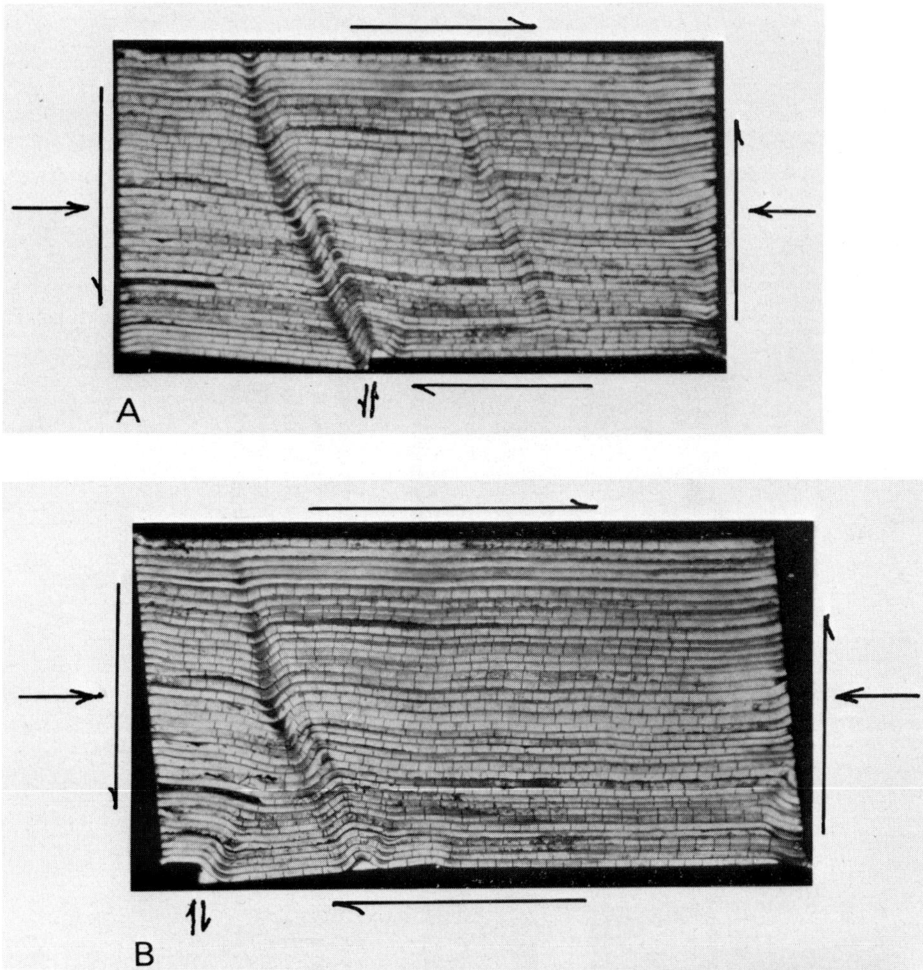

Fig. 8. Multilayer of thicker rubber strips subjected to same stress state as that for multilayer shown in Fig. 7A. Before kink bands appeared, edge of apparatus was "faulted". A. Shear induced by faulting of same sense as that induced by overall simple shear. B. Shear induced by faulting opposite to that induced by overall simple shear.

shear. Thus, the sense of asymmetry for the asymmetric, rounded folds shown in Fig. 9A4 is the sense one would expect where a pattern of initially sinusoidal lines were deformed passively in homogeneous, simple shear. The amount of asymmetry of folded layers, however, is less than those of passive lines, as can be recognized by comparing the tilt of the lines drawn on the model, initially normal to layering, and the tilt of traces of axial planes of the folds themselves. We showed in Part V that, for high-amplitude folds formed in lubricated multilayers under conditions of, first, layer-parallel

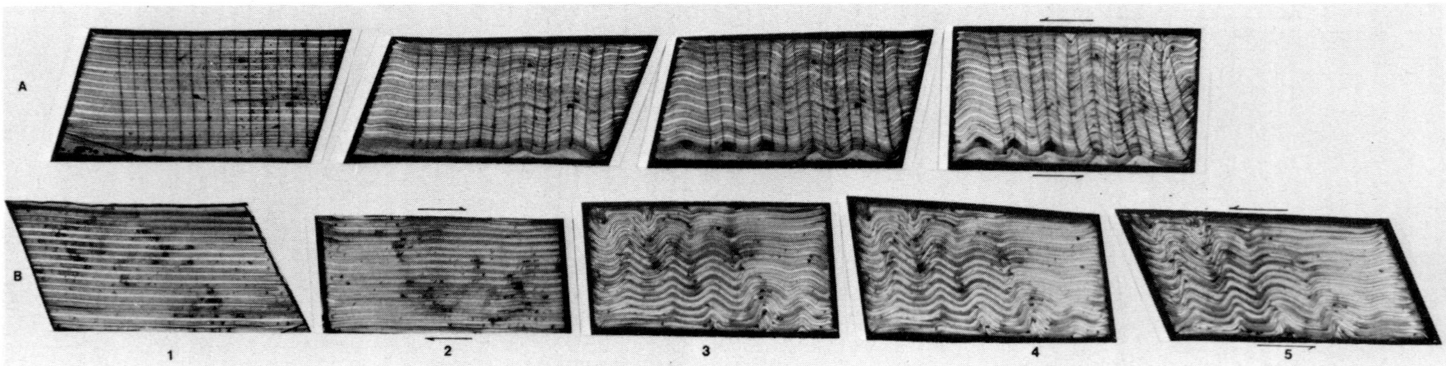

Fig. 9. Effect of shear on folding of multilayer of interbedded rubber and gelatin. A. Multilayer subjected to simultaneous layer-parallel shear and shortening. A1. Initial shape of multilayer. A2. First-formed folds are symmetric. A4. Final folds are asymmetric. Sense of asymmetry consistent with that of rule of thumb for drag folds. B. Multilayer first subjected to layer-parallel shear, then layer-parallel shortening, then simultaneous layer-parallel shear of opposite sense and layer-parallel shortening. B1. Initial shape of multilayer. B2. Multilayer after layer-parallel shear but before shortening. B3. First-formed folds essentially symmetric. B5. Final folds asymmetric.

shortening and, then, layer-parallel shear, the tilt of traces of axial planes and layers correspond closely to the tilt of passive markers parallel to the axial planes (fig. 14B, Part V). We also showed there that, the higher the amplitude of folds at the time of layer-parallel shear, the more nearly the layers and axial planes deformed as passive markers in response to the shear. Here we have shown that the tilt of traces of axial planes underestimates the total amount of layer-parallel shear if the folds are subjected simultaneously to layer-parallel shortening and shear.

The effect of initial shear on folding of multilayers of interbedded soft and stiff layers bonded tightly together is also shown in Fig. 9B. Figure 9B1 shows the parallelogram shape of the multilayer before loading. First the multilayer was deformed into the rectangular shape shown in Fig. 9B2. Thus, relatively large shear stresses were set up within the multilayer before layer-parallel shortening commenced. However, layer-parallel shortening produced essentially symmetric folds (Fig. 9B3), so that the shear is not reflected by the fold pattern. During subsequent shear, the sense of shear tended to *relieve* shear stresses set up during the first loading (Figs. 9B1 and 9B2), yet the folds transformed into asymmetric, rounded forms (Fig. 9B5). This experiment clearly illustrates that it is the gross shear strain of the fold pattern, not the shear stress between layers, that produces the asymmetry.

THEORY

The experimental results and field observations present several theoretical problems. One problem is whether the theory predicts that first-formed folds in multilayers of interlayered stiff and soft materials will tend to be symmetrical, sinusoidal shapes. Another is whether the existence and sense of monoclinal kinking can be understood theoretically. In the course of trying to solve these and other problems, we have developed further insights into possible processes of locking and propagation of kink bands, which apply to both monoclinal and symmetric conjugate kinking. As in Part IV, we deal only with reverse kink folds.

Sinusoidal folding of interbedded stiff and soft layers subjected to layer-parallel shear

Let us consider the folding of interlayered stiff and soft material tightly bonded together and subjected to initial layer-parallel shear stress, much as in the experiments shown in Figs. 9A and 9B. The theory for such conditions was presented in eqs. IV,37 and IV,38 of Part IV, except that the shear strength, τ, in eq. IV,37 is now replaced by the initial shear stress. If we select the origin of coordinates where amplitudes of folds are zero, eq. IV,38a for the deflection form becomes:

$$v = [C_1 \sin(\alpha x) + C_2 \sin(\beta x)] \cos(\pi y/T) \tag{1}$$

where α and β are defined in eq. IV,38b and T is the total thickness of the multilayer. We could eliminate the arbitrary constants in eq. 1 by means of boundary conditions, as in eq. IV,39, but the operation is not necessary for the problem considered here. Although eq. 1 can describe an asymmetric waveform which is the sum of two sine terms as shown in Fig. 10A, the asymmetric waveform probably cannot develop. The parameters α and β in eq. 1 generally cannot be real and distinct because this would require an axial load greater than the critical load for symmetric, *Biot* waves. Thus, as the axial load increases, the first waveform that becomes unstable is the symmetric, *Biot* waveform, which developes at the critical load P_B shown in Fig. 10B. The critical load for the pair of waves shown in Fig. 10A is higher, as indicated in Fig. 10B.

The theory would predict, therefore, that symmetric, sinusoidal folds

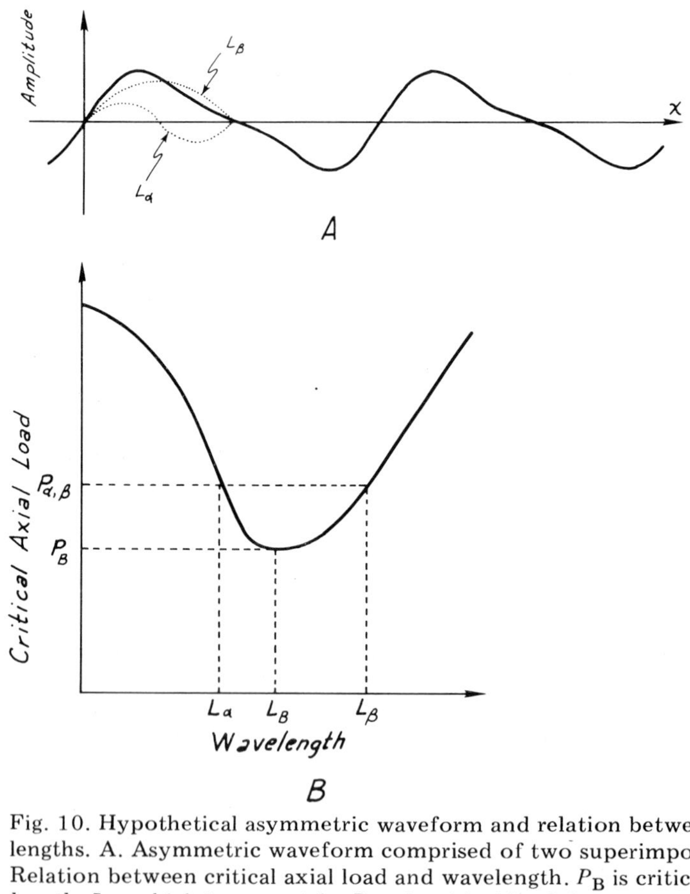

Fig. 10. Hypothetical asymmetric waveform and relation between axial load and wavelengths. A. Asymmetric waveform comprised of two superimposed sinusoidal waves. B. Relation between critical axial load and wavelength. P_B is critical load for *Biot wavelength*, L_B, which is symmetric. $P_{\alpha,\beta}$ is critical load for two wavelengths shown in A.

should develop in multilayers of stiff and soft materials bonded tightly together, subjected to layer-parallel shear and shortening. This prediction is in accord with our experiments, where the first-formed folds were essentially symmetric, whether shear was applied during (Fig. 9A) or before (Fig. 9B) layer-parallel shortening of the multilayer.

Treagus (1973) analyzed the folding of a single viscous layer in a viscous medium subjected to principal stresses oblique to the contacts. She found that the waveform is symmetric sinusoidal and that its dominant wavelength is independent of the inclination of the principal stresses. She suggested that the progressive development of folds in layers oblique to the principal stresses can produce asymmetry of the final fold form, however. Our results are consistent with hers, although we have dealt solely with materials with memory.

Elementary theory of monoclinal kink folding

A theory for the development of an ideal *kink form* in multilayers with contact strength was presented in Part IV, eqs. 7—25. There an initial shear stress was incorporated in the analyses but its effect was unclear. We can understand its effect and, further, explain why kink bands tend to be monoclinal by considering Fig. 11. Figure 11A shows part of a multilayer which

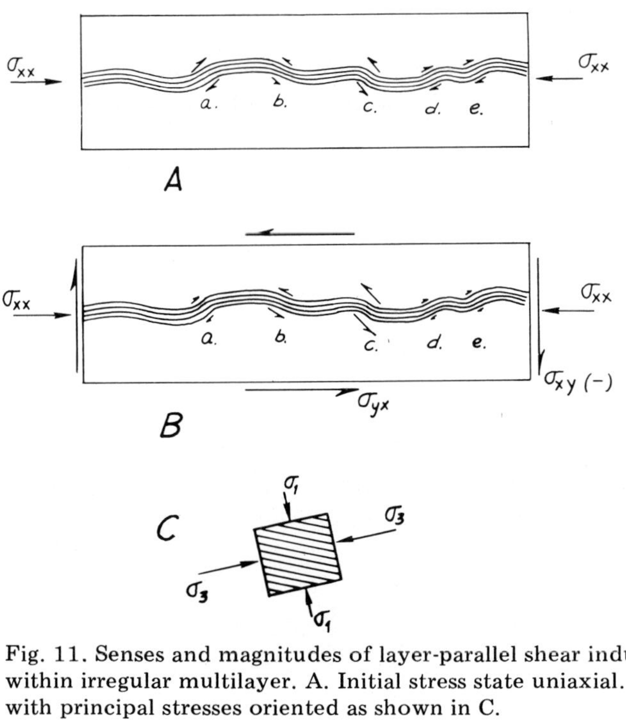

Fig. 11. Senses and magnitudes of layer-parallel shear induced by local slopes of layers within irregular multilayer. A. Initial stress state uniaxial. B. Initial stress state biaxial, with principal stresses oriented as shown in C.

contains random irregularities in slope. The multilayer is subjected solely to layer-parallel compression, σ_{xx}. As a result of the initial irregularities, there is shear between layers. The shear is zero where slopes are zero and maximal where slopes are maximal, regardless of the sign of the slope. Points *a*—*e* represent places where initial slopes are relatively large, and the small arrows indicate the relative magnitudes and senses of layer-parallel shear stresses. The shear stresses are largest at points *a* and *c*, but the senses are opposite. That shown at *a* is positive and that at *c* is negative according to the usual sign convention for shear stresses. If the multilayer were subjected solely to axial stress, σ_{xx}, Fig. 11A, we would expect yielding to occur simultaneously at points *a* and *c*. We would expect the resulting kink bands to be conjugate and symmetric because nothing is introduced to favor one sense of kinking over the other. This is the situation analyzed in Part IV.

Suppose, however, that the outer edges of the same multilayer are subjected to both initial axial compression, σ_{xx}, and negative shear, σ_{xy}, stresses, as shown in Fig. 11B. The equivalent, principal stresses are shown in Fig. 11C; the direction of maximum compression is rotated counterclockwise relative to the x-direction. As a result of the initial shear, σ_{xy}, the layer-parallel shear stresses are *increased* at points *b* and *c*, but *decreased* at points *a*, *d*, and *e*. Thus, the negative state of initial shear stress increases the layer-parallel shear stress at points where the layers are tilted in the clockwise sense relative to the x-direction. Accordingly, we would now expect yielding to occur first at point *c*, then at point *b*. Yielding will not occur at points *a*, *d* or *e* for this state of stress. For this reason, a negative state of initial shear stress (Fig. 11B) favors initiation of right-lateral kink folding. This conclusion is in accord with the observations of folding of multilayers of rubber strips with contact strength, Fig. 7, and with the sense of shear deduced for the field examples of monoclinal kinking shown in Figs. 3 and 4.

Now we see how to interpret the effect of initial shear stress in eqs. 7—25 of Part IV. The Terzaghi-Coulomb shear strength, $\pm\tau$, becomes the *effective* shear strength, τ_0:

$$\tau_0 = \tau - \sigma_{xy}; \qquad (dv/dx)_0 > 0 \tag{2a}$$

where initial slopes, $(dv/dx)_0$, are positive, and the critical slope required for initiation of kink folding (eq. 23, Part IV) becomes:

$$(dv/dx)_0 > (\tau_0 bt/BI)(x_0/\pi)^2; \qquad (dv/dx)_0 > 0 \tag{2b}$$

where b is unit width, t is thickness and BI is bending rigidity of layers and x_0 is the half-width of the *kink form*. Where slopes are negative, eq. 2a becomes:

$$\tau_0 = -\tau - \sigma_{xy}; \qquad (dv/dx)_0 < 0 \tag{3a}$$

and eq. 2b becomes:

$$-(dv/dx)_0 < (-\tau_0 bt/BI)(x_0/\pi)^2; \qquad (dv/dx)_0 < 0 \tag{3b}$$

The state of initial stress, σ_{xy}, is negative if as shown in Fig. 11B and positive if opposite.

Further, the various derivations concerning widths and other properties of kink bands presented in Part IV are valid if there is initial shear stress. Indeed, the effective strength τ_0, was used in most of the derivations, so we need not repeat them here.

The first index of kinking, K_1, eq. IV,33b, was derived assuming that initial shear stress was zero. In light of the conclusion that initial shear does not affect sinusoidal folding, we can merely replace the shear strength in eq. IV, 23b with the effective shear strength, or, in general:

$$K_1 = |\partial v_0/\partial x|[G_a/(\tau - |\sigma_{xy}|)] \quad (4)$$

Here G_a is the effective shear modulus of the multilayer, eq. IV,29b, and $\partial v_0/\partial x$ is the maximum slope of initial, *Biot* waves.

We showed in Part IV, eq. 33a, that if $K_1 = 0$, the axial load must equal the critical load for sinusoidal, Biot waves before yielding between layers could occur. The same conclusion applies if there is initial, layer-parallel shear stress. Therefore, we can generalize the conclusions presented in Part IV and state that if K_1 defined by eq. 4 is very small:

$$K_1 \simeq 0$$

sinusoidal folds and concentric-like folds would be expected. On the other hand, if K_1 is significantly greater than zero:

$$K_1 > 0$$

kink folds would be expected. Accordingly, small initial slopes, $\partial v_0/\partial x$, and small average shear moduli, that is, thick, soft interbeds, both favor sinusoidal folding. Further, zero average shear modulus, G_a (which requires zero contact strength and zero initial, layer-parallel shear stresses), favors sinusoidal and concentric-like folding. However, high average shear moduli, that is, thin, stiff interbeds, as well as moderate contact strength, τ, and high initial, layer-parallel shear stress all favor kink folding. Indeed, if the initial shear stress approaches the shear strength, K_1 becomes boundlessly large unless G_a is identically zero.

Accordingly, the only modification of the first criterion we need to make where initial shear stress is important is that initial shear stress tends to favor kink folding rather than sinusoidal folding of a multilayer with contact strength.

Kinking as a process of yielding instability

We can gain further insights into the development of monoclinal kink bands by ignoring temporarily the bending within layers and by considering the state of stress within the multilayer to be uniform. Considerations of bending are useful because they allow us to determine factors that affect the widths of kink bands and to derive criteria to distinguish between conditions favorable to kink folding and to sinusoidal folding.

In following pages we will consider only multilayers comprised of layers with identical rheological properties and identical contact strengths. For such materials we can more accurately determine the stress state required for initiation of yielding of contacts than was possible with the elementary theory in Part IV, because bending is zero until the contacts between layers have yielded. We can assume here that the multilayer behaves as an isotropic body with planes of weakness so that the stress state of the multilayer is approximately uniform.

Further, we will ignore effects of bending even after yielding between layers has occurred. This will allow us to determine conditions of locking of kink bands, which we cannot analyze with buckling theory because slopes of layers far exceed the limits of that approximate theory. This assumption probably is justified for some multilayers. We probably can ignore effects of bending after individual layers internally have yielded elastically or plastically in hinges of kink folds, or after kink bands have widened significantly. We suggest that this assumption of no bending resistance is approximately valid for predicting conditions of locking of kink bands for the thin rubber strips used in our experiments and for the paper cards used in Weiss' (1968) experiments. However, it is only approximate for predicting locking of kink bands in the thick rubber strips used in our experiments, because their bending resistance is large until they yield elastically. With these qualifications firmly in mind, let us proceed with the analysis.

Initiation of conjugate kink bands. Let us consider a body of layered material between which contact strength is entirely frictional, with a friction angle of ϕ. Locally the layers are inclined at an angle θ to the direction of flat layering. θ is positive if it is a counterclockwise angle from the positive x-direction (Fig. 12B). The principal stresses are parallel and normal to layering, where σ_3 is the *maximum compression*, acting parallel to flat layering, and σ_1 is the minimum compression, acting normal to flat layering (Fig. 12A). As in previous derivations, tensile normal stresses are positive. We will consider only compressive stresses here so both the principal stresses are negative and the magnitude of σ_3 is greater than the magnitude of σ_1. If the state of stress within the multilayer is homogeneous, we can use Mohr's circle or equations for equilibrium of forces on various planes to derive expressions for normal and shear stresses on bedding planes in the multilayer. If the principal stresses are sufficiently large so that slippage is impending locally where the layers are inclined by an angle θ, we can show that:

$$(\sigma_3/\sigma_1) = \tan(\theta \pm \phi)/\tan(\theta) \tag{5}$$

where the sign of ϕ is selected to be the same as the sign of θ. (This equation can be derived from eq. 12, p. 66, Jaeger and Cook, 1969, assuming zero cohesion.)

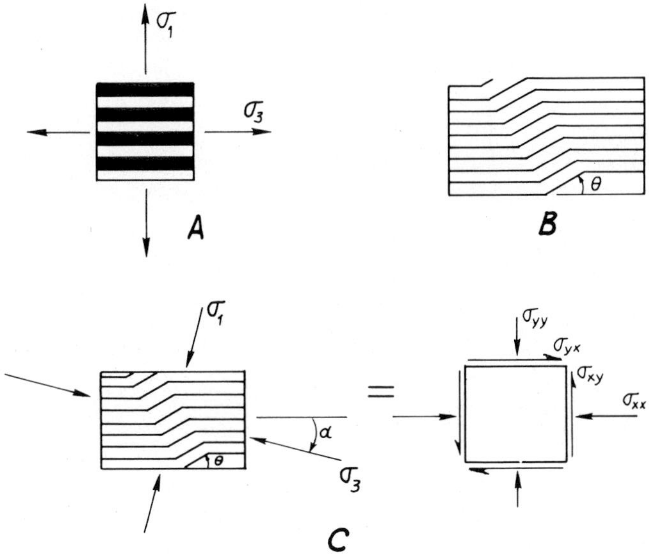

Fig. 12. A. Multilayer subjected to standard positive state of principal stresses (tensile stresses positive). Principal stresses parallel or normal to layering. B. Positive slope angle of layering within kink band. C. Maximum compression σ_3, inclined to layering at a positive angle, α. θ is angle between nonfolded layering and layering within kink band. Inclination of principal stresses shown in C produces *positive* shear, σ_{xy}, normal to layers and σ_{yx}, parallel to layers.

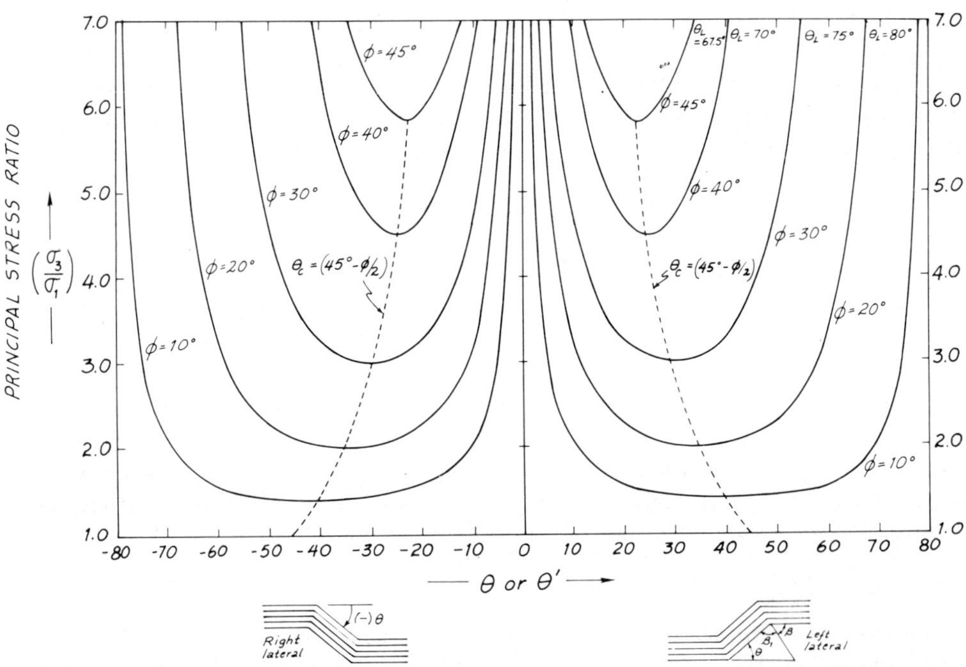

Fig. 13. Relations between ratio of principal stresses, σ_3/σ_1, and slope angle of layers locally, θ, for various friction angles, ϕ, for multilayers with purely frictional contact strengths.

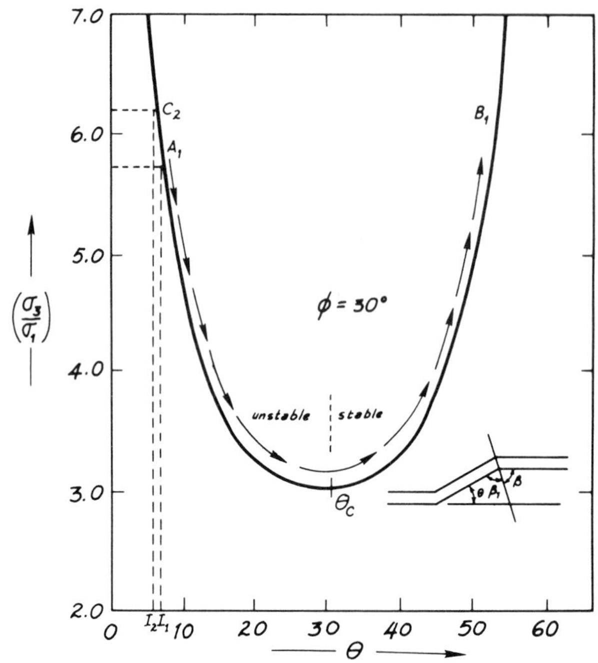

Fig. 14. Relation between ratio of principal stresses and slope angle of layers in kink band for foliated materials with friction angle of 30°. Arrows indicate stress path as layers within kink band rotate.

Figure 13 shows eq. 5 plotted for layering inclined in positive and negative directions and for friction angles ranging from 10° to 45°. In order to illustrate the relations, we will consider the curve for $\phi = 30°$ and $\theta > 0$ (Fig. 14). Suppose that a multilayer contains two areas of positive initial slope, one with layers inclined at about 7° (I_1, Fig. 14) and another at 6° (I_2, Fig. 14). According to Fig. 14, slippage cannot occur until the stress parallel to flat layers is about $\sigma_3 \simeq 5.7\ \sigma_1$, corresponding to point A_1. Then the layers become *unstable* in the sense that an increase in slope locally decreases the stress that can be supported there, and, presumably, a kink band begins to develop by buckling as discussed in earlier paragraphs and in Part IV. We call this instability *unstable yielding of contacts.* As the layers within the kink band steepen, the amount of stress they can resist decreases, according to the arrows shown in Fig. 14, until the slope angle reaches a certain critical value, θ_c, which is 30° for this example. The critical angle is in general is:

$$\theta_c = \pm(45° - \phi/2) \tag{6}$$

For greater slope angles of layering within the kink band, the stress ratio must increase, so the process of increasing the slope angle is *stable.*

At least two things can happen after the first initial irregularity originates a kink band. The layers within the first kink band might continue to rotate as the stress ratio is increased until the second initial irregularity originates a new kink band. In general, however, the first kink band will reflect near a stiff boundary or at an unusually thick layer in the multilayer and originate a kink band of the conjugate set. We showed many examples of conjugate kink bands developing by reflection in Part IV. The first kink band forms a pucker at the boundary and the pucker provides an initial slope of layers which can originate a kink band of the conjugate set. Accordingly, the stress state might not follow the path indicated by arrows shown in Fig. 14A, rather it might end slightly above the minimum, corresponding with θ_c, because the first kink band provides relatively large slopes of layers near the boundary for initiating other kink bands, and when these, in turn, reflect, they provide initial slopes for further kink bands. In this way a complete pattern of conjugate kink bands can develop at a stress state only slightly above that represented by the minimum stress ratio shown in Fig. 14; the complete pattern of kink bands can be initiated at a single place within the multilayer.

In some situations the first-formed kink band may be unable to reflect. For example, the confining medium might be very soft or the multilayer might be very thick. In these cases, the second irregularity in the multilayer can originate a second kink band. When the stress ratio reaches the critical value of $\sigma_3 \simeq 6.1\,\sigma_1$ for initiation of the second kink band with an initial slope angle of about 6° ($C2$, Fig. 14), the layers in the first kink band stop rotating at an angle corresponding to point B_1 in Fig. 14, about 53°, and the second band develops unstably, under decreasing layer-parallel stress.

Locking of kink bands. An interesting feature of Fig. 14 is that it provides an estimate of the angle of layers in fully developed kink bands. The layers for a material with a friction angle of 30° are stable at angles greater than 30°. The larger the angle above this value, the higher the stability. If the stress ratio is limited, the layering will stop rotating at an angle corresponding with the limiting stress ratio. *In this sense*, the kink band "locks". A locked kink band can always be unlocked by increasing the stress ratio, however. Thus, the final locking of a kink band reflects the stress state as well as the strength properties of the multilayer.

If we assume that there is no dilatation in the kink band ($\beta = \beta_1$ in Fig. 14), the axial traces of the hinges of the kink band are inclined at an angle β which is related to the slope of the layers, as shown by Paterson and Weiss (I, 1966):

$$\beta = \pm(90° - \theta/2) \tag{7a}$$

As shown by substituting values of $\theta > \theta_c$ from Fig. 14 into eq. 7a, a material with a friction angle of 30° should lock with directions of kink bands inclined about 60—75°. Weiss (I, 1968, p. 313) reported that directions of kink bands in experiments with paper cards, which have a friction angle of about 30°, range from about 57° to 73°, depending upong the magnitude of

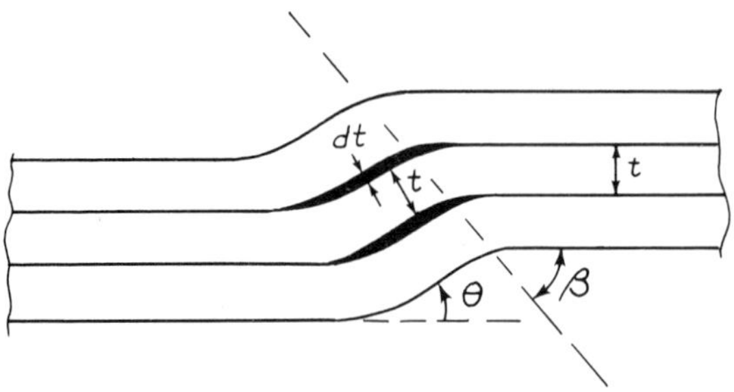

Fig. 15. Idealized kink band, showing dilatation of layers within band. β is slope angle between edge of kink band and nonfolded layering. t is thickness of layer. dt is thickening of layer or space between layers. Dilatation, Δ, is defined as dt/t. Angle θ is measured between layers within kink band and nonfolded layers.

confinement. Thus, the theory helps to explain conditions of "locking" as well as initiation of kink bands and it provides an indication of one type of condition which might control the orientation of kink bands in foliated materials in which bending resistance is relatively low.

Effect of dilatation on locking of kink bands. Paterson and Weiss (I, 1966) showed that the amount of thickening or thinning of layers within a kink band depends upon the slope angle, θ, of layering within the kink band and the angle β between the nonfolded layering and the boundary of the kink band (Fig. 15). The amount of thickening or thinning, or dilatation, Δ, is (e.g., Ramsay, I, 1967, eq. 7-53):

$$\Delta = \mathrm{d}t/t = \sin(\theta)\,\mathrm{ctn}(\beta) + \cos(\theta) - 1 \tag{7b}$$

where dt is the change in thickness or spacing between layers and t is the original thickness of layers. Dilatation has at least two important effects on kinking. It can markedly affect initiation of kink bands; an increase in thickness of layers within incipient kink bands releases normal stress between layers, thereby encouraging unstable yielding locally. We shall discuss this process more completely in following pages when we consider deformations near the tip of a propagating band.

Dilatation also can markedly affect the locking angle of layers within a kink band. As kinking proceeds, the bands become locked due to an increase in compression normal to layers within the bands, as discussed above. However, if β is relatively small, that is, smaller than the value defined by eq. 7a, there will be thickening of layers within the band which can compensate for part of the increase in normal stress due to rotation. Thereby the locking angle of layers will be increased. If β is larger than the value defined by eq. 7a, on the other hand, there will be thinning of layers within the band, thereby causing the layers within the band to lock at lower angles than would be predicted by the theory presented in earlier pages.

Thus, the possibility of dilatation within kink bands is one further reason it is difficult to predict locking angles of kink bands.

Monoclinal kink bands. Now let us consider the initiation and locking of kink bands in response to principal stresses inclined at an angle α to the layers (Fig. 12C). A positive (clockwise) angle α produces a positive state of shear stress if the principal stresses are compressive, as shown in Fig. 12C. We merely replace eq. 5 with:

$$(\sigma_3/\sigma_1) = \tan(\theta' \pm \phi)/\tan(\theta') \tag{8a}$$

where:

$$\theta' = \theta + \alpha \tag{8b}$$

$$|\alpha| < 45° - \phi/2 \tag{8c}$$

and where the sign of ϕ in eq. 8a is the same as the sign of θ'. Thus Fig. 13 remains valid if we replace θ with θ'.

In order to evaluate the effect of inclination of the principal stresses to layering, it will be convenient to consider three conditions: General slippage and kinking; general slippage, alone; and slippage and kinking in places where the local slope of layering is different from the general slope of layering. First let us suppose that the layers are initially inclined uniformly at an angle, α, to the direction of maximum compression and consider conditions of stable and unstable general slippage. Let us consider two multilayers, inclined at equal angles, δ, to the critical direction, α_c, requiring the minimum stress ratio for slippage (Figs. 16A and 16B). The layers in one multilayer, Fig. 16A, are inclined at α_1, which is less than α_c, and the layers of the other, Fig. 16B, are inclined at α_2, which is greater than α_c. Suppose that the Mohr circle in Fig. 16C represents the stress state for arbitrary compressive stresses, σ_1 and σ_3. The envelopes represent the strength of the bedding; failure across bedding is excluded from the analysis. Slip might occur on a plane oriented parallel to that with an angle α_c which is $45° - \phi/2$ in physical space. However, no bedding planes are available with this orientation. The stresses on bedding planes inclined at angles α_1 and α_2 are too small to cause yielding. Thus, slip is impossible for this state of stress.

Let us assume that the maximum compression is increased, keeping the minimum compression constant (Fig. 16D) until the shear stresses on bedding planes shown in Figs. 16A and 16B equal the contact shear strength. Yielding occurs on bedding with both orientations. The results of yielding, however, are quite different. Figs. 16E and 16F show the two multilayers before and after yielding along layers. In both cases the layers steepen in response to yielding along contacts. An argument for steepening of layering is that slippage is always resisted by contact strength, regardless of the sense of slippage, and a virtual increase in α has no resistance because the shear stress is equal to the shear strength but a virtual decrease in α has large resistance because that sense of layer-parallel shear is resisted both by the strength and by the applied shear stress along contacts between layers.

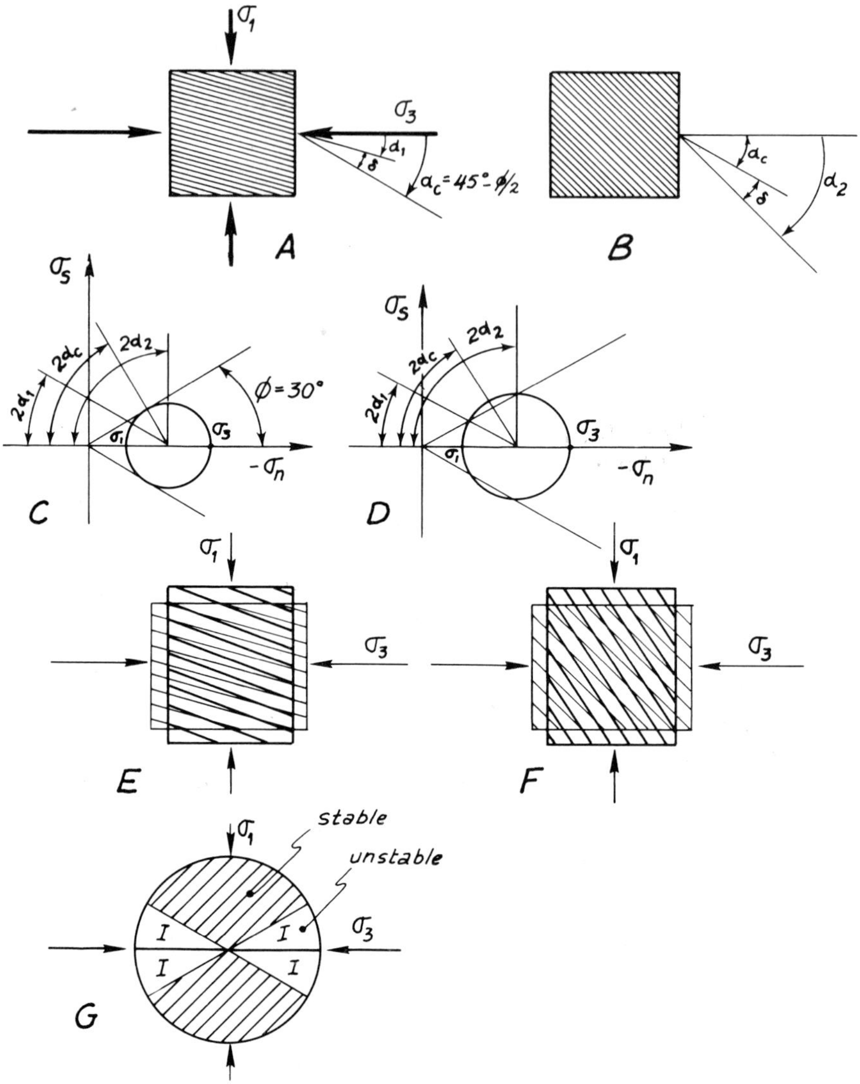

Fig. 16. Concept of yielding instability of foliated materials. If foliation is inclined at small angles to direction of maximum compression, as in A, yielding is unstable. If foliation is inclined at large angles, as in B, yielding is stable. C and D are Mohr circle representations of orientations of foliation shown in A and B for two different stress states.

Therefore, yielding of the multilayer with layers inclined as shown in Fig. 16B is *stable*, that is, yielding is self-equilibrating, so that the larger the rotation, the larger the stress ratio required to maintain slippage. In contrast, yielding of the multilayer with layers inclined as shown in Fig. 16A is *unstable*; yielding and a resulting increase in α requires the stress ratio to drop for equilibrium to be maintained.

According to this analysis, there is solely general slippage along layering if the ratio of stresses satisfies the relations:

$$\sigma_3/\sigma_1 = \tan(\alpha \pm \phi)/\tan(\alpha) \quad (9a)$$

$$|\alpha| \geqslant (45° - \phi/2) \quad (9b)$$

where the sign of ϕ in eq. 9a is the same as the sign of α. Yielding is stable (Fig. 16G). Conversely, there is unstable yielding if the general slope angle of layering satisfies relation 9a and:

$$|\alpha| < (45° - \phi/2) \quad (9c)$$

Therefore, the conditions of stable and unstable yielding of layers inclined to the direction of maximum compression are the same as the conditions of stable and unstable yielding of short segments of layers described in earlier pages and shown in Fig. 13.

There will be kinking if the compressive normal stress parallel to layering is greater than that normal to layering and inequality 9c is satisfied, because there is both yielding and buckling instability. Limits can be placed on the widths of the initial kink bands by considering bending resistance, as explained in detail in Part IV.

For a given state of stress, conditions of slippage and stability of layers with different orientations can be represented as shown in Fig. 17A. Layers

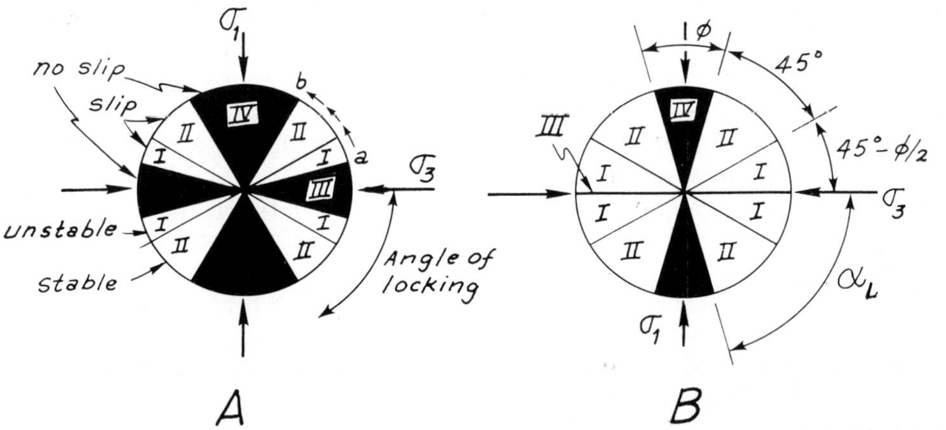

Fig. 17. Zones of yielding and no yielding for certain states of stress. A. If orientation of foliation falls within zones I or II, yielding is possible. Zone I is unstable yielding, zone II is stable yielding. Orientation of foliation must fall within zone I for there to be kink folding. Zones III and IV represent orientations for which yielding is not possible for this state of stress. B. Limits of boundaries of zones where maximum compression is much larger than minimum compression.

with orientations corresponding with zones III and IV cannot slip. The shear stresses on those with orientations corresponding with zones I and II could be equal to the shear strength of the contacts. Zone I is the zone of possible unstable slippage and zone II is the zone of possible stable slippage. Those with orientations corresponding with point *a* are unstable and they will tend to assume the orientation *b*, which is stable, if the state of stress returns to its original value.

As the initial stresses assume extreme values, where the maximum compression is much larger than the minimum compression, $|\sigma_3| \gg |\sigma_1|$, the four zones reduce or expand into those shown in Fig. 17B. Zone IV of no slip can never subtend an angle smaller than ϕ, the friction angle. Its boundary is the *ultimate* locking angle:

$$\alpha_L = \pm(90° - \phi/2) \tag{10}$$

because as $|\sigma_3| \to \infty$, $\alpha \to \alpha_L$. Slippage is possible for angles greater than those defined in eq. 10 only if σ_1 is tensile, a possibility we have excluded from the analysis. Zone III of no slip can reduce to nearly zero. Zones I can never subtend an angle greater than $45° - \phi/2$.

Finally we will consider slippage and kinking only at places where the local slope of layering is different from that of the general layering. The stresses and angular relations are assumed to satisfy eqs. 8, with $\theta \neq 0$. Let us consider a specific example in which there are *local* initial slopes, θ, of +10° and −10° in a multilayer with a contact friction angle of 30°. The maximum compression is inclined +5° (counterclockwise) to flat layering. Thus, $\alpha = 5°$ and $\theta' = +15°$ for one set of initial slopes and $\theta' = -5°$ for the other set. As the axial compression is increased relative to the confinement, the parts of layers that initially slope at +10° become unstable, at a stress ratio of about 3.7 (Fig. 13). The places where local slopes are opposite require a much higher stress ratio, greater than 7 according to Fig. 13. Only a single direction of kink bands initiates. For this example, the local initial slope of layers would have to be −20° ($\theta' = -15°$) and +10° in order for kink bands of both senses to be initiated simultaneously.

In this way we can understand why monoclinal kink bands may not reflect to produce conjugate kink bands in multilayers subjected to principal stresses inclined to layering. The more the principal stresses are inclined to layering, the more is the tendency for monoclinal rather than conjugate kink bands to develop.

For the hypothetical example we have been considering, the magnitude of the local slope angle for the second set would need to be increased only ten degrees, to a total of −20°, for the second set to become unstable. Thus, the folds will be truly monoclinal if the principal stresses are relatively steeply inclined to layering. If the magnitude of the angle of locking of the first-formed kink band is less than the magnitude of the angle required for the conjugate kink band to initiate, the kinks will be monoclinal because they will be unable to reflect.

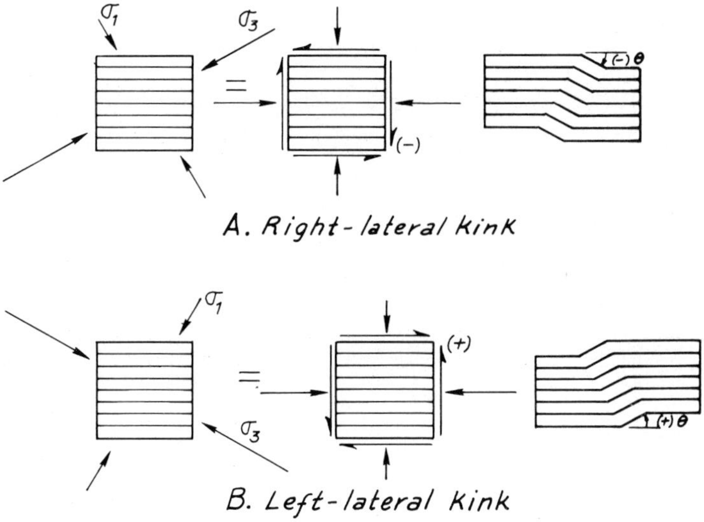

Fig. 18. Stress states for right-lateral and left-lateral monoclinal kink folds.

Another conclusion that we can draw from the analysis is that, where there are monoclinal folds, the principal stresses were inclined to layering in the senses shown in Fig. 18. Negative shear (Fig. 18A) relative to layering favors right-lateral kink bands, and positive shear (Fig. 18B) favors left-lateral kink bands. These conclusions are in accord with experimental observations reported in earlier pages (Figs. 7A and 7D).

One further conclusion we can draw from the analysis is that the angle of locking of one kink band in a conjugate, asymmetric set is lower than the angle of locking of the other. The angles are shown in Fig. 19. This form is strictly theoretical and has not been observed experimentally. Nevertheless, in the theoretical example discussed above, if the stress ratio is not allowed to exceed the value required to initiate the right-lateral kink band, the angle of locking for that band would be $\theta_R = +39°$ ($\theta' = +44°$, $\theta' = \theta_R - \alpha$); and the angle of locking for the left-lateral kink band would be $\theta_L = -49°$.

Experiments by Gay and Weiss

The theoretical predictions concerning the effect of inclination of the principal stresses to layering are consistent with the results of Gay and Weiss (1974), who experimented with slate and paper cards. Figure 20 shows theoretical relations between the ratio of maximum to minimum compression, σ_3/σ_1, and the inclination angle, α, of layering to the direction of maximum compression, according to eq. 9a. The relations are based on the assumption that slippage is impending between the layers. The relations are for friction angles of 20°, 30° and 40°. Also shown in Fig. 20 are points

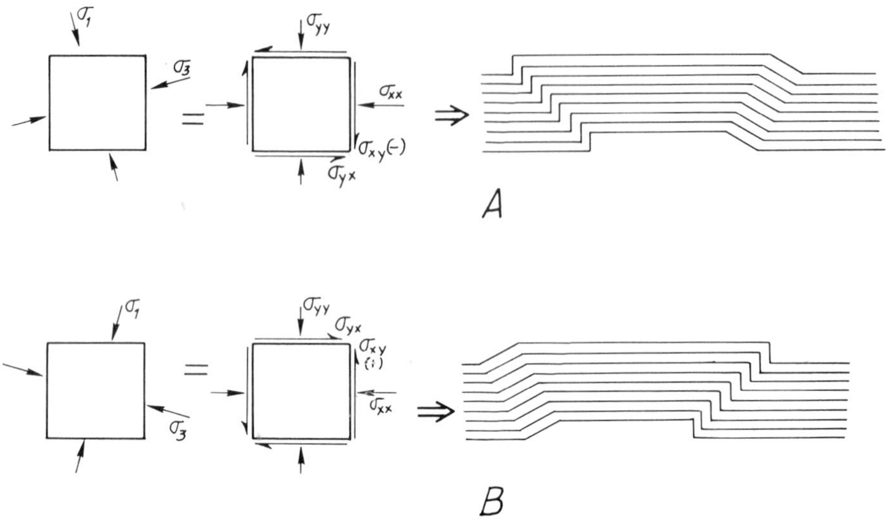

Fig. 19. Stress states for asymmetric, conjugate kink folds. Sense of asymmetry is consistent with rule of thumb for drag folds.

derived from data presented by Gay and Weiss (1974, fig. 4) for the stress state and the general angle of layering for initiation of kinking in decks of paper cards. According to Weiss (I, 1968), the static friction angle of the cards is about 31° and the dynamic friction angle is about 27°. The figure shows that, for inclinations of layers greater than about five degrees, the relation between the stress ratio required for initiation of kinking and the inclination of the layering is compatible with kinking and general slippage along layers. Gay and Weiss report that slippage before kinking was noticeable for angles of inclination greater than $\alpha = 15°$. The data for $\alpha < 5°$ fall well below the theoretical curves but if there were initial local slope angles of layers of about five or less degrees, to initiate slippage, the data points would be shifted to the theoretical curves. Thus, the theoretical relations between ratio of maximum to minimum compression required to initiate kinking and the angle of inclination of layering are consistent with the data by Gay and Weiss.

Gay and Weiss do not present relations between angle of locking of layers in kink bands and ratio of principal stresses, so we cannot compare theoretical and experimental results for the remainder of the curves shown in Figs. 13 and 20. Nevertheless, the data Gay and Weiss present for Delabole slate (shown in their fig. 6; 1974, p. 295) are consistent with predictions. The mean of angles of layers in kink bands in the slate, relative to the direction of maximum compression, is about 47° for slope angles ranging from zero to 30°. The data range from about 35° to 67°, and presumably depend upon the ratio of principal stresses during locking of the kink bands. According to

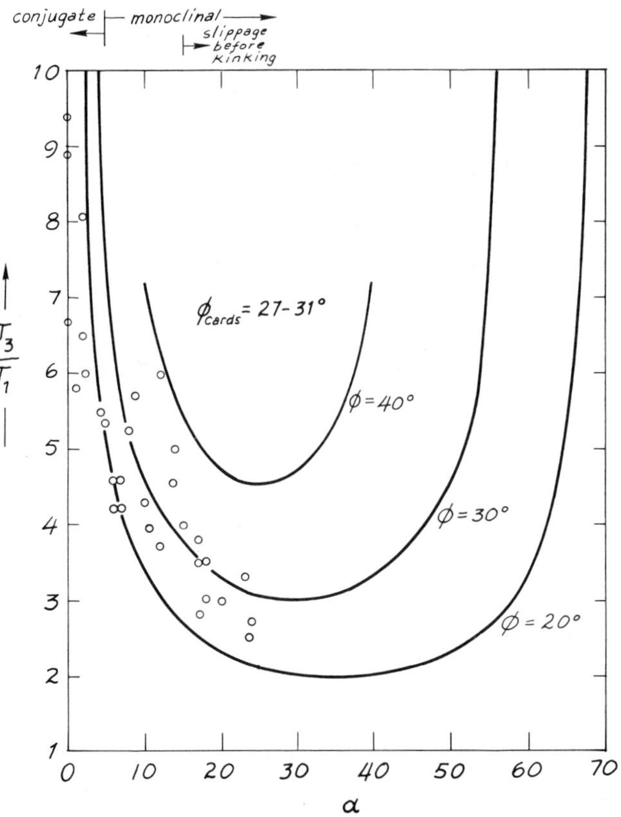

Fig. 20. Theoretical relations between stress ratio and orientation of direction of maximum compression, assuming general yielding between layers, for friction angles of 20°, 30° and 40°. Dots represent approximate stress ratios and orientations of principal stresses when kink folding began in paper cards tested by Gay and Weiss (1974, fig. 4).

Fig. 14, if $\phi = 30°$, the range of locking angles should be 30—75° and normally 40—70°, depending upon the stress ratio. Similarly, the angle of inclination of the edge of the locked kink band relative to the direction of maximum compression, assuming zero dilatation within the band, theoretically would be about 52—75°. According to Gay and Weiss, the angle ranged from about 48° to 73°. Thus, the theoretical predictions agree closely with the experimental results of Gay and Weiss.

Spacing of monoclinal kink bands

We noted in the description of the experiments with rubber strips that monoclinal kink bands appear to have distinct spacings (e.g., Figs. 7A3 and 7D3). There seems to be a tendency for a series of monoclinal kink bands to

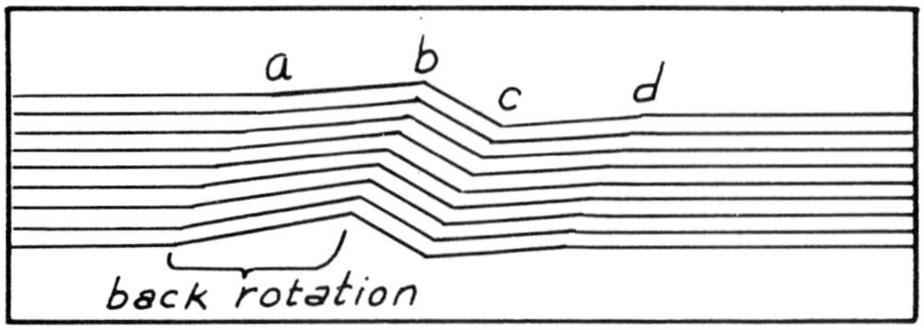

Fig. 21. Back rotation of layers near a monoclinal kink band.

develop at relatively uniform distances from each other along the length of the multilayers. Then, as shortening continues, monoclinal kink bands of a second series tend to develop roughly equidistantly from those of the first series. Weiss (I, 1968), Gay and Weiss (1974) and Johnson (I, 1970, p. 319) show similar features in different experimental materials.

At one time we thought that the spacing was a type of "wavelength", but we could derive no expression for it. Now, however, we believe that the tendency for monoclinal kink bands to be spaced apart can be explained in terms of the theory presented in earlier pages. Figure 21 shows a single, idealized, monoclinal kink band. The layers between *b* and *c*, within the kink band, have slipped relative to each other. The layers have rotated in the clockwise direction. Layers between *a* and *b* and between *c* and *d* have rotated counterclockwise, that is, they have "back rotated" (Gay and Weiss, 1974). Between *a* and the left-hand end and between *d* and the right-hand end of the multilayer, the layers have not rotated. We measured slippage between layers in one experiment and found that there was slippage between layers within the kink band. There was no slippage between layers on either side, but layers were tilted there. Thus, the back rotation is presumably produced by strain and rigid-body rotation. The back rotation apparently is the way the multilayer adjusts to slippage and rotation of layers within the kink band.

The back rotation has important consequences. Because of the rotation, slippage within zones *a*—*b* and *c*—*d* (Fig. 21) requires higher applied stresses than slippage within zones closer to the ends of the specimen. Therefore, the next monoclinal kink band will occur between *a* and the left-hand end and *d* and the right-hand end of the multilayer, where the slopes of layers were unaffected by formation of the kink band. It is for this reason, we believe, that the monoclinal kink bands tend to be separated somewhat, rather than developing side-by-side.

Effect of cohesive strength and pore pressure on kinking

All the derivations we have presented thus far have been restricted to contact strength which is purely frictional and we have ignored effects of pore-water pressure. We have used this simplified model of contact strength between layers because it seems to describe the contacts between rubber strips, with which we have experimented, and the contacts between paper cards, with which Weiss has experimented. For some foliated materials, which have been kinked, however, including phyllite and slate (e.g., Paterson and Weiss, I, 1966; Donath, I, 1968; Gay and Weiss, 1974), contact strength probably has a cohesive component, independent of normal stress, and for crystals such as gypsum and brucite (e.g., Turner, I, 1962; Turner and Weiss, I, 1965), as well as for potter's clay (Ellen, I, 1971) and plasticene (Ghosh, I, 1968; Cobbold et al., I, 1971), the contact strength may be more nearly cohesive than frictional. Further, natural kink bands might form under high pore-water pressures. Thus, we will generalize the model of contact strength and include effects of cohesion and pore-water pressure. The Terzaghi-Coulomb model of shear strength, which we now consider, was used in Part IV where we derived relations for conjugate kink folding.

Effects of pore-water pressure are especially easily incorporated into the analysis in previous pages. Instead of relating the total stress ratio, σ_3/σ_1, to friction angle, ϕ, and slope angles, θ and α, of layering, we merely relate the *effective* stress ratio, σ'_3/σ'_1, to these variables. Thus, eq. 5, for example, becomes:

$$\sigma'_3/\sigma'_1 = (\sigma_3 - p_w)/(\sigma_1 - p_w) = \tan(\theta \pm \phi)/\tan(\theta) \tag{11}$$

in which p_w is pore-water pressure, which is negative if compressive, as are σ_3 and σ_1. The effective stress ratio is larger than the total stress ratio if pore pressures are compressive, so the effect of pore-water pressure generally is to decrease the magnitude of the maximum compression required to initiate slippage between layers. If the slopes of layers are very small, and the pore-water pressure is equal to the stress σ_1, normal to layers, the contact strength would be zero and we would expect sinusoidal folding rather than kink folding, as discussed at some length in Part IV.

The relations for shear of materials in which contact strength is purely cohesive are not fundamentally different either. Equation 5, for example, becomes:

$$\sigma_3/\sigma_1 = [(2C/|\sigma_1|)/\sin(2|\theta|)] + 1; \qquad \phi = 0 \tag{12a}$$

or:

$$|\sigma_3 - \sigma_1| = 2C/\sin(2|\theta|) \tag{12b}$$

where C is cohesive strength.

According to eq. 12a, we could draft a figure similar to Fig. 13, but for each value of confinement, σ_1, there would be only one curve for positive angles of layering and one for negative angles. Each curve would be symmetric about $\theta = \pm 45°$. Thus, the general results derived from Fig. 13 remain valid for simple cohesive contacts. For example, the angle for which the stress ratio is minimized is given by eq. 6; the ultimate angle of locking is given by eq. 10; the slope angle of the kink band is given by eq. 7; and the angles describing conditions of stable, eq. 9b, and unstable, eq. 9c, orientations of layers all remain the same. We merely set $\phi = 0$ in those equations.

The ratio of effective principal stresses for materials with both cohesive and frictional contact strengths between layers is defined by a more complex relation:

$$|\sigma_3'/\sigma_1'| = \tan(|\theta| + \phi)/\tan(|\theta|)$$

$$+ [2C/|\sigma_1'|]/[\sin(2|\theta|)(1 + \tan|\theta|\tan\phi)] \tag{13}$$

but, again, the basic conclusions are unchanged. One difference is that the principal stresses may be tensile; they are not restricted to compressive values if there is cohesive strength. However, none of the conclusions depends upon the sign of the principal stresses. The orientations of layering for stable and unstable yielding, for example, are the same and the possibility of buckling is the same, whether the normal stresses normal and parallel to layering are compressive or tensile. As shown by the relations between critical initial axial stress and wavelength ratio for compressive and tensile stresses normal to layers in fig. 2, Part III, and as shown by Biot (I, 1965a), it is the stress difference, e.g., $\sigma_3 - \sigma_1$, which must reach a certain critical value for buckling.

Therefore, cohesive strength of contacts between layers and pore-water pressure quantitatively affect conditions of kinking and yielding instability, but not qualitatively.

Propagation of kink bands

In Part IV we described the propagation of kink bands in multilayers of rubber strips with which we have experimented. We indicated that the propagation seems to involve dilatation and partial release of confinement on layers within the terminal part of the kink band, thereby allowing layers there to buckle into a kink form and become part of the body of the kink band (figs. 2 and 4, Part IV). Here we will present a theory that approximately describes phenomena near the tip of a propagating kink band in elastic materials. The theory ignores the buckling aspect of the process, so it is known to be incomplete.

The shear stress, σ_s, acting on interfaces between gently sloping layers is approximately:

$$\sigma_s = -(\sigma_{xx} - \sigma_{yy})(\partial v/\partial x) \tag{14a}$$

where σ_{xx} and σ_{yy} are the initial normal stresses parallel and normal to flat layers, respectively, and $\partial v/\partial x$ is the local slope of layers. At critical equilibrium, the shear stress equals the shear strength of contacts, which we will assume to be described in terms of Coulomb properties:

$$|\sigma_s| = -\left[\lambda\frac{\partial u}{\partial x} + (2G + \lambda)\frac{\partial v}{\partial y} + \sigma_{yy}\right]\tan\phi + C \tag{14b}$$

where C is cohesion, σ_{yy} is the stress normal to layers to the first approximation and $(2G + \lambda)(\partial v/\partial y) + \lambda(\partial u/\partial x)$ is the normal stress induced by local normal strains, the dilatation, $\partial v/\partial y$, normal to layers, and the strain, $\partial u/\partial x$, parallel to layers, associated with kinking. Here λ and G are Lamé constants. As before, normal stresses and strains are negative if compressive. Equating (14a) and (14b):

$$[(2G + \lambda) \tan \phi](\partial v/\partial y) - (\sigma_{xx} - \sigma_{yy})|\partial v/\partial x| = C - \left[\sigma_{yy} + \lambda\frac{\partial u}{\partial x} \tan \phi\right] \quad (14c)$$

We interpret eq. 14c as follows. Within the body of the kink band, the normal strain $\partial v/\partial y$, is zero or constant, so that the slope of layers, $\partial v/\partial x$, is constant. Equation 14c only crudely describes the slopes of layers there because the slope angles typically are larger than the limit of 10—15° for which eq. 14a is a close approximation to the shear stress between layers. The exact expression was used to derive eq. 5 in previous pages. Equation 14c provides some insight into conditions near the termination of a kink band, however, where slopes of layers are small. The coefficient $(\sigma_{xx} - \sigma_{yy})$ of the slope, $\partial v/\partial x$, in eq. 14c is negative for conditions of kinking because both σ_{xx} and σ_{yy} are negative and the value of σ_{xx} is larger than that of σ_{yy}. Therefore, eq. 14c indicates clearly that critical equilibrium can be maintained in the terminal regions of kink bands if the dilatation, $\partial v/\partial y$, increases as the slope, $\partial v/\partial x$, decreases. The effect of an increase in $\partial v/\partial y$ locally is to decrease the confinement on layers locally, thereby allowing slippage.

One further piece of information may be extracted from eq. 14c. That equation is hyperbolic so that it can be used to define real characteristic directions, which were discussed in some detail in Part I. The characteristics are derived with eq. 14c in combination with the definition of the total derivative:

$$dv = dy(\partial v/\partial y) + dx(\partial v/\partial x) \quad (14d)$$

by methods explained in Part I. The slope of one characteristic is:

$$(dy/dx)_\beta = +(2G + \lambda) \tan \phi/(\sigma_{xx} - \sigma_{yy}); \qquad \partial v/\partial x > 0 \quad (15a)$$

which is defined only where slopes or layers are positive. There is a second characteristic direction where slopes are negative, because of the absolute value sign in eq. 14c, and its slope is:

$$(dy/dx)_\alpha = -(2G + \lambda) \tan \phi/(\sigma_{xx} - \sigma_{yy}); \qquad \partial v/\partial x < 0 \quad (15b)$$

The characteristics thus derived, unfortunately, do not define the slopes of boundaries of the main body of kink bands because the slopes of layers there are large. The equations do provide some information about boundaries of kink bands near their terminations, however. According to eq. 15a, the slope of the β-characteristic direction, associated with positive slopes of layers, is negative because $\sigma_{xx} - \sigma_{yy}$ is negative. Further, the slopes of the characteristics are constant if the state of stress, $\sigma_{xx} - \sigma_{yy}$, is constant. Or, conversely, if the characteristics change in slope, the state of stress must be changing. Further, the equations indicate that the slope of the characteristics decreases as the axial compression, σ_{xx}, increases relative to the confinement. These theoretical conclusions are consistent with many of our observations of terminal parts of propagating kink bands.

We can perform the same analysis for conditions where there is initial shear stress parallel to layers. The initial shear stress, σ_{xy}, is added to the right-hand sides of eqs. 14a and 14c. The effect of initial shear is as expected. If the initial shear is positive, it effectively decreases the Coulomb strength of contacts in terminal parts of left-lateral kink bands and effectively increases the Coulomb strength in terminal parts of right-lateral kink bands. Further, if the slopes of layers are of the same magnitude somewhere within the terminal parts of left- and right-lateral kink bands, the dilatation must be larger within the right-lateral kink band than within the left-lateral kink band. These statements concerning the effect of shear are merely reversed if the initial shear is negative rather than positive.

Summary of kinking process and conditions of asymmetric folding

Our concept of the kinking process has two essential parts; buckling and yielding instabilities. The two types of instabilities may augment each other if the rotation due to buckling decreases the resistance of contacts to slippage. On the other hand, the resistance of contacts to shear may inhibit the buckling of layers, forcing wavelengths to be shorter than they would be in materials with perfectly lubricated contacts, as discussed in Part IV.

Bending is an essential element or our concept of the kinking processes. Bending resistance of layers forces widths of kink bands to be finite; if bending resistance were zero, widths would tend toward zero. Further, consideration of the buckling aspect of kinking led to two criteria of kink folding, presented in Part IV, which seem to differentiate conditions favorable to kink folding on the one hand from conditions favorable to sinusoidal, concentric-like and chevron folding on the other.

Shear strength of contacts between layers is an essential element of our concept of kinking processes also. It provides an estimate of conditions required for initiation of kink bands, for the propagation in terminal regions of kink bands, for the locking of kink bands, for the development of monoclinal and conjugate kink bands, and for the inclination of boundaries of kink bands to nonfolded layering and to the direction of maximum compression.

Let us consider only multilayers for which the criteria presented in Part IV and summarized in Part V indicate that kink folding rather than sinusoidal folding is favored. Then, one of the conclusions from the theoretical analyses is that yielding between layers is unstable and, therefore, kinks will tend to develop if the slope angle, α, between the general layering and the direction of maximum compression, σ_3, or if the local slope angle, $\theta + \alpha$, between layering and σ_3 is within the range:

$$-(45^\circ - \phi/2) < \alpha < (45^\circ - \phi/2); \qquad \alpha \neq 0$$

or:

$$-(45^\circ - \phi/2) < \theta + \alpha < (45^\circ - \phi/2); \qquad \theta + \alpha \neq 0$$

where θ is the angle of inclination of the layers within the kink band to nonfolded layers. Thus, kinking can initiate only if layering, at least locally, is inclined at less than 45° to the direction of maximum compression. Further, the analysis indicates that the kink bands may lock when layering within kink bands becomes inclined at angles, $\theta + \alpha$, in the range:

$$\pm(45^\circ - \phi/2) \text{ to } \pm(90^\circ - \phi/2)$$

depending upon the stress ratio, σ_3/σ_1. That is, the layers within the kink band lock with slope angles that are acute relative to the direction of maximum compression, if dilatation within the kink band is compressive or negligible. If there is zero dilatation within the locked kink band, the slope angle, β, of the edges of the kink band relative to nonfolded layering is within the range:

$$\pm(67.5^\circ + \phi/4 - |\alpha|/2) \text{ to } \pm(45^\circ + \phi/4 - |\alpha|/2)$$

The angle α must fall in the range $\pm$ (45° $- \phi/2$) for there to be instability in shear. The minimum is:

$$\beta_m = \pm(45^\circ + \phi/4 - |\alpha|/2)$$

which is consistent with experimental results reported by Gay and Weiss (1974). Also, it is consistent with the monoclinal kink bands observed in interbedded limestone and shale near Cody, Wyoming (Fig. 2), which are inclined at an angle of about 45° to nonfolded layering.

The analysis presented here indicates that monoclinal kink bands tend to develop where there is significant layer-parallel shear stress. Most kinking probably begins with a single, monoclinal band, but for a variety of reasons a fully developed pattern of conjugate kink bands may be developed subsequently. If layer-parallel shear is zero, kink bands with both left-lateral and right-lateral senses may initiate simultaneously, at the same stress ratio and local initial slopes of layering. Also, an original monoclinal kink band may reflect at relatively stiff boundaries or at unusually thick layers to produce a conjugate pattern. Truly monoclinal kink bands theoretically will develop where the layering in the reflected part of the band slopes in the same sense relative to the direction of maximum compression as the layering in the main part of the kink band, or where the layering in the reflected part of the band requires a higher stress ratio for unstable growth than is applied to the multilayer. Thus, the occurrence of monoclinal kink bands indicates that the principal stresses were inclined to nonfolded layering, with the senses indicated in Fig. 18, but the actual directions of the principal stresses cannot be deduced accurately without knowledge of the stress ratio and the strength properties of contacts between layers at the time of folding.

The left-lateral, monoclinal kink bands within the limb of a syncline in siliceous shale of the Monterey Formation, Fig. 4, are consistent with the right-lateral, layer-parallel shear which we normally would associate with development of the syncline. Further, the right-lateral monoclinal kink bands in chert with thin shale interbeds of the Franciscan Complex, Fig. 3, are consistent with left-lateral, layer-parallel shear deduced from the sense of asymmetry of chevron-like folds nearby in the chert with thicker shale interbeds, as was suggested in Part V.

The experiments and theory reported here and in Parts IV and V indicate that the *sense* of asymmetry of folds in a multilayer depends upon the properties of the multilayer, the confinement normal to layers, the sense of layer-parallel shear, and the sequence of deformations.

Monoclinal and conjugate kink bands. The experiments described in Part V (figs. 3 and 4) suggest that there is a transition from symmetric conjugate kink bands, where initial shear stresses are zero, to monoclinal kink bands, where initial shear stresses are high. The magnitudes of the initial shear stresses were controlled indirectly in the experiments by changing the contact strength and the magnitude of the shear deformation of the loading frame, so our information is necessarily qualitative. Forms intermediate between the two extremes theoretically are possible, and apparently depend upon the magnitude of the initial shear stresses (Fig. 22A4), that is, the inclination of the direction of maximum compression to layering.

Gay and Weiss (1974) report that kink bands produced in decks of paper cards are conjugate if the cards are inclined at angles less than 5° to the direction of maximum compression, whereas monoclinal kink bands tend to form if the angle is between about 5° and 30°. The angle of internal friction of the paper cards is 27—31° according to Weiss (I, 1968).

Simultaneous layer-parallel shear and shortening. For conditions that favor kink folding (explained in Parts IV and V), the first-formed kink bands theoretically may be monoclinal (Fig. 22A3) or conjugate asymmetric (Fig. 22A4), depending upon the magnitude of the initial shear stress. The sense of asymmetry of the conjugate kink bands is the same as that predicted by the standard rule of thumb of drag folding; the sense of asymmetry shown in Fig. 22A4 indicates the positive state of shear stress shown in Fig. 22A1. The sense of monoclinal kinking is left-lateral for positive states of shear stress and right-lateral for negative states of shear stress. The one shown in Fig. 22A3 is left-lateral and the state of shear stress shown in Fig. 22A1 is positive.

For conditions that favor sinusoidal folding (explained in Parts IV and V), the first-formed folds are symmetric (Fig. 22A5) and the folds become asymmetric according to the rule of thumb of drag folds as their amplitudes grow (Fig. 22A6).

First layer-parallel shortening, then shear. For conditions that favor kink

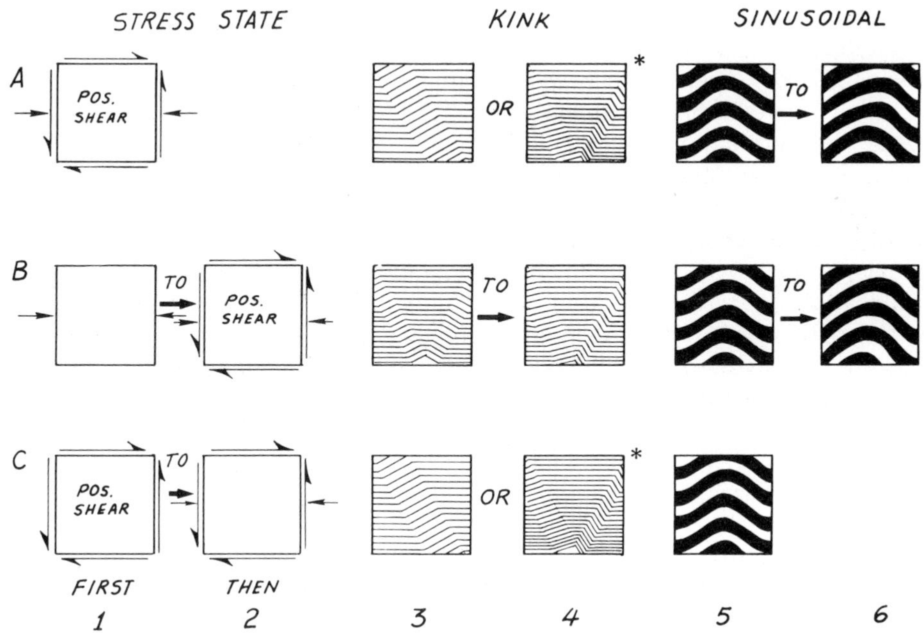

Fig. 22. Ideal fold patterns developed in different kinds of multilayers subjected to different states of stress and different deformation histories. Patterns deduced from experiments and theory. A. Simultaneous layer-parallel shear and shortening. B. First layer-parallel shortening, then shear. C. First layer-parallel shear, then shortening. Fold patterns marked (*) are theoretical; they have not been observed experimentally.

folding, the kink bands are first conjugate symmetric (Fig. 22B3). Then, one kink band is steepened and the other is flattened (Fig. 22B4). Sinusoidal folds are first symmetric (Fig. 22B5) and then become asymmetric (Fig. 22B6). The sense of asymmetry of both sheared conjugate kink folds and sinusoidal folds is as predicted by the rule of thumb of drag folds.

First layer-parallel shear, then shortening. For conditions that favor kink folding, the sense of asymmetry is the same as that for simultaneous shear and shortening, at least for the rubber with which we have experimented. For conditions that favor sinusoidal folding in rubber and gelatin there is no asymmetry, at least for low-amplitude folds (Fig. 22C5).

Local fault displacement can trigger kinking. In some experiments, fault-like displacements of edges of the experimental apparatus apparently initiated kinking. However, the development of monoclinal kink bands with the same senses of displacement near faults with different senses of displacement suggests that the kinking was controlled by the *general* conditions of the multilayer rather than by the conditions induced by *local* faulting. Presumably the fault was a trigger for the kinking, rather than a cause, in the experiments.

ACKNOWLEDGEMENT

Some of the ideas presented here were developed during discussions with Dr. Elmont Honea, Department of Geology, Humboldt State College, California. We thank Elmont for this assistance. Dr. David Pollard, U.S. Geological Survey in Menlo Park, California, and Dr. Raymond Fletcher, Department of Geology, Stanford, reviewed the manuscript and made many useful suggestions for improving it. We wish to express our gratitude to these individuals, while we absolve them from responsibility for errors the manuscript contains.

We wish to thank Peter Gordon for constructing the apparatus used to perform the experiments, and Perfecto Mari for drafting the figures.

The research reported in this part was supported by the National Science Foundation, Grant nos. GA-36917 and EAR 76-03273.

REFERENCES NOT CITED IN CHAPTER 3

Aydin, A., 1973. Field study and theoretical analysis of some small faults in Montana, Wyoming and Utah: M.S. Report, Branner Library, Stanford Univ., Stanford, Calif., 51 p.

Billings, M.P., 1972. Structural Geology. Prentice-Hall, Englewood Cliffs, N.J., 3rd edition, 606 p.

Dennis, J.G., 1972. Structural Geology, Ronald Press, New York, N.Y., 532 p.

Dewey, J.F., 1965. Nature and origin of kink-bands. Tectonophysics, 1: 459—494.

Gay, N.C. and Weiss, L.E., 1974. The relationship between principal stress directions and the geometry of kinks in foliated rocks. Tectonophysics, 21: 287—300.

Honea, E. and Johnson, A.M., 1976. A theory of concentric, kink and sinusoidal folding and of monoclinal flexuring of compressible, elastic multilayers. IV. Development of sinusoidal and kink folds in multilayers confined by rigid boundaries. Tectonophysics, 30: 197—239.

Jaeger, J.C. and Cook, N.G.W., 1969. Fundamentals of Rock Mechanics. Methuen, London, 515 p.

Johnson, A.M. and Ellen, S.D., 1974. A theory of concentric, kink and sinusoidal folding and of monoclinal flexuring of compressible, elastic multilayers. I. Introduction. Tectonophysics, 21: 301—339.

Johnson, A.M. and Honea, E., 1975a. A theory of concentric, kink and sinusoidal folding and of monoclinal flexuring of compressible, elastic multilayers. II. Initial stress and nonlinear equations of equilibrium. Tectonophysics, 25: 261—280.

Johnson, A.M. and Honea, E., 1975b. A theory of concentric, kink and sinusoidal folding and of monoclinal flexuring of compressible, elastic multilayers. III. Transition from sinusoidal to concentric-like to chevron folds. Tectonophysics, 27: 1—38.

Johnson, A.M. and Page, B.M., 1976. A theory of concentric, kink and sinusoidal folding and of monoclinal flexuring of compressible, elastic multilayers. VII. Development of folds within Huasna syncline, San Luis Obispo County, California. Tectonophysics, 33: 97—143.

Ramberg, H., 1963. Evolution of drag folds. Geol. Mag., 100: 97—106.

Ramberg, I.B. and Johnson, A.M., 1976. A theory of concentric, kink, and sinusoidal folding and of monoclinal flexuring of compressible, elastic multilayers. V. Asymmetric folding in interbedded chert and shale of the Franciscan Complex, San Francisco Bay area, California. Tectonophysics, 32: 295—320.

Treagus, S.H., 1973. Buckling stability of a viscous single-layer system, oblique to the principal compression. Tectonophysics, 19: 271—289.

Chapter 9

DEVELOPMENT OF CONCENTRIC-LIKE AND CHEVRON FOLDS AND DOUBLY PLUNGING FOLDS IN TERTIARY ROCKS IN PART OF COAST RANGES, CALIFORNIA *

SUMMARY

Folds in the Huasna area of the southern Coast Ranges of California provide an opportunity to study different fold forms and to estimate dimensional and relative rheological properties of rocks at the time of folding. Plunging, concentric-like and chevron-like folds with wavelengths ranging from about 0.1 to 1 km are clearly visible in natural exposures at the south end of the Huasna syncline, which has a wavelength of 12—16 km. Examination of two fresh roadcut exposures in the Miocene Monterey Formation suggests that folding within part of the Monterey was accommodated primarily by layer-parallel slip between structural layers with thicknesses ranging from 30 to 43 m, even though lithologic layers range from a few mm to a few dm in thickness. This part of the Monterey is folded into a series of concentric-like folds, with chevron-like folds at their cores and with a ratio of wavelength to total thickness of layers of about $L/T_T = 0.42$. Theoretical analysis of multilayers, comprised of identical, elastic or elastic—plastic layers with frictionless contacts, indicates that the effective, or weighted-average thickness of structural layers corresponding with an L/T_T ratio of 0.42 is about 41 m. Thus, the theoretical predictions are roughly in agreement with available data concerning these folds.

Thicknesses of structural units in other folds of this area are inadequately known to closely check theoretical predictions, but most of the data are consistent with predictions. An exception is the Huasna syncline which has a larger wavelength than we would predict. There are several likely explanations for this discrepancy. Layers in the underlying Franciscan complex may have taken part in the folding, making our estimates of total thickness too small. The basement rocks may have been much softer, relative to the overlying sedimentary rocks, than we assumed. The Huasna syncline could be partly a result of gravitational instability of relatively low density, Miocene siliceous and porcelaneous shales, overlain by relatively high density, Pliocene sandstones.

The Huasna syncline and some of the smaller folds in the Miocene rocks are doubly plunging; that is, they are elongated domes or basins. Theoretical analysis of doubly plunging sinusoidal folds indicates that plan shapes of such folds should be related to the ratio of principal stresses acting along the axes, such that the larger the compression, the

---

* Originally published as: Johnson, A.M. and Page, B.M., 1976. Part VII, Development of folds within Huasna syncline, San Luis Obispo County, California. Tectonophysics, 33: 97—143.

shorter the corresponding axis. Differences in plan shape should be sensitive to differences in stress ratios for ratios of long to short axes of folds ranging from about 1 : 10 to 1 : 1. Folds which form under uniaxial compression, normal to the fold axis, should be cylindrical but may readily range from cylindrical to doubly plunging with axis ratios up to about 1 : 10. Dome-shaped folds, however, with an axis ratio approaching unity, should not form unless the principal horizontal stresses parallel to the axes are equal. The Huasna syncline has an axis ratio of about 1 : 5 and a smaller fold has an axis ratio of about 1 : 4. These axis ratios correspond to a ratio of principal regional stresses of about 1 : 10. Long axes of folds are oriented northwest—southeast, roughly parallel to traces of the major faults in the area, suggesting that the maximum compression was approximately normal to the traces of the major faults.

INTRODUCTION

The Coast Ranges of central and southern California are characterized by large faults, many of which are of the strike-slip variety (Fig. 1A), and by anticlines and synclines with wavelengths on the order of 5—20 km (Fig. 1B). One of the folds, the Huasna syncline, is centered about 15 km north of Santa Maria and about 20 km southeast of San Luis Obispo, California (Fig. 1A). It is doubly plunging, so one can observe internal structures at each end of the syncline. This seventh paper in our series on folding * is a result of a study of structures exposed near the southern end of the syncline, where it plunges about 30 degrees northwestward. We found excellent exposures of many small folds in this area, the wavelengths of which range from a few centimeters to hundreds of meters.

The region around the Huasna syncline has been mapped by several investigators, including Taliaferro (1943) and King (1943). Hall and Corbató (1967) mapped the Nipomo quadrangle in detail and showed the Huasna syncline and some of the smaller folds in their map and cross-sections. However, none of these investigators has studied the folds or possible mechanisms of folding in the manner that we will report herein. The region provides a place to study the development of concentric-like and chevron-like folds in rocks, and thus the study complements our theoretical and experimental researches into origins of different fold forms, reported in other parts of this series (Parts III, IV, V, and VI).

The purposes of this paper are to describe the folds; to further develop theory to describe possible processes of folding in the area; to combine the theoretical and field observations in order to explain some features of the smaller folds; and to speculate about the origin of the Huasna syncline itself. The folds are described by means of a geologic map, a composite structural

---

*** Part I: Johnson and Ellen, 1974; Part II: Johnson and Honea, 1975a; Part III: Johnson and Honea, 1975b; Part IV: Honea and Johnson, 1976; Part V: Ramberg and Johnson, 1976; Part VI: Reches and Johnson, in preparation.**

Note: Part I = Chapter 3, Part II = Chapter 4, Part III = Chapter 5, Part IV = Chapter 6, Part V = Chapter 7, Part VI = Chapter 8.

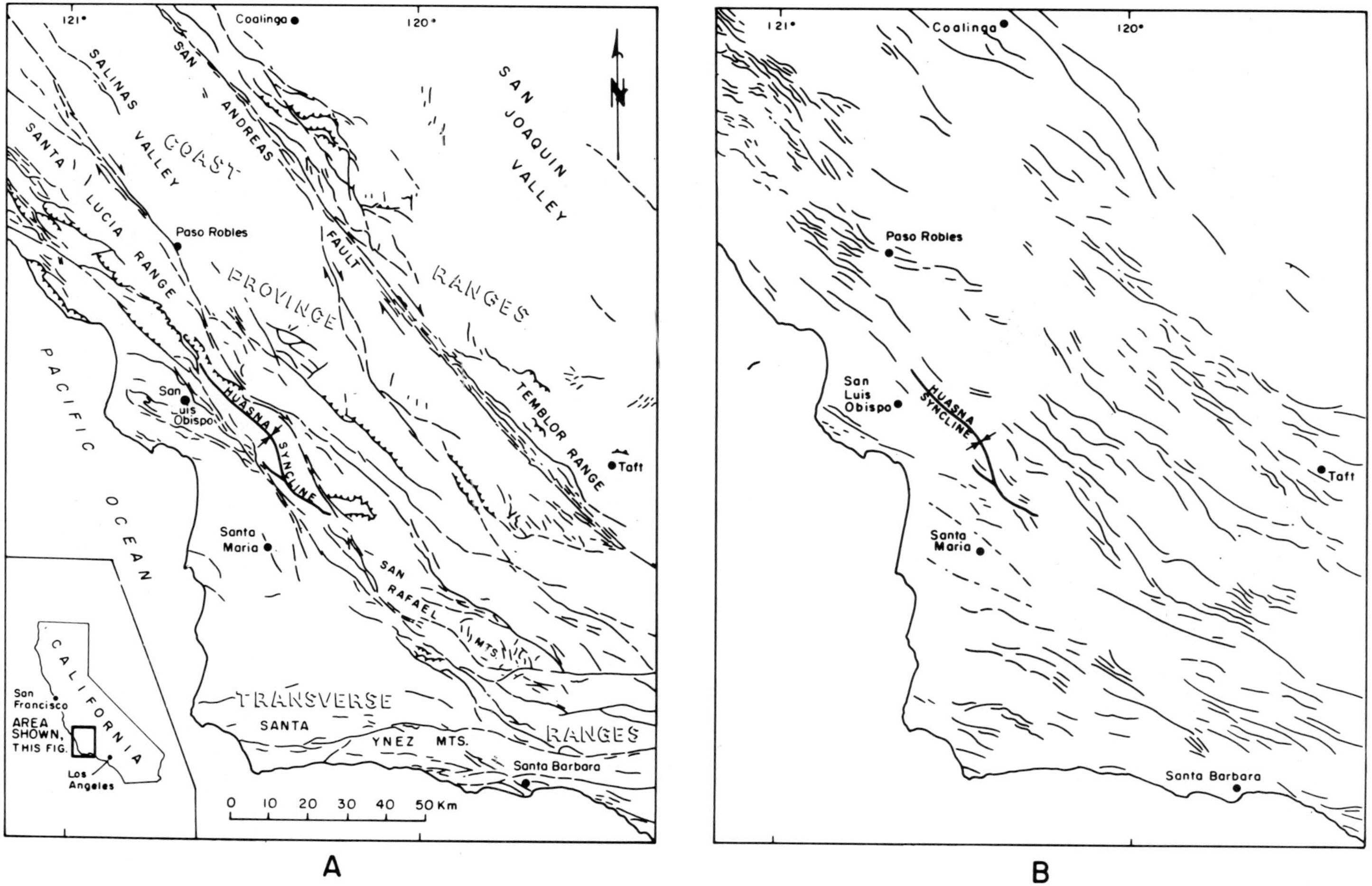

Fig. 1. Structural setting of Huasna syncline. A. Traces of major faults (after Jennings, 1973). B. Axes of major folds. (Compiled largely from Open File maps of U.S. Geol. Survey, T.W. Dibblee, Jr.)

section of the southern part of the Huasna syncline, a controlled sketch of minor folds exposed in the side of a hill, and photographs. The theory deals with the folding of elastic multilayers confined by elastic media, the simultaneous development of several orders of folds in elastic material, and the formation of three-dimensional, doubly plunging folds. The theory of doubly plunging folds for single or multiple layers includes the solution for the reaction of an infinite half-space to the three-dimensional deformation of one surface, a solution which should be useful for other problems.

THE HUASNA SYNCLINE

Regional geologic setting

The Huasna syncline (Fig. 1) is in the southern California Coast Ranges. The areal geologic setting is shown in the San Luis Obispo Sheet of the Geologic Map of California (Jennings, 1958). The "basement" rocks in the area are part of the Franciscan Complex (Bailey et al., 1964; Berkland et al., 1972). Where exposed to the east and west of the Huasna syncline, the Franciscan is largely mélange, consisting of sheared, argillaceous matrix containing lumps, blocks, and slabs of rocks such as graywacke, greenstone, radiolarian chert, and serpentinite. The Franciscan is overlain by Late Jurassic and Cretaceous marine sandstone, mudstone, and conglomerate, which are equivalent to the Great Valley Sequence. These rocks are in tectonic contact with the Franciscan. Tertiary formations commonly overlie the Late Mesozoic rocks (Fig. 2). Except for some Oligocene fluviatile sediments, most of the Tertiary formations are marine sedimentary rocks of Late Oligocene, Miocene, and Pliocene ages. All are described in detail by Hall and Corbató (1967) and those exposed in the area of our geologic map are described in following pages.

There are several important faults in the region, generally trending northwest, but they do not transect the main part of the Huasna syncline (Fig. 1A). The imprecisely defined Sur—Nacimiento fault zone (Page, 1970) is at the northeastern side of the Huasna syncline (Fig. 1A). The zone is marked approximately by Late Cenozoic, high-angle faults such as the one labeled "Nacimiento Fault" on the San Luis Obispo Sheet of the Geologic Map of California (Jennings, 1958). This particular fault is 5—15 km from, and subparallel with, the axial trace of the Huasna syncline. Vedder and Brown (1968) find evidence for large vertical movements along this fault in the interval from Early Tertiary to Miocene, and do not rule out the possibility of some strike-slip.

The East Huasna and West Huasna faults are on either side of the Huasna syncline, within 1—7 km of the axial trace (Fig. 3). They are both steeply dipping, and may be steep thrust, strike-slip, or oblique-slip faults. Probably they have undergone both thrusting and right-hand strike-slip movements. In any case, Hall and Corbató (1967) find differences in thicknesses and

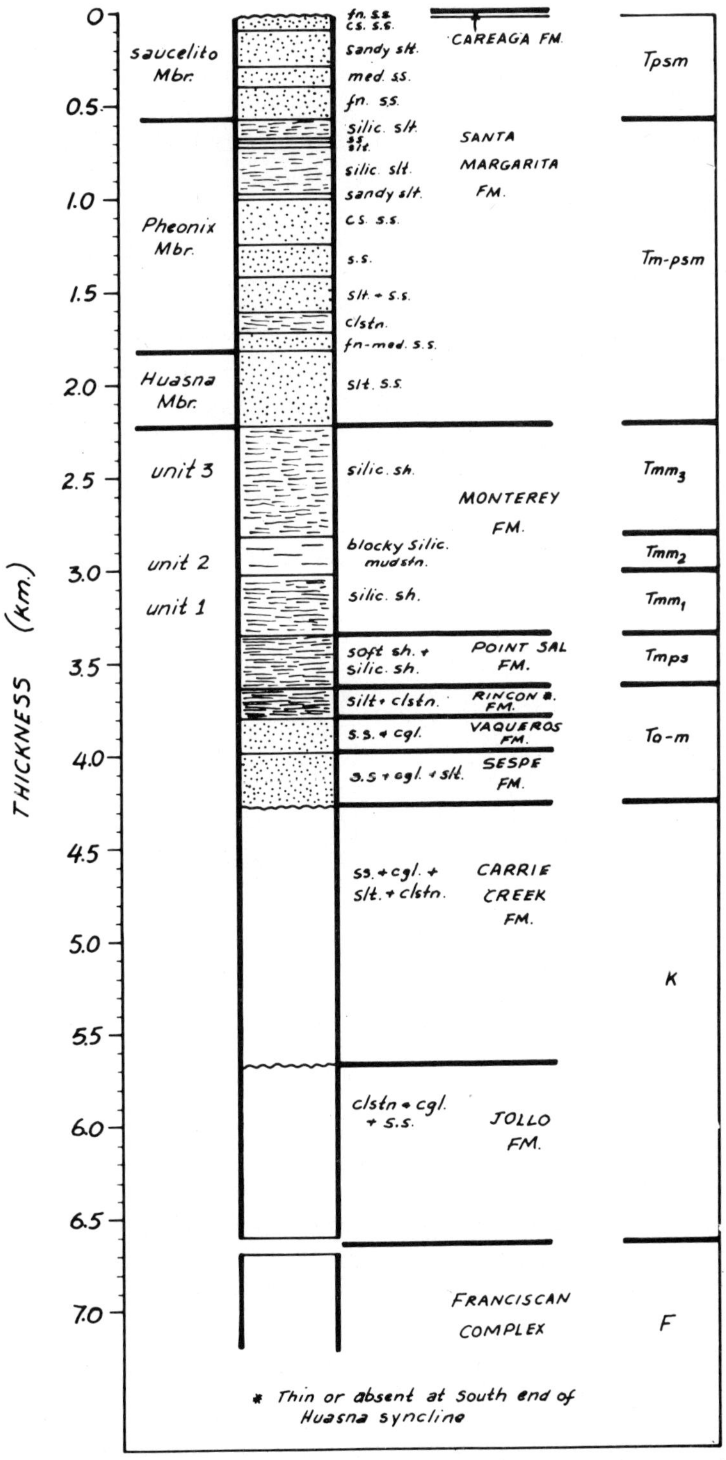

Fig. 2. Stratigraphic units in area of Huasna syncline. Maximum thicknesses shown except for Point Sal Formation and three units of Monterey Formation, for which thicknesses are those at south end of Huasna syncline (largely after Hall and Corbató, 1967, fig. 2).

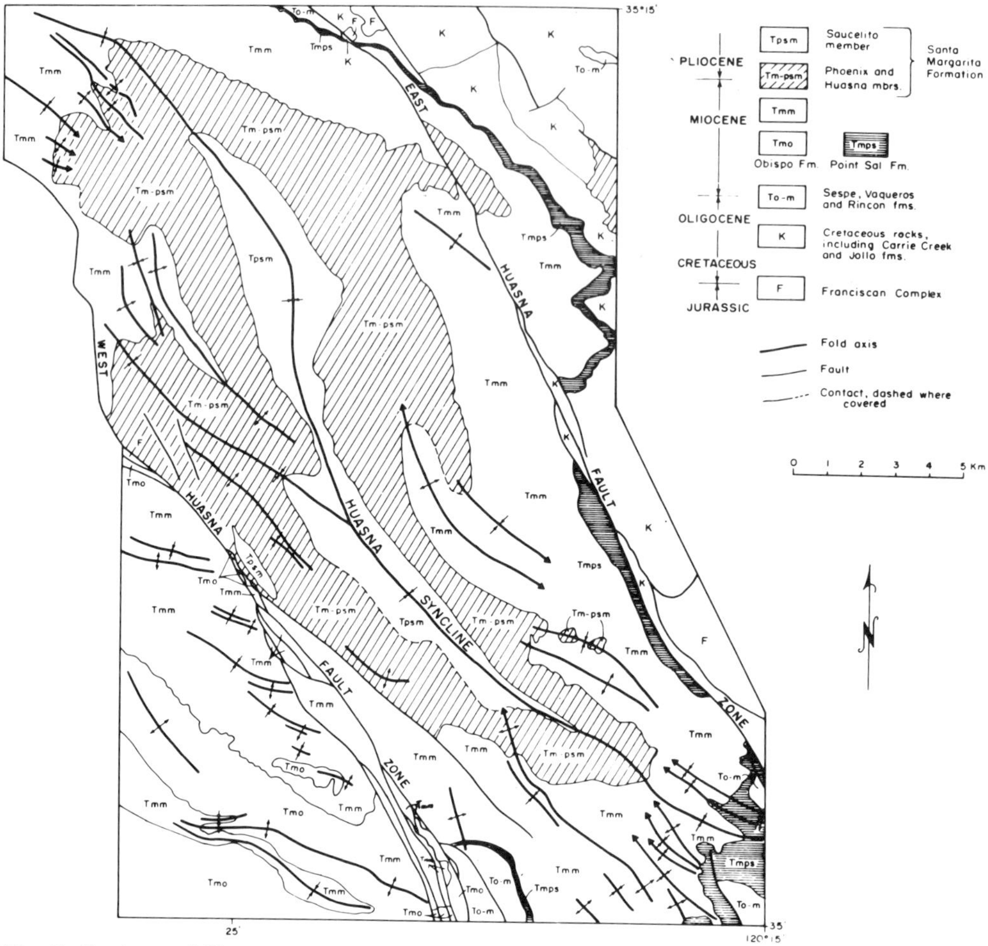

Fig. 3. Geology of Huasna syncline (after Hall and Corbató, 1967, pl. 1).

facies of equivalent units on opposite sides of each of these faults.

Folds are prevalent in the region, and most of them trend northwestward, more or less parallel to the faults. In some other parts of California the folds are reported to show an en-echelon arrangement, with axes at an acute angle to the trend of the San Andreas fault (Wilcox et al., 1973), but this arrangement of folds is conspicuously lacking in the region of our study (Fig. 1B).

Form of the Huasna syncline

The areal configuration of the Huasna syncline is well shown by Hall and Corbató (1967, pl. 1); our Fig. 3 is a synoptic adaptation of their detailed map. The main fold is slightly curvilinear or sigmoidal in plan, trending northwest at the ends and almost north—south near the middle sector. The more northerly trending part of the fold is within Late Miocene and Pliocene rocks. There the fold is broad and open, with gently dipping limbs. The flanks and ends of the syncline expose Middle Miocene rocks which are folded to produce a synclinorium. The map pattern shows that the folds generally plunge. The north part of the Huasna syncline plunges 20—35 degrees southeast, and the south part plunges 20—30 degrees northwest.

The cross-sectional shapes of the folds can be surmised by the dips of various parts of the limbs, crests, and troughs, and by the map pattern as well. Because of the relatively uniform dips of some limbs in map traverses normal to axial traces, it is inferred that these limbs are planar as in chevron folds. This inference is supported by angular map traces of beds where they pass around axes. Down-plunge views of the geologic map of Hall and Corbató (1967, pl. 1) also suggest the presence of chevron folds, and these authors show chevron forms locally in their cross-sections (op. cit., pl. 2). In other places, dips and map patterns indicate gradual curvature of beds around axes, and commonly a particular fold shows both chevron and concentric-like forms, the former being nested within the latter. Detailed structural sections made near the north end of the Huasna syncline by Hill (1976) and by Scofield (1975) substantiate these conclusions.

Times of folding

The folding in the area of the Huasna syncline is young, because the syncline involves rocks as young as Pliocene, and some other Coast Range folds involve Pleistocene sediments. There probably has been a long history of folding in the area, however. There is some indication that folding of the Huasna syncline began at least as early as Late Miocene, during the time of deposition of the Huasna member of the Santa Margarita Formation. A distinctive coarse-grained sandstone facies of the Huasna member occurs only along the axis of the syncline and fingers out within about 500 m on either side of the axis (Hall and Corbató, 1967, pl. 1, sections 10 and 15, T 31 S, R 14 E). Folding occurred earlier in the general region of the syncline, however, in rocks of Late Mesozoic to Early Tertiary ages. Where these rocks are overlain by Late Tertiary rocks, there is invariably an unconformity, generally angular. Late Mesozoic rocks are not exposed in the area of the Huasna syncline, but they crop out in terrains on either side, where they are overlain by Tertiary units with angular unconformity. Thus we presume that the exposed Tertiary formations of the Huasna syncline rest on previously deformed Mesozoic rocks. There is no way of knowing how this circumstance affected the geometry of the syncline.

The following summary describes what is believed to be the regional tectonic setting, according to current concepts of plate tectonics. In this model, the Franciscan "basement" is an assemblage formed in a subduction zone west of the continental margin during Late Jurassic, Cretaceous and Early

Tertiary (e.g., Hamilton, 1969). The Great Valley-type rocks overlying the Franciscan accumulated in an arc—trench gap (Dickinson, 1970) and, in the area of this study, they have been underthrust by the Fransican during subduction (Page, 1970). The Sur—Nacimiento fault zone roughly corresponds with the former continental edge as it existed during part of the time of subduction, the fault zone approximately marking a tectonic boundary between sialic continental basement rocks and the Franciscan Complex. Subduction ceased 20—25 m.y. B.P. and was followed by NW—SE, right-hand transform motion between the Pacific and the North American plates (Atwater, 1970).

The post-Oligocene rocks and folds in the area of our study are younger than the time of subduction but are contemporaneous with the time of transform motions. In other areas where Coast Range folds are en echelon and oblique to the San Andreas fault, the folds commonly are explained in terms of "wrench tectonics" (e.g., Wilcox et al., 1973). This view is consistent with the foregoing model of plate tectonics; however, the folds we studied, although formed during the time of transform motions, are not oriented in such a way that their origin can be simply explained by "wrench tectonics". Indeed, fold axes generally are parallel to the major faults, as shown in Fig. 1. Thus, the current model of plate tectonics may not accurately describe the tectonic history of this region.

FOLDS NEAR SOUTH END OF HUASNA SYNCLINE

Rock units at south end of Huasna syncline

The southern part of the Huasna syncline is shown in an oblique aerial photograph in Fig. 4. There are six large folds in the area, one of which is the Huasna syncline itself. The rock units are the Oligocene to Miocene Vaqueros Formation (To-mva), the Miocene Point Sal Formation (Tmps), well exposed in The Eagle, the Miocene Monterey Formation (Tmm), and the overlying, Miocene to Pliocene, Santa Margarita Formation (Tm-psm). The Santa Margarita is not exposed within our area, but it immediately overlies rocks we mapped. According to Hall and Corbató (1967, p. 572), it is up to about 2.2 km thick, and consists of quartz or arkosic sandstones with silty sandstone, siltstone and siliceous claystone (Fig. 2).

The rocks in the area of detailed study are all Tertiary, and are of marine sedimentary origin. Except for the Vaqueros Formation, most of them are thin-bedded, shaly and siliceous rocks, including siliceous shale and porcelanite. For stratigraphic details, the reader is referred to Hall and Corbató (1967).

Vaqueros Formation (Oligocene—Lower Miocene). The Vaqueros Formation is mainly massive sandstone. It occurs in beds 3—15 m thick without internal partings or visible layering except for a few ill-defined pebbly lenses. The sandstone is medium to coarse-grained and is locally cemented with calcite.

Fig. 4. Oblique aerial view of south end of Huasna syncline, showing axes of large folds. View northwest.

In addition to sandstone, the Vaqueros includes a few shale members 1—10 m thick. The thickness of the formation in the area studied appears to exceed 100 m. According to Hall and Corbató (1967), it ranges from 16 to 200 m in the surrounding region.

Point Sal Formation (Middle Miocene). The Point Sal Formation, which rests conformably upon the Vaqueros, is a rather distinctive shaly unit of both organic and clastic composition. Much of the rock is fissile, "punky" shale which may be diatomaceous; it is soft, light, porous, and rich in organic components.

The bulk of the shale is microscopically nearly isotropic matrix material with fine opaline particles, brown organic matter, and silt-size detrital mineral grains, including mica. The Point Sal Formation also includes sequences (5—15 m thick) of flaggy, hard, siliceous shale, like parts of the overlying Monterey Formation. In addition, there are prominent white- or yellowish-weathering dolomite beds 0.5—1 m thick; these commonly have films of dried asphalt adhering to joint surfaces. The total thickness of the Point Sal Formation is about 330 m in the area we mapped but, according to Hall and

Corbató (1967), in the surrounding region it ranges from less than 85 to 600 m.

Monterey Formation (Middle and Upper Miocene). The Monterey Formation rests with gradational contact upon the Point Sal. For mapping purposes, we divided the formation informally into three local units, the middle one being slightly thicker-bedded and less porcelaneous than the other two.

The lower unit of the Monterey is thin-bedded, with rhythmic alternations of porcelanite and less dense siliceous shale, each type of layer commonly ranging from 0.5 to 8 cm in thickness. In places opaline chert rather than porcelanite forms the denser layers — or it may occur as thin central portions within the porcelanite beds. All these materials are hard and brittle in varying degrees, but the siliceous shale is porous. The sequences of siliceous beds are interrupted at intervals by dolomite layers and lenses 0.2—1 m thick, like the dolomite in the underlying and overlying rock units. Although lithic layering is extremely conspicuous in this part of the Monterey, bedding joints do not appear until weathering commences; cross-joints also become visible during weathering. The lowest unit of the Monterey is 260—330 m thick.

The middle unit of the Monterey Formation is siliceous mudstone of somewhat blocky appearance as compared with adjacent units. Whereas lithic layers in the lower and upper units are generally 0.5—8 cm thick, in the middle unit they are 15—30 cm thick; moreover, bedding fissility and prismatic jointing are not as marked as in the middle unit, and the weathered rock tends to form small chunks rather than thin plates or prisms. Porcelanite and chert are lacking in the middle unit. Although the siliceous mudstone is quite hard, it is very porous. It is composed of sharply angular silt-size quartz and feldspar grains, minute opaline or (volcanic?) vitreous particles, brown organic specks, and obscure ultra-fine shreds or flakes. The middle unit is not exclusively siliceous mudstone, and at intervals it contains sequences (1—10 m thick) of thin-bedded porcelanite like that of adjacent units. It contains rather uniformly spaced dolomite beds 0.3—1 cm thick. The thickness of the middle unit is more than 235 m.

The upper unit of the Monterey in the area of Fig. 5 is nearly identical with the lower unit, consisting mainly of thin-bedded siliceous shale and porcelanite. It contains dolomite beds 0.3—1 m thick, and a few sandstone beds ranging from 5 cm to perhaps 2 m in thickness. This unit is 380—450 m thick.

The foregoing descriptions do not quite span the entire stratigraphic column of the Monterey, as the uppermost part of the formation lies beyond the area we mapped. In the surrounding region, the thickness of the entire Monterey Formation ranges from 760 to more than 1830 m (Hall and Corbató, 1967).

Forms of the folds

In order to estimate shapes and relative positions of folds in our area, we mapped attitudes of layers at many places (Fig. 5*) and constructed a down-plunge, or composite, structural cross-section (Fig. 6*). Plunges of the folds were determined by the usual method, utilizing a stereographic net, and are indicated along traces of axial planes in Fig. 5. Most of the data were projected to the structural section by assuming a plunge of 30 degrees northwest. However, the data for the northwestern part of the area were projected along local plunge directions to points of intersection with the cross-section. The resulting composite section is only an approximate representation of the structure of the area; in constructing such a section one assumes that the structures are cylindroidal, which they are not. Nevertheless, it provides an opportunity to examine fold patterns through a stratigraphic section of about 1.8 km thickness.

The fold in the center of the composite section is the Huasna syncline. The syncline, at middepth in the section, within the Monterey Formation, is a chevron form, with a sharp hinge and an interlimb angle of about 90 degrees (Loc. *B*, Fig. 4). It grades upward into a more open chevron form, with a sharp hinge but an apical angle of about 120 degrees near the top of the Monterey and the base of the sandstones of the Santa Margarita (Loc *A*, Fig. 4). Most of the syncline within the Santa Margarita Formation is an open fold with a rounded hinge. It appears to grade downward into a more rounded hinge within the massive sandstone of the Vaqueros; however, control of the fold form within the Vaqueros is poor. Thus, the Huasna syncline appears to grade from a relatively rounded fold at depth to a sharp-hinged, chevron fold at middepth, to a relatively rounded fold near the surface.

On either side of the Huasna syncline, within the Monterey, are smaller synclines and anticlines with wavelengths on the order of 500—750 m (Fig. 6). The folds shown in the center left of the composite section are clearly exposed in a cliff along the Cuyama River (Loc. 2c, Fig. 5; Fig. 7). The spacing between crests of the two anticlines is about 500 m and the folds extend through a section of about 1.2 km, so the ratio of wavelength to total thickness is on the order of 0.42. The folds are nearly classic, concentric-like folds, with sharply bent layers in the cores and open outer edges. The sharpness is more prominent in the core of the two anticlines than in the syncline (Figs. 6 and 7). The folds grade upward, toward the contact with the Santa Margarita, into smooth, broad warps. The gradation of fold forms downward into the Point Sal is not clearly visible in most places. However, throughout the area the Point Sal and the lower part of the Monterey are folded intensely, as indicated by discordant attitudes of bedding within and between small outcrops.

The intense folding within the Point Sal and the lower part of the Monterey

---

See fold-out pages 293—296.

Fig. 7. Oblique aerial view of concentric-like and chevron-like folds within Monterey Formation in riverbank exposure, along Cuyama River. View northwest. Concentrations of brush mark layering. Small chevron fold at *A*; asymmetric fold at *B*.

is clearly exposed in the bare, eastern face of The Eagle (Loc. 4f, Fig. 5; Figs. 4 and 8). We mapped folded layers in the rock face, using detailed photographs of various parts of the face, and compiling the results on an oblique aerial photograph. Figure 8 shows traces of the layering on the photograph, roughly oriented so that the view is down the plunge of the larger anticline, the core of which is The Eagle. Thus, Fig. 8 provides an indication of the fold shapes but it incorporates distortions inherent to the imperfect processes used to construct it.

The folds in The Eagle are complex, but certain patterns are recognizable. Those near the center of the exposure are combinations of symmetric, chevron forms and rounded, concentric-like forms (Fig. 9A). Those on either side are asymmetric. On the southwest side (left in Fig. 8), the sense of asymmetry is that one would expect if layers above moved toward the axis of the anticline relative to layers below, after chevron folds had developed (e.g. Ramberg and Johnson, Part V). On the northeast side (right in Fig. 8), the sense of asymmetry is opposite.

A rather symmetric chevron fold (Fig. 9B) is exposed in the lower part of the Monterey Formation along the Cuyama River (Loc. *A*, Fig. 7). An asym-

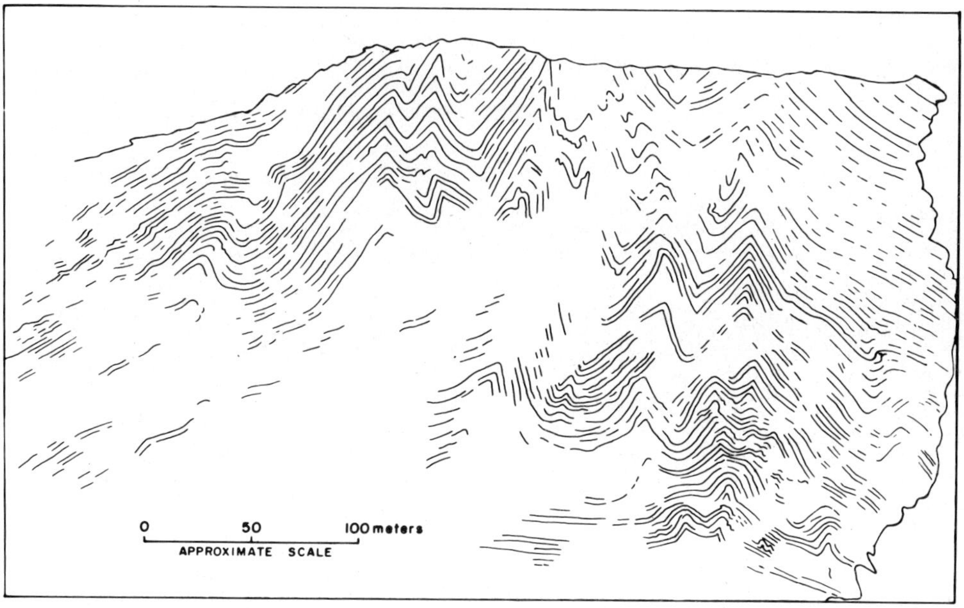

Fig. 8. Controlled sketch of fold patterns at The Eagle (Loc. 4f, Fig. 5). View approximately down plunge of large fold axis. Folds are within upper part of Point Sal and lower part of Monterey formations.

metric chevron fold or kink band is exposed nearby, on one limb of a larger anticline (Loc. *B*, Fig. 7; also, Fig. 6).

Tiny kink bands and faults are common within siliceous shales of the Monterey Formation in the *riverbank exposure* (Loc. 2c, Fig. 5; Fig. 7). The kink bands range in width from a few mm to a few cm and the layers involved are paper-thin laminae of shale. Such structures are absent in the large roadcut exposures of the blocky and thin-bedded units of the Monterey Formation.

Field evidence for shear in folded rocks

Slip parallel with bedding. One of the structural problems of the area is how the rocks were deformed during folding. Possibilities include relatively homogeneous strain, slip between mechanical layers, and slip along small faults and joints. All the rock units exposed in the area of detailed study, with the exception of the Vaqueros Formation, are conspicuously stratified. One of the well-known features of the Monterey Formation is its predominantly thin, rhythmic bedding. From a mechanical point of view, one might instinctively equate the formation with a deck of playing cards. Yet, surprisingly, there is no visible evidence of slip between most of the strata; in fact, we have concluded that slip surfaces are few and widely spaced. The

Fig. 9. Chevron-like folds. A. Upper part of The Eagle. Dark tree in left foreground is about 4 m high. B. Small fold within lower part of Monterey Formation at location *A*, Fig. 7, within the riverbank exposure.

most instructive exposures are in unweathered rock in a man-made channel along the Cuyama River (Loc. 2-3f, Fig. 5) and in deep highway cuts in the west part of the area (Locs. 3b and 3c, Fig. 5). These fresh exposures show thin bedding delineated by rhythmic lithic variations. With few exceptions, the beds are bonded together; most shaly intercalations show no evidence of shear (Fig. 10). No slickensides, slip surfaces, or cataclasis are visible except at widely spaced intervals normal to bedding. Even bedding joints are lacking in fresh exposures.

The few surfaces along which bedding-plane slip was observed in highway cuts are, on the average, 30—35 m apart measured normal to bedding. These surfaces are quite different from the predominant tightly bonded bedding planes, being marked by slickensides, 1—3 cm of gouge, and fracturing of the adjacent rock. In addition to these slip surfaces, within the same exposures there is about an equal number of small faults oriented at an acute angle to the bedding. The faults may have played almost the same role as the surfaces of bedding-plane slip, or they may have formed during growth of the folds. These small faults each show 4—10 cm of apparent slip in the appropriate sense for contemporaneity with folding; that is, the strata above the fault surface in each case moved toward the axis of an anticline.

Faulting and jointing. In addition to the small faults which make acute angles with bedding, there are a few small faults oriented at large angles to the bedding. The dip separation on individual faults ranges from 1 cm to 3 m. Two larger faults were seen or inferred, and are shown in Fig. 5. They are steeply dipping, and have displacements of tens or hundreds of meters. It is unlikely that the few larger faults, or the smaller faults at large angles to bedding, are related to folding, as the sense of displacement is generally inappropriate.

Bedding joints are conspicuous in natural outcrops of all rock units except the Vaqueros Formation, and they might be supposed to play a part in slip parallel with stratification. However, they are not seen in fresh artificial exposures and become visible only after the onset of weathering. Because of their virtual absence in fresh rock, and because slickensides are lacking, these joints are interpreted to be unrelated to the strain accompanying folding.

Cross-joints normal to bedding are characteristic of natural outcrops of the Monterey Formation. Some occur in conjugate sets, each joint transecting several beds, the spacing being on the order of 2—20 cm. Other more closely spaced joints are restricted to the most siliceous rock layers; in some places they develop in chert or porcelanite without extending into adjacent siliceous shale intercalations. The various kinds of cross-joints tend to subdivide the rock into prisms whose sides are normal to the bedding. Although we have seen instances elsewhere where joint-bounded prisms or blocks have been collectively tilted by shear parallel with bedding, such is not the case in the study area. Not only are the prisms undisturbed, but they, like the bedding joints, are non-existent or non-apparent prior to weathering.

Fig. 10. Characteristics of bedding within siliceous shale of the Monterey Formation exposed in large roadcuts west of riverbank exposure. A. Thin-bedded siliceous shale in most of exposure. Blocky siliceous mudstone in lower right of exposure. B. Layers and joints in thin-bedded siliceous shale. Dark layers are porcelaneous; light layers are siliceous shale.

We conclude that the faults, except perhaps those at acute angles to bedding, and virtually all the joints, are of no significance to the folding in the area of this study. The widely separated surfaces of bedding-plane slip, and perhaps the faults at acute angles to bedding, were the loci of significant slippage during the folding process.

ANALYSIS OF FOLDS AT SOUTH END OF HUASNA SYNCLINE

The field observations reported in previous pages pose the following problems which we have chosen to analyze: The penecontemporaneous development of several scales of folds; the relative scarcity of surfaces of slip or of marked differences of properties of layers within the Monterey Formation; and the common occurrence of doubly plunging folds. Most of the theory required to analyze the problems has been presented in Parts III, IV and V, but the remainder will be developed here. First we will consider the folding of a multilayer confined by soft media. The theory is an extension of that developed for multilayers confined by rigid boundaries (Biot, I *, 1964; Johnson, I, 1969). Then we will use that theory to develop a method of predicting the simultaneous formation of folds at different scales.

Ramberg wavelength for multilayer in confining media

Soft Interbeds. Let us consider the folding of a multilayer of interbedded stiff (1) and soft (2) layers confined above and below (Fig. 11A) or merely below (Fig. 11B) by infinitely thick confining media with modulus, B_0. The basic differential equation for the displacement, v, normal to layers, is (e.g., Johnson, I, 1969; I, 1970; or Part IV, eq. 35):

$$B_1 I_1(\partial^4 v_1/\partial x^4) + (P - tbG_a)(\partial^2 v_1/\partial x^2) - tbB_v(\partial^2 v_1/\partial y^2) = -P(\partial^2 v_0/\partial x^2) \quad (1a)$$

where v_0 is the initial sinusoidal deflection, before axial loading,

$$v_0 = \delta_0 \sin(lx) \cos(\pi y/nT) \quad (1b)$$

$$l = 2\pi/L \quad (1c)$$

L is wavelength and δ_0 is amplitude of initial waves,

$$I_1 = bt_1^3/12$$

$$t = t_1 + t_2$$

$$G_a = tG_1G_2/t_1G_2/(t_1G_2 + t_2G_1)$$

$$B_v = tB_1B_2/(t_1B_2 + t_2B_1)$$

$$B_v' = B_v(t/t_1)^3 \quad (1d)$$

---

* References marked by an I, followed by a date, are cited in Part I and will not be repeated in the list of references of this Part.

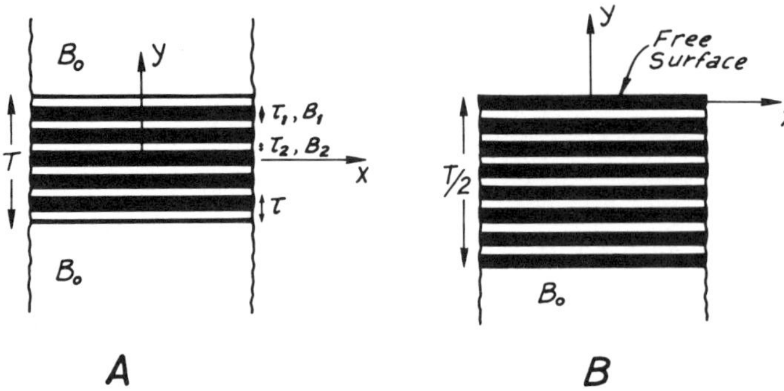

Fig. 11. Multilayers confined by elastic media. A. Multilayer confined above and below by media. B. Multilayer confined below by medium but with free surface above. Thickness, T, used in calculations is twice the actual thickness of the multilayer.

G_i is shear modulus, t_i is thickness of stiff or soft layers, b is unit width,

$$B_i = E_i/(1 - \nu_i^2) \tag{1e}$$

and E_i is Young's modulus and ν_i is Poisson's ratio for stiff, $i = 1$, or soft, $i = 2$, layers, or for the medium, $i = 0$.

The relevant sinusoidal solution of eq. 1a is (e.g., Johnson, I, 1970, chapter 6):

$$v = v_0 + v_1 = [\delta_0/\{1 - (P/P_{\text{crit}})\}] \sin(lx) \cos(\pi y/nT) \tag{2a}$$

where P_{crit} is the critical axial load,

$$P_{\text{crit}} = tbG_a + B_1 I_1 l^2 + tbB_v(\pi/nTl)^2 \tag{2b}$$

in which l is the wavenumber defined in terms of wavelength, L, in eq. 1c.

Here n is a factor that depends upon the rigidity of the confining media and upon the wavelength of the fold. It will be derived in following paragraphs. It ranges from 1 for rigid media to infinity for no media. Here T is the total thickness of the multilayer confined above and below by media (fig. 11A) or it is *twice* the thickness of a multilayer confined below by a medium and having a free upper surface (Fig. 11B).

As usual in folding theory of elastic materials, we compute the wavelength, L, that requires the minimum critical axial load by differentiating eq. 2b with respect to L, and by setting the result equal to zero. The resulting, *Ramberg wavelength* is:

$$L_R = 2T\sqrt{\pi n/N_p}\,\sqrt[4]{(B_1/12B_v')/(1 - D)} \tag{3a}$$

where N_p is the number of pairs of stiff and soft layers and:

$$D = 1/[1 + (\pi/2n)\{\text{ctn}(\pi/2n) + \tan(\pi/2n)\}] \tag{3b}$$

which is related to the factor n, derived by equating the normal stresses at the frictionless interface between the multilayer and the medium:

$$\underset{\text{multilayer}}{(\partial v_1/\partial y)B_v} = \underset{\text{medium}}{-(B_0 l/2)v_1} \quad {}_{y=\pm T/2}$$

Thus:

$$n \operatorname{ctn}(\pi/2n) = (B_v/B_0)(L_R/T) \quad (3c)$$

The amplitude of folds at the edge, δ_e, of the multilayer is proportional to the amplitude of folds at middepth, δ:

$$\delta_e = \delta\cos(\pi/2n) \quad (3d)$$

The *Ramberg wavelength* is derived by solving eqs. 3a—3c iteratively, because the equations are nonlinear. Equation 3a is the most general, approximate solution we have derived thus far for wavelengths of folds in elastic multilayers. It should be accurate for a wide range of conditions for multilayers containing at least ten layers (see Part IV, fig. 5). That equation and another, for the folding of widely separated stiff layers (eq. 26, Part III), should be adequate for predicting wavelengths for most simple multilayers.

The *Ramberg wavelength* reduces to the *Biot wavelength* (Johnson, I, 1969):

$$L_B = 2T\sqrt{\pi/N_p}\sqrt[4]{B_1/12B_v'} \quad (4)$$

if the media are rigid, that is, is the modulus of the medium approaches infinity. Thus, if $B_0 \to \infty$, $n \to 1$, and $D \to 0$, eq. 3b, so that eq. 4 is a special case of eq. 3a.

Frictionless contacts. Where folds in multilayered rock form under conditions of high pore pressures, effective strengths of contacts between layers are very low. Evidence is available that suggests that rocks comprised of interbedded sandstone and shale, or of interbedded porcelaneous shale and siliceous shale, or of interbedded porcelaneous shale and siliceous shale, behave much as homogeneous layers with slippery contacts (e.g., Honea, 1975; Ramberg and Johnson, Part V). The theory of folding of multilayers with zero contact strength can be derived readily from the theory in preceding paragraphs. If the multilayer consists of homogeneous layers with frictionless contacts, the thickness, t_2, of soft layers vanishes so that $G_a = 0$, and $B_v = B_1$. Thus, the *Ramberg wavelength*, eq. 3a, becomes:

$$L_R = 2T\sqrt{\pi n/N}\sqrt[4]{1/[12(1-D)]} \quad (5a)$$

and the critical axial stress, $P_{crit}/t_1 b = \sigma_{crit}$, becomes (eq. 2b):

$$\sigma_{crit} = B_1[(t_1^2 l^2/12) + (\pi/nTl)^2] \quad (5b)$$

The factor, n, (eq. 3c) becomes:

$$n \operatorname{ctn}(\pi/2n) = (B_1/B_0)(L_R/T) \tag{5c}$$

but the factor D is unchanged (eq. 3b). Here N is the number of layers.

We have solved eqs. 5a, 5c and 3b simultaneously for values of relative stiffness, B_0/B, of medium to layers, ranging from 0.1 to infinity, and for numbers, N, of individual stiff layers ranging from 10 to 1000 (Figs. 12A and 12B). The results are plotted in terms of ratios of *Ramberg wavelength* to total thickness of the multilayer, L_R/T, but can be converted readily to ratios of wavelength to thickness for individual stiff layers:

$$L_R/t_1 = N(L_R/T) \tag{5d}$$

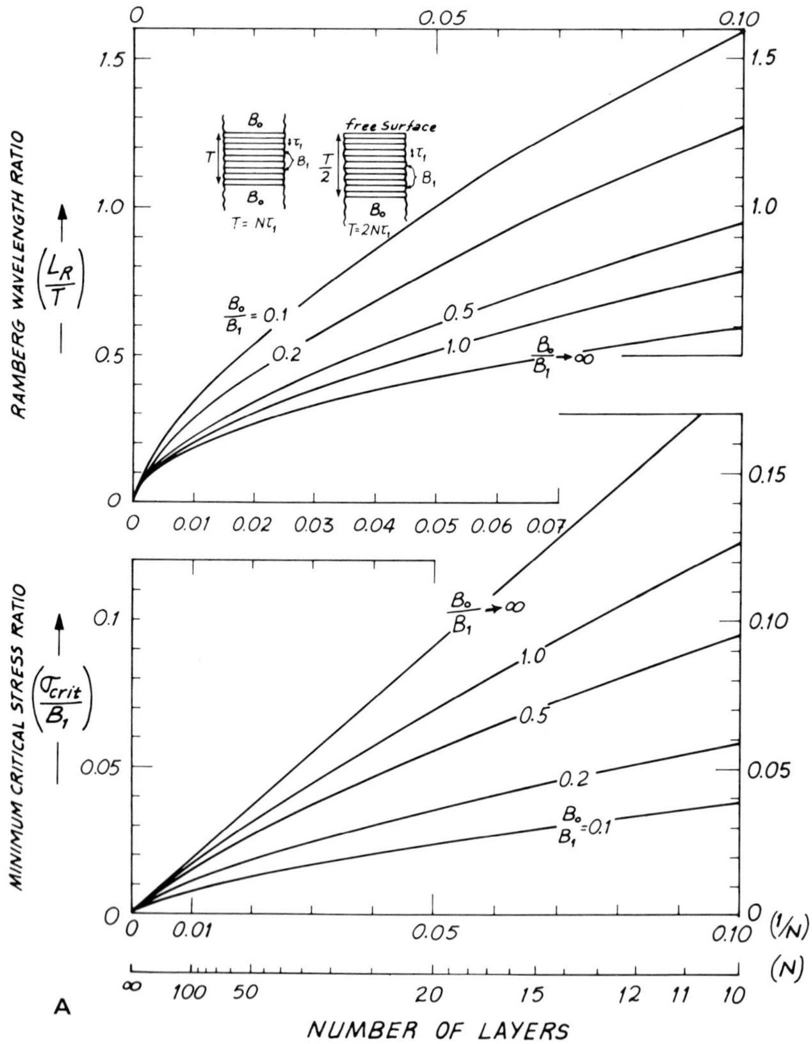

Fig. 12A.

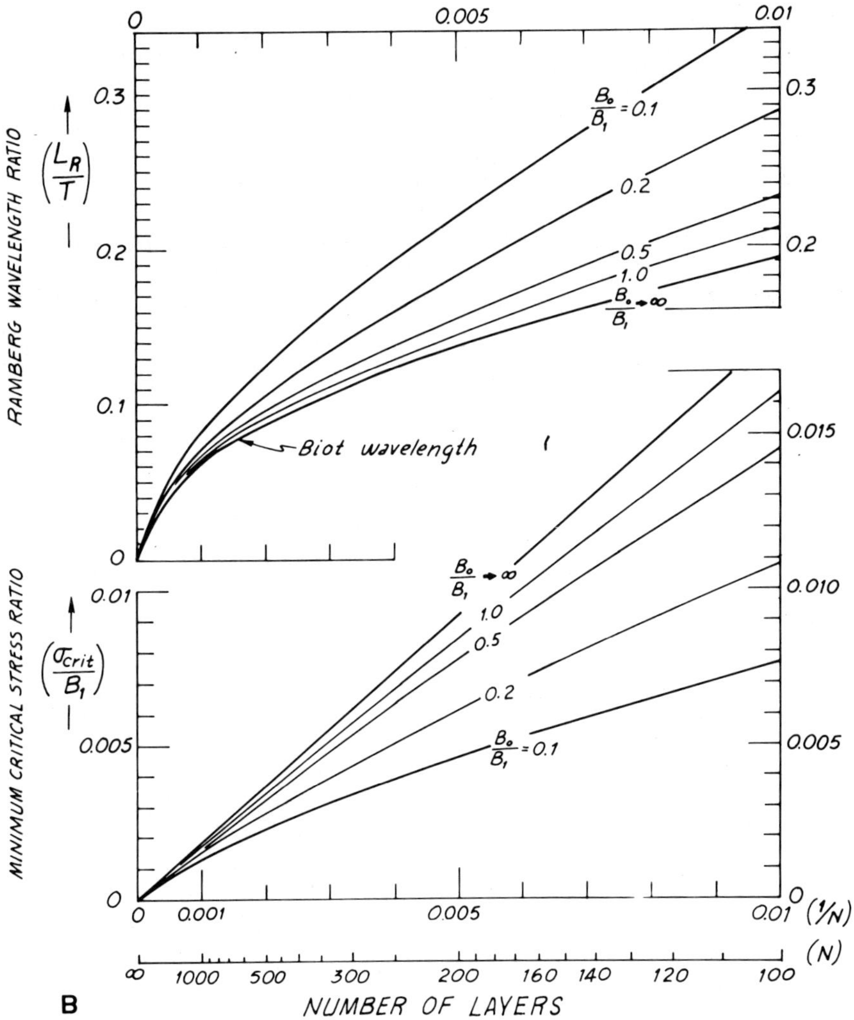

Fig. 12B.

Fig. 12. Relations between ratio of *Ramberg wavelength* and total thickness, and number of layers, and between minimum critical stress ratio and number of layers. A. Relations for number of layers between 10 and 100. B. Relations for number of layers between 100 and 1000.

for a multilayer stack confined above and below by media, or:

$$L_R/t_1 = 2N(L_R/T) \tag{5e}$$

for a multilayer with a free upper surface, but confined below by a medium.

Relations between minimum critical axial compression and numbers of layers, for various modulus ratios, computed by substituting eqs. 5a, 5c and 3b into eq. 5b, are also shown in Figs. 12A and 12B. These figures will be useful in following paragraphs for the discussion of different scales of folds developed simultaneously. The figures show that the wavelength ratio is quite sensitive to the modulus ratio of the layers and medium if the medium is softer than the layers, but that it is insensitive to the modulus ratio if the medium is stiffer than the layers. For example, if there are twenty layers within the multilayer, $1/N = 0.05$, and according to Fig. 12A, if the modulus ratio is unity, the wavelength ratio is 0.52. If the modulus of the medium approaches infinity the wavelength ratio decreases about 20% to a value of 0.42, and if the modulus of the medium is reduced to one-tenth of that of the layers, the wavelength ratio nearly doubles to a value of 1.0.

Simultaneous folding at different scales

Simple multilayers. Folds with different wavelengths can develop simultaneously in complex multilayers comprised of layers with various thicknesses and rheologic properties (Ramberg, I, 1964; Ramberg and Strömgård, I, 1972). In viscous materials, as shown by Ramberg (op. cit.), several orders or wavelengths of folds can develop if folds with different wavelengths grow at about the same rate. In elastic materials, several wavelengths of folds can develop if they require about the same critical axial stress. For example, let us consider the solutions for the buckling of a multilayer confined by rigid boundaries. The wavelength is given by eq. 4 and the critical axial load is (Part IV, eq. 32b):

$$P_B = (2b\pi/T)\sqrt{B_v B_1 (t_1 + t_2) t_1^3/12} + b(t_1 + t_2) G_a \tag{6}$$

The moduli, B_v and G_a, depend upon the thicknesses and elastic properties of the stiff and soft layers. Now, if there were a series of multilayers separated by very stiff media, the different multilayers could fold simultaneously, with different wavelengths, if the dimensional and rheological properties of the multilayers satisfy certain conditions. The conditions are more clear, albeit more restrictive, for multilayers of homogeneous materials with frictionless contacts. In this case the wavelength and critical axial stress depend on only three parameters:

$$L_B \simeq 1.9\, T\sqrt{1/N} \simeq 1.9\, t_1\sqrt{N} \tag{7a}$$

$$\sigma_B = P_B/(bt_1) = [\pi/\sqrt{3N}]B_1 \tag{7b}$$

That is, the critical axial stress is inversely proportional to the number, N, of layers in the multilayer and directly proportional to the modulus of the layers. The wavelength is proportional to the thickness of individual layers and to the square root of the number of layers. Thus, folds could form simultaneously in several multilayers if the multilayers contained the same number of layers with identical elastic properties, because the critical stress would be the same for all the multilayers. The wavelengths would be different if the thicknesses, t_1, of individual layers were different in the adjacent multilayers.

The same general conclusions apply for multilayers with frictionless contacts confined by media with various properties (Figs. 12). As shown in Figs. 12, for a certain critical axial stress, multilayers with different numbers of layers and different modulus ratios, B_0/B, can fold simultaneously, and the folds can have a variety of wavelengths. For example, suppose a multilayer containing thirty frictionless layers and confined by media with the same properties ($B_0/B = 1.0$) was folded, and we wish to determine the properties of other multilayers which would fold simultaneously. According to Fig. 12A, the critical axial compression ratio is 0.048, and multilayers containing thirteen layers with a modulus ratio of 0.2, and those containing 24 layers with a modulus ratio of 0.5 should fold simultaneously. The ratios of wavelength to total thickness and wavelength to individual layer thickness for these three multilayers should be:

L_R/T	B_0/B	N	L_R/t_1
0.4	1.0	30	12
0.53	0.5	24	13
1.08	0.2	13	14

If the modulus ratio for several multilayers were the same, the multilayers containing the same number of layers would fold under the same state of stress. The ratio of wavelength to total thickness in this case should be the same but the wavelengths could be different for the multilayers if the thicknesses of individual layers were different. Thus, such a multilayer with thirty layers, each 1 mm thick, should fold with a wavelength of about 12 mm and another multilayer with 30 layers, each 10 m thick, should fold with a wavelength of 120 m, under the same state of stress.

Multilayers containing layers of different thicknesses. The theory developed thus far helps us to understand conditions of simultaneous development of folds of different wavelengths but it is restricted to multilayers within which each stiff and each soft layer is identical. The layers involved in folding of the Huasna syncline, however, are definitely of different thicknesses, so now we will develop an approximate theory for such multilayers. We will simplify the problem and consider multilayers of homogeneous materials with frictionless contacts. In this case the differential equation for an element of a stiff layer becomes:

$$B_1\bar{I}_1(\partial^4 v_1/\partial x^4) + \sigma_a(\partial^2 v_1/\partial x^2) - B_1(\partial^2 v_1/\partial y^2) = -\sigma_a(\partial^2 v_0/\partial x^2) \quad (8a)$$

where:

$$B_1\bar{I}_1 = B_1(1/12N)\sum_N t_i^2 \quad (8b)$$

is the average bending rigidity and N is the number of layers. We note that the bending rigidity depends upon the square of the thickness of layers, not upon the cube, as assumed by Currie et al. (I, 1962) and by Johnson (I, 1970, p. 152). They assumed that each layer is folded with the same amplitude, an approximation which requires that the media be much softer than the multilayer. The theory derived here should be approximately correct for multilayers confined by stiff or by soft media.

The critical axial stress is the same as computed before, eq. 5b, except that t_1^2 is replaced by the average of the squares of the thickness of stiff layers:

$$\sigma_{\text{crit}} = B_1[(l^2/12N)\sum_N t_1^2 + (\pi/ln\sum_N t_i)^2] \tag{9a}$$

The *Ramberg wavelength*, which requires the least critical stress, is derived as before and is:

$$L_{\text{R}} = 2\sqrt{\pi n(\sum_N t_i)}\sqrt[4]{(1/N)(\sum_N t_1^2)/[12(1-D)]} \tag{9b}$$

where D is defined as in eq. 3b and n is defined by:

$$n\ \text{ctn}(\pi/2n) = (B/B_0)(L_{\text{R}}/\sum_N t_i) \tag{9c}$$

The wavelength defined by eq. 9b is necessarily approximate, because it assumes symmetry of fold forms above and below middepth in the multilayer. Nevertheless, eq. 9b should provide first estimates of wavelengths in complex multilayers within which the distribution of thicknesses of layers above and below middepth is the same. Similar equations can be derived for multilayers of interbedded stiff and soft materials by inspection of eqs. 9a—9c and eqs. 2b, 3a and 3c.

For some purposes the lengthy procedure of solving eqs. 3b, 9b and 9c can be avoided. We define the *effective number* of layers as the number of layers of equal thickness, t_1 = constant, in a multilayer which develops the same wavelength as a multilayer with layers of unequal thickness, t_i = variable. We also specify that the total thickness, T_{T}, of the multilayer, with variable t, and of the multilayer with constant t are the same. For example:

$$T_{\text{T}} = N_1 t_1 + N_2 t_2$$

if there are only two thicknesses of layers in the complex multilayer. The equivalent thickness squared for individual layers in the multilayer with variable t is:

$$(T_{\text{T}}/\overline{N})^2 = (N_1 t_1^2 + N_2 t_2^2)/(N_1 + N_2)$$

where $\overline{N}$ is the effective number of layers within the multilayer of constant t which folds with the same wavelength as the multilayer with variable t. Solving these two equations for $1/\overline{N}$:

$$1/\overline{N} = [\sqrt{(N_1 t_1^2 + N_2 t_2^2)/(N_1 + N_2)}]/(N_1 t_1 + N_2 t_2) \tag{9d}$$

Thus, in order to solve for the ratio of wavelength to total thickness for the multilayer with variable t, we would compute the effective number of layers with eq. 9d and use Figs. 12.

For multilayers containing several thicknesses of layers, we would use the equations:

$$(1/\bar{N}) = [\sqrt{(\Sigma_N t_i^2)/N}]/T_{\mathrm{T}}; \qquad T_{\mathrm{T}} = \Sigma_N t_i \tag{9e}$$

where N is the actual number of layers and $\bar{N}$ is the effective number of layers in the multilayer containing layers of uniform thickness.

Figure 13 shows five multilayers, confined by media with the same properties as the layers, which would fold with the same wavelength and would require the same magnitude of axial stress for folding. Fig. 13A is a simple multilayer, containing 20 layers of uniform thickness. The effective number, $\bar{N}$ of layers is equal to the actual number. Figure 13B shows a multilayer which would fold with the same wavelength, but the layers are of different thicknesses. One layer is three times as thick as those in the simple multilayer, and the remaining layers are somewhat thinner than those in the simple multilayer. The actual number of layers is 27 but the effective number of layers is 20. The multilayer shown in Fig. 13E contains two thick layers. The

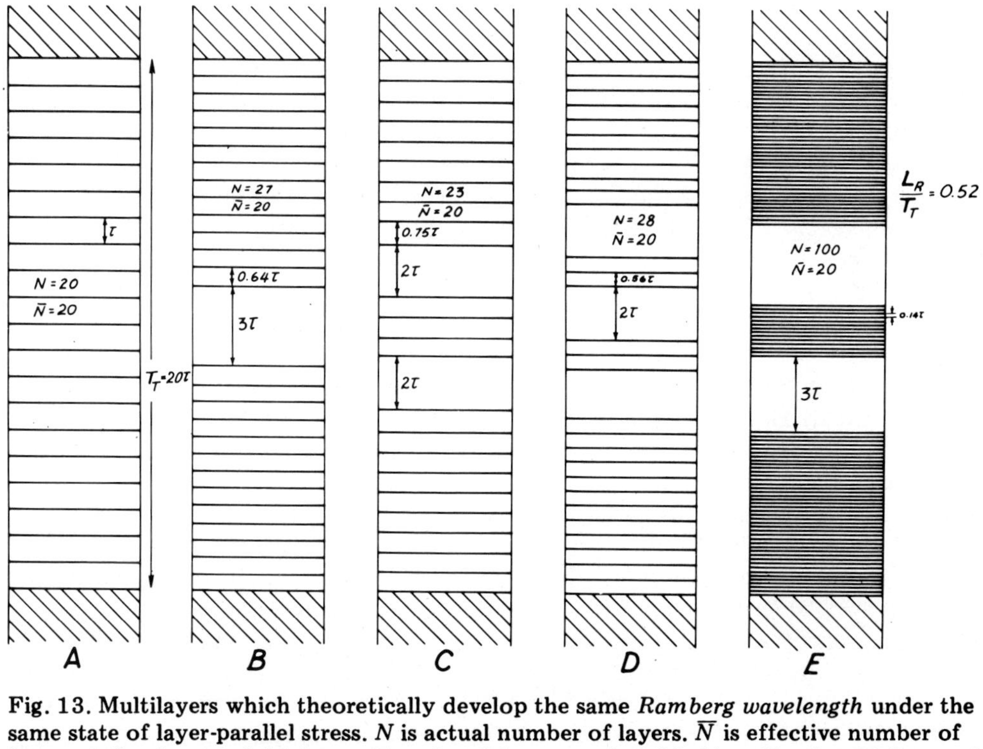

Fig. 13. Multilayers which theoretically develop the same *Ramberg wavelength* under the same state of layer-parallel stress. N is actual number of layers. $\bar{N}$ is effective number of layers, defined as total thickness, T_{T}, of multilayer stack divided by effective thickness of individual layers. For the examples shown here, the effective thickness of layers in multilayers B—E is equal to the thickness of individual layers in the simple multilayer shown in A.

actual number of layers is 100, but the effective number is 20. Thus, merely by changing the dimensional properties of a few layers in a multilayer one can markedly change the effective number of layers.

This analysis emphasizes the remarkable variability of relations among wavelength, layer thickness, and total thickness for a multilayer. Clearly, selection of suitable values for the variables in a field situation demands excellent exposures, detailed observations, and a set of reasonable criteria for identifying wavelength, total thickness and thicknesses of individual layers. We will discuss a particular application in the following section.

Folds in riverbank exposure

The folds at the south end of the Huasna syncline range widely in scale, with wavelengths of a few mm within certain units of the Monterey Formation, to about 1 km for the large fold at The Eagle. We have sufficient information to make some estimates of the dimensional properties of the layers involved in some of these folds, and for others we can place limits on possible dimensional properties.

The forms of the folds in the riverbank exposure (Loc. 2c, Fig. 5; Fig. 6) are the best known. These folds are concentric-like, and include one syncline and two anticlines, involving all three units of the Monterey Formation. They die out downward within the top of the Point Sal Formation, and upwards within the top of the Monterey or the bottom of the overlying Santa Margarita (Figs. 6 and 7). Their wavelength, at middepth, is about 500 m, and the thickness of rock involved in folding is estimated to be about 1200 m so that the ratio of wavelength to total thickness is about:

$$L/T_{\mathrm{T}} = 0.42$$

Let us assume that the rocks involved in the folding can be represented approximately in terms of homogeneous, elastic layers with frictionless contacts, and that the rocks above and below behaved as homogeneous, elastic media, with thicknesses somewhat greater than the wavelength of the folds. The thickness of overburden at the time of folding was about 2 km or about four times the wavelength of the folds. We presume that pore-water pressures were high at the time of folding so that strength of contacts was zero for certain structural layers within the Monterey Formation. In order to apply the theory we have developed, we must assume, further, that the folds were a response to layer-parallel shortening of the rocks, and that the wavelengths observed are those requiring the least critical shortening, or axial stress. Then, according to the theory, a ratio of wavelength to total thickness of 0.42 corresponds with about $\overline{N}$ = 29 layers of equal thickness (Fig. 12A). That is, the effective thickness, t_1, of individual layers was:

$$t_1 = T_{\mathrm{T}}/N = 41 \text{ m}$$

As explained in earlier pages, however, a multilayer with layers with an effective thickness of 41 m may contain many layers thicker or thinner than this value (e.g., Fig. 13).

We measured average spacings between zones of bedding-plane slip in a large roadcut immediately west of the riverbank exposure, within limbs of one of the folds being considered here (Loc. 3b, Fig. 5). Within 150 m of section of siliceous shale, part of the uppermost unit of the Monterey, we counted five zones of bedding-plane slip, providing an average thickness of structural layers of about 30 m. However, there are also five faults inclined at an acute angle to the layering within this unit, so the structural layers may have been as thin as 15 m for this unit. Within 95 m of section of the blocky, siliceous mudstone, part of the middle unit of the Monterey, we noted only two zones of bedding-plane slip, suggesting a thickness of structural layering of 43 m. If we include all the section studied, 245 m, but only the zones of bedding-plane slip, we calculate a thickness of 35 m, which is close to the theoretical, equivalent thickness of 41 m. The estimate can be made closer by postulating a few thick layers. Thus, although the data are incomplete, we conclude that the field data are consistent with the theoretical predictions for these folds.

Significance of the ratio of stress to elastic modulus

During our field work on the folds at the southern end of the Huasna syncline we noted that faulting is uncommon. There are a few small faults inclined at low angles to bedding and a few inclined steeply to bedding, but the layering in large exposures is strikingly continuous, especially for rocks in California. Thus, we suggest that the rocks generally folded under a stress state that was below that required for development of large faults, which we tentatively define as the faulting strength. Thus, if the rocks were subjected to stresses equal to their faulting strength, they would fault rather than fold, so the magnitude of the critical stress required for folding, relative to the faulting strength of the rock, is an important parameter for evaluating the application of our theory of folding to folds in rock. It is also important to evaluate the relation between faulting strength and laboratory measurements of compressive strength.

According to Fig. 12A, folds in homogeneous layers with a ratio of wavelength to total thickness of 0.42 require a critical stress ratio of:

$$(\sigma_{crit}/B_1) = 0.051$$

where σ_{crit} is the minimal critical stress for folding and B_1 is a modulus which is essentially equal to Young's modulus, E_1 (eq. 13). If the material is incompressible, $B_1 = (4/3)E_1$ and if it is perfectly compressible, $B_1 = E_1$. Therefore, a critical stress ratio of 0.051 would seem to imply stresses in excess of the compressive strength of rock, because ratios of compressive strength to secant Young's modulus for most rock range from 0.003 to 0.006. However, this conclusion is incorrect for folding of many rocks be-

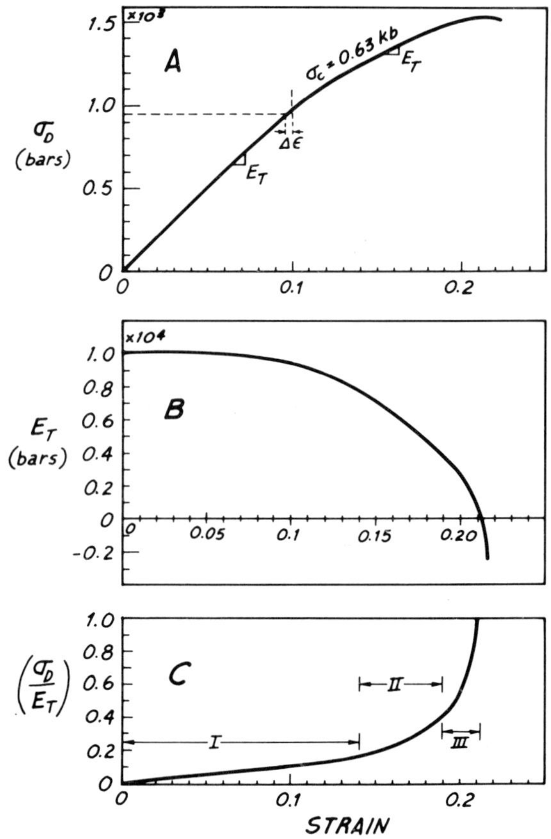

Fig. 14. Relations between axial strain and: A. differential axial stress; B. tangent Young's modulus; C. ratio of differential axial stress and tangent Young's modulus. Test for Chattahoochee sand under confining pressure of 0.63 kbar. Data for A from Vesic and Clough (1968). The stress ratio in C is the value one uses to compute the critical stress for folding of elastic and elastic—plastic layers. It is the stress ratio shown in Figs. 12.

cause the theory considers the *incremental* or *tangent* moduli of materials being folded, not the total or secant moduli (e.g., Part II, fig. 5). The tangent modulus is the local slope of a stress—strain curve for a material, and as the yield point of an elastic—plastic or nonlinear material is approached, the stress—strain curve may flatten, producing a tangent modulus of zero (Fig. 14A). Therefore, the ratio σ_{crit}/B_1 for many rocks can range from a value of 0.003—0.006 to nearly infinity.

Let us examine the ratios of stress to Young's modulus for a variety of rocks. The ratios are derived as follows. Figure 14A shows a stress—strain curve for highly compacted Chattahoochee sand, tested by Vesić and Clough (1968). Much of the deformation of this material is permanent, but our analyses of folding assume merely that small, incremental strains, $\Delta\epsilon$, and incremental stresses associated with the folding, are linearly related for each value of initial stress, σ_D (Fig. 14A and Fig. 5, Part II). The total strains need not be recoverable. The incremental, tangent modulus, E_T, is the local slope of the stress—strain curve and, as shown in Fig. 14B, it ranges from a maximum for low differential stresses, to zero at the peak strength, to even negative values for larger strains. Finally, Fig. 14C shows the ratio of the differential axial stress to the tangent modulus (called the *stress ratio* in following paragraphs) for various values of axial strain of the sample. The curve rises towards infinity as the peak strength is approached. The relation between stress ratio and strain can be characterized by a gently rising curve, zone I, between strain values of zero and about 0.15 for the compact sand (Fig. 14C), by a steeply rising curve, zone III, between strain values of 0.19 and 0.26 for the compact sand, and by a transition zone II. We assume for the moment that the faulting strength and compressive strength are the same. Then, we suggest that if the critical stress ratio for folding falls within zones I or II, the folds can develop. If it falls within zone II, there may be faulting associated even with low-amplitude folding. However, we suggest that if the critical stress ratio for folding falls within zone III, faulting rather than folding will develop. This is a tentative classification of the stress ratio/strain curve, proposed here only for purposes of general discussion of curves for a variety of rock types.

Figures 15 show curves derived similarly from data for several rocks sub-

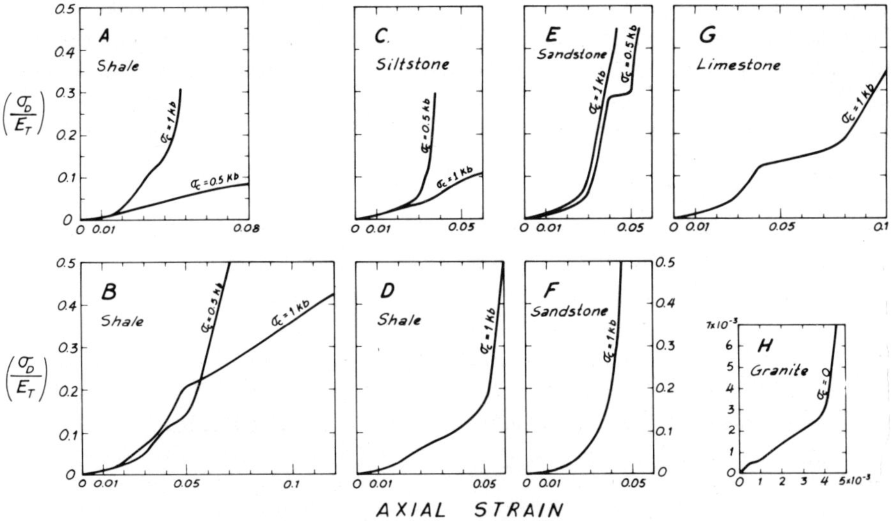

Fig. 15. Relations between axial strain and ratio of differential axial stress and tangent Young's modulus for a variety of rock types. Data for A—G from Handin and Hager, 1957 (H.H.). A. Green River shale shortened normal to bedding (H.H., fig. 26). C. Repetto siltstone (H.H., fig. 30). D. Pennsylvanian shale (H.H., fig. 25). E. Weeks Island sandstone (H.H., fig. 24). F. Barnes sandstone (H.H., fig. 19). G. Marianne limestone (H.H., fig. 14). H. Chelmsford granite (Johnson, I, 1970, fig. 5.15).

jected to different confining pressures. Figures 15A and B are for Green River shale with maximum compression parallel (15A) or normal (15B) to layering (data from Handin and Hager, 1957). The curves show that, even for 0.5 kbar confining pressure, stress ratios greater than 0.1 can be reached before the axial strain approaches the value corresponding with the peak strength, that is, for regions I and II. A stress ratio of 0.051 was computed for the folds in the Monterey Formation in the riverbank exposure at the south end of the Huasna syncline. The other curves in Figs. 15 show that the same conclusion applies to a Pennsylvania shale (15D), the Barnes sandstone (15F) and the Marianne limestone (15G). The Weeks Island sandstone (15E) appears to enter zone III at strains of about 0.028 for 1 kbar confining pressure, and the stress ratio is about 0.75 at that point. Values for Repetto siltstone (15C) enter zone III at stress ratios on the order 0.6 to 1.2 at 0.5 kbar confining pressure. There are, of course, rock types and test conditions that provide stress—strain curves that do not satisfy the requirement of large stress ratios. For example, Fig. 15H shows results for a sample of Chelmsford granite subjected to uniaxial loading (data from Johnson, I, 1970, p. 204). The peak strength corresponds with a strain of about $4 \cdot 10^{-3}$ for this material, and the corresponding stress ratio is about 0.0025, which is more than an order of magnitude smaller than the value of 0.051 calculated for the folds.

Folds at The Eagle

Comparison of theoretical and measured thicknesses of structural layers within the folds in the riverbank exposure at the south end of the Huasna syncline, and comparison of theoretical critical stress ratios with stress ratios for a variety of sedimentary rocks, indicate that the theoretical predictions appear to be consistent with available data. On this basis we will use the theory to learn something about other folds within the Huasna syncline, for which data are quite incomplete.

The folds at The Eagle (Loc. 4f, Fig. 5 and Fig. 8) are within the upper part of the Point Sal and the lower part of the Monterey formations. Those in the central part of The Eagle are chevron-like, with sharp crests and straight limbs (Figs. 8 and 9A). The ratio of wavelength (measured along limbs) to total thickness of layers involved in folding is about $L/T_T = 0.32$ for one series of folds and 0.35 for another. According to Fig. 12A, the effective number of layers and the stress ratio for the two sets of folds are:

L/T_T	N	σ_{crit}/B_1
0.32	48	0.032
0.35	39	0.039

If we assume that the mechanical layers are defined in terms of layers mapped in Fig. 7, the number of layers would be about 85 instead of 48 for one group of folds and about 50 instead of 39 for the other group. Thus, the equivalent mechanical layers at the time of folding apparently were 1.2 to 1.7 as thick as the layers we mapped in the field.

The critical stress ratio for the folds in The Eagle is 0.032—0.039 and that for the folds in the riverbank exposure is about 0.042. We are uncertain whether this difference is significant, but it might be a result of different properties of the rocks involved or a result of different amounts of shortening required for folding. Without knowing the properties of the rock, in terms of the amount of shortening required to produce the folds, we cannot evaluate the stress ratios. For example, suppose that the rocks in The Eagle behaved similarly to Green River shale loaded parallel to layering and confined by 0.5 kbar (Fig. 15A), and that the Monterey rocks in the riverbank exposure behaved as the Repetto siltsone under the same conditions (Fig. 15C). Then, the folds in The Eagle would require layer-parallel shortening of about 0.036, and the folds in the riverbank exposure would require shortening of about 0.026. That is, the folds in the riverbank exposure in this case would develop before those in The Eagle. The relative magnitudes of critical stress ratios, alone, therefore, do not indicate the order in which folds should form unless the properties of the various multilayers are the same.

The chevron-like folds in The Eagle apparently formed after the large anticline containing them began to develop because they are most highly developed near the axial plane of the larger fold. However, folds similar to those in The Eagle were observed at many other places within the Point Sal Formation, both on the limbs as well as near the axial planes of larger folds.

We suggest that the asymmetric folds on either side of The Eagle (Fig. 8) formed as a result of layer-parallel shear of symmetric, chevron-like folds, such as those near the center of The Eagle. The layer-parallel shear, presumably, occurred during growth of the large anticline containing The Eagle. The axial planes of most of the folds within The Eagle, both symmetric and asymmetric, are nearly vertical, which supports our interpretation. Details of the process of producing such asymmetric folds were discussed in Parts V and VI.

Larger folds

The large anticline within which the folds at The Eagle formed has a wavelength of about 980 m. The total thickness of rocks involved in this fold is not known, but it includes the Vaqueros, the Point Sal, the Monterey and, perhaps, part of the Santa Margarita formations. Thus, the minimum thickness involved in the folding is 1.7 km. so the maximum ratio of wavelength to total thickness is 0.58. According to Fig. 12A, the effective number of layers is greater than 16, so the maximum effective thickness of individual layers is about 106 m. There are layers much thinner than this in the folded section, so there must be some relatively thick structural layers that forced the wavelength to be so large. Perhaps the massive sandstone of the Vaqueros is one of these. Its thickness is about 200 m. but we do not know how many structural layers it contains. Alternatively, the fold may extend upwards into the Santa Margarita Formation, so that the total thickness involved in folding is greater than 1.7 km.

Hall and Corbató (1967, pl. 2) show folds with a variety of wavelengths in their cross-sections of the Huasna syncline, but they state that the folds shown at depth are largely speculative. We doubt that the cross-sections are accurate. We would expect the folds within the Monterey Formation beneath the entire syncline to be similar to those at both ends of the syncline. Thus, we will not analyze any of the folds shown by Hall and Corbató (1967, pl. 2) except the Huasna syncline itself. The Huasna syncline is roughly canoe-shaped, with narrow widths at each end and maximal width near midlength. The wavelength at midlength is on the order of 12—16 km. depending upon where one selects the edges of the syncline to be. The fold involves all the Tertiary formations described in earlier pages, with a total thickness of about 2.4 km (Hall and Corbató, 1967, pl. 2), plus at least 1.5 km of Upper Cretaceous, Carrie Creek Formation (Hall and Corbató, 1967, p. 565). The Carrie Creek consists of siltstone, arkosic sandstone and conglomerate. The fold may also involve the Early Cretaceous, Jolo Formation, which is about 900 m thick in the Huasna area. The Jolo consists of mudstone, siltstone, and shale interbedded with sandstone. Franciscan rocks underlie the Jolo. Thus, the total thickness of rocks above the Franciscan which may be involved in the folding is about 4.8 km.

The Huasna fold apparently involves the ground surface, so we use twice the thickness of section in computing the wavelength-to-thickness ratio (Fig. 11B). The L/T_{T} ratio is estimated to be:

$L/T_{\mathrm{T}} = 1.3$ to 1.7

According to Fig. 12A, such a large ratio is not possible for a multilayer confined by a medium with the same properties as the layers themselves, unless the number of layers is substantially less than ten, for which the theory presented here is invalid, as shown in Part IV. Even if we assume the maximum thickness of units, shown in Fig. 2, we obtain wavelength ratios of 0.9—1.2, which are still large.

The large wavelength ratio could be a result of several factors, two of which are:

(1) The medium, the Franciscan, might behave as a much softer material than the overlying rocks. Thus, if the Franciscan were 0.1 times as stiff as the overlying sedimentary rocks, $B_0/B = 0.1$ and the number of layers would be about 11 according to Fig. 12A. However, we see no reason to presuppose the Franciscan to have a much lower modulus than the overlying rocks.

(2) The fold might be a result of a combination of gravity instability as well as of layer-parallel shortening. Many units within the Point Sal and the Monterey have low specific gravities, on the order of 1.4 (C. Isaacs, personal communication, 1975), whereas the sandstones of the Santa Margarita probably have specific gravities of 2.2 to 2.5. The effect of such density inversion,

where there are low-density rocks beneath high-density rocks, is to increase the wavelengths of folds produced by shortening (e.g., Ramberg, I, 1970). The effect of the density contrast between the air and the rocks at the ground surface during folding would be to decrease the wavelength (e.g., Biot, I, 1961; Johnson, I, 1970, p. 158). Therefore, without doing a detailed analysis and without data on thicknesses of units, we are unable to evaluate the magnitudes of the effects of the density inversion on the wavelength.

It is possible, of course, that the Huasna syncline itself is primarily a result of gravity instability or of processes other than layer-parallel shortening, as we have assumed in all our analyses. The smaller folds, however, appear to be a result of layer-parallel shortening.

DEVELOPMENT OF DOUBLY PLUNGING FOLDS

In general in folding theory we assume that folds are two-dimensional, cylindroidal structures. Actual folds terminate, of course, and many folds are known to terminate by plunging at each end; that is, they are *doubly plunging*. Most of the folds in the Huasna area are doubly plunging, so we will develop a theory to investigate the significance of such structures in elastic materials. First we will consider the reaction of a medium to the deflection of a layer resting upon it. The problem is three-dimensional, but it is analogous to the two-dimensional problem solved by Biot (e.g., Johnson, I, 1970, p. 258—266). Then we will consider the development of doubly plunging folds in a multilayer, or in a single layer resting upon the medium, or sandwiched between two media.

Reaction of infinite medium

The problem of the reaction of a medium to sinusoidal folding of a layer on its surface is solved as follows: We assume sinusoidal displacements of one surface of an infinite half-space and compute normal stresses required to maintain those displacements. We begin with the three-dimensional equations of elasticity (e.g., Timoshenko and Goodier, I, 1970, p. 241):

$$G\nabla^2 u + (G + \lambda)(\partial e/\partial x) = 0 \tag{10a}$$

$$G\nabla^2 v + (G + \lambda)(\partial e/\partial y) = 0 \tag{10b}$$

$$G\nabla^2 w + (G + \lambda)(\partial e/\partial z) = 0 \tag{10c}$$

where ∇^2 is the Laplacian operator, e is volumetric strain, and G and λ are Lame's elasticity constants.

Let z be vertical and x and y be in the plane of the surface of the medium (Fig. 16A). Let us assume that the vertical displacement, w, of the surface of the medium can be described in terms of a sine in the x-direction and a sine in the y-direction. Then, the solutions of eqs. 10 must be in the form:

$$u = F(z)\cos(l_x x)\sin(l_y y) \tag{11a}$$

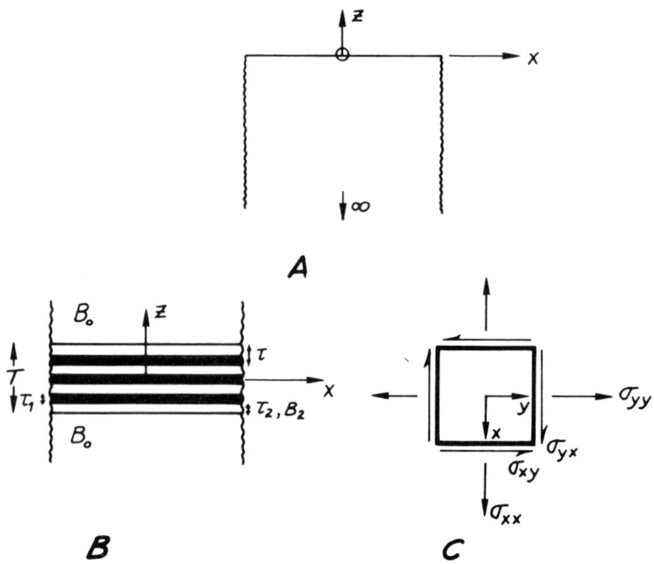

Fig. 16. Definitions of coordinates and stress directions for three-dimensional folding. A. Coordinates for a medium. B. Multilayer confined by media above and below. C. Positive state of stress as viewed down z-axis.

$$v = g(z)\sin(l_x x)\cos(l_y y) \tag{11b}$$

$$w = f(z)\sin(l_x x)\sin(l_y y) \tag{11c}$$

in which f, g and F are functions of z, and l_x and l_y are the wave numbers:

$$l_x = 2\pi/L_x \tag{12a}$$

$$l_y = 2\pi/L_y \tag{12b}$$

L_x is the wavelength of the deflection form for the x-direction and L_y is the wavelength for the y-direction.

Substitution of eqs. 11 into eqs. 10a and 10b, and eliminating $\mathrm{d}f/\mathrm{d}z$ between the resulting equations provide the result:

$$F(z) = (l_x/l_y)g(z) \tag{13}$$

Substituting eqs. 11 into eqs. 10b and 10c, and using eq. 13, we derive:

$$(\mathrm{d}^2 f/\mathrm{d}z^2)(1+\chi) - \bar{l}^2 f - (\chi \bar{l}^2/l_y)(\mathrm{d}g/\mathrm{d}z) = 0 \tag{14a}$$

$$(\mathrm{d}^2 g/\mathrm{d}z^2) - \bar{l}^2(1+\chi)g + \chi l_y(\mathrm{d}f/\mathrm{d}z) = 0 \tag{14b}$$

where:

$$\chi = 1 + \lambda/G \tag{14c}$$

$$\bar{l} = \sqrt{l^2{}_x + l^2{}_y} \tag{14d}$$

Differentiating eq. 14b with respect to z, solving eq. 14a for $\mathrm{d}g/\mathrm{d}z$, and eliminating g between the two resulting equations:

$$(d^4f/dz^4) - 2\bar{l}^2(d^2f/dz^2) + \bar{l}^4 f(z) = 0 \tag{15}$$

The general solution of eq. 15 is:

$$f(z) = (C_1 + C_2 z)\exp(-\bar{l}z) + (C_3 + C_4 z)\exp(\bar{l}z) \tag{16}$$

The displacements are required to vanish as $z \to -\infty$, so $C_1 = C_2 = 0$ in eq. 16, and the relevant solution is:

$$f(z) = (C_3 + C_4 z)\exp(\bar{l}z) \tag{17}$$

Substitution of eq. 17 into eq. 14a, and integrating:

$$g(z) = (l_y/\bar{l}^2)[C_3\bar{l} + C_4(2 + \chi)/\chi + C_4\bar{l}z]\exp(\bar{l}z) \tag{18}$$

and, according to eq. 13, $F(z)$ is directly related to $g(z)$ thus derived. There was an arbitrary constant of integration in the solution for $g(z)$, but it is of no interest to the problem here, so it was set equal to zero.

By substituting eqs. 13, 18 and 17 into eqs. 11, we have the solution for the displacements within the medium. The remaining problems are to evaluate the arbitrary constants and to compute the magnitudes of the stresses. According to elasticity theory:

$$\sigma_{zz} = 2G(\partial w/\partial z) + \lambda e \tag{19a}$$

$$\sigma_{zx} = G(\partial w/\partial x + \partial u/\partial z) \tag{19b}$$

At the surface, $z = 0$:

$$\sigma_{zz} = (\sigma_{zz})_0 \sin(l_x x) \sin(l_y y) \tag{20a}$$

$$\sigma_{zx} = 0 = \sigma_{zy} \tag{20b}$$

where $(\sigma_{zz})_0$ is the amplitude of the normal stress. Substituting eqs. 11a, 11c, 13, 17 and 18 into eqs. 19b, and using 20b, we derive:

$$C_3 = -C_4(\chi + 1)/(\bar{l}\chi) \tag{21a}$$

Substituting eqs. 11, 13, 17 and 18 into eqs. 19a, equating the result to eq. 20a, and using eq. 21a:

$$C_4 = -(\sigma_{zz})_0/2G \tag{21b}$$

Thus, at the surface, $z = 0$, eq. 11c becomes:

$$w = [(\chi + 1)/(2G\bar{l})](\sigma_{zz})_0 \sin(l_x x) \sin(l_y y) \tag{22}$$

Substituting eqs. 20a and 14 into eq. 22, we derive the relation between the normal stress and the deflection of the surface of the infinite half space:

$$\sigma_{zz} = (\bar{l}B_0/2)w;\ \text{at } z = 0 \tag{23}$$

where $B_0 = E_0/(1-\nu_0^2)$, and the zero subscript indicates the medium. This is the expression for the reaction of an infinite half-space to the sinusoidal displacement of its upper surface. Equation 23 reduces to that for a two-dimensional medium in plane strain (Johnson, I, 1970, eq. 7.54) because, for the two-dimensional case, $l_y = 0$ and $\bar{l} = l_x = 2\pi/L_x$.

Folding of single layer within soft media

The differential equations for the three-dimensional buckling of plates are derived and explained in detail by Timoshenko and Gere (I, 1961). The theory is quite similar to that for two-dimensional buckling, so we will not explain it here. Instead, we will proceed directly to the differential equations and their solution. Let us consider a stiff layer that folds in response to principal horizontal stresses, σ_{xx} and σ_{yy}, where compressive normal stresses are negative (Fig. 16B). The differential equation for the deflection of the layer within a medium is:

$$BI'[(\partial^4 w_1/\partial x^4) + 2(\partial^4 w_1/\partial x^2 \partial y^2) + (\partial^4 w_1/\partial y^4)] - \sigma_{xx}[\{\partial^2(w_1 + w_0)/\partial x^2\} + \sigma^*\{\partial^2(w_1 + w_0)/\partial y^2\}]t + \bar{l}B_0 w_1 = 0 \quad (24a)$$

where:

$$I' = t^3/12 \quad (24b)$$

and

$$\sigma^* = \sigma_{yy}/\sigma_{xx} \quad (24c)$$

and the other symbols are defined in previous equations. The initial waveform is assumed to be:

$$w_0 = \delta_0 \sin(l_x x)\sin(l_y y) \quad (24d)$$

The last term in eq. 24a is the reaction of the medium, eq. 23, but it is doubled here because we have assumed that the medium is above and below the stiff layer. The first term is the resistance due to bending of the layer and the middle term is the driving term.

The relevant solution of eqs. 24a and 24d is:

$$w = w_1 + w_0 = \delta_0 \sin(l_x x)\sin(l_y y)/(1 + \bar{\sigma}/\bar{\sigma}_{crit}) \quad (25a)$$

where*:

$$\bar{\sigma} = \sigma_{xx}[(l_x/\bar{l})^2 + \sigma^*(l_y/\bar{l})^2] \quad (25b)$$

$$\bar{\sigma}_{crit} = [BI'(\bar{l}) + B_0/\bar{l}]/t \quad (25c)$$

$\bar{l}$ is a wavenumber defined in eq. 14d:

$$\bar{l}^2 = l_x^2 + l_y^2$$

---

* R.C. Fletcher of Stanford University suggested this form for the variables, which greatly simplifies the analysis and emphasizes the similarity of cylindrical folds and doubly plunging folds.

and $\overline{\sigma}$ is a driving stress. The driving stress is negative, if compressive. Thus, eq. 25a is identical in form to that for cylindrical folding, which is the special case where $\bar{l} = l_x$ and $\sigma^* = 0$.

The wavelength, $\bar{L}m$, where:

$$\bar{L}m \equiv (2\pi/\bar{l}_m) \equiv (2\pi/\sqrt{l_x^2 + l_y^2}) \equiv L_x/\sqrt{1 + (L_x/L_y)^2} \tag{26a}$$

requiring the minimum driving stress, $\bar{\sigma}$, is calculated in the usual way, and is equal to the wavelength for cylindrical folding:

$$\bar{L}m = Lm \equiv 2\pi t(B/6B_0)^{1/3} \tag{26b}$$

The wavelength in the x-direction, which becomes equivalent to the wavelength for cylindrical folding if the wavelength in the y-direction approaches infinity, is:

$$L_x = Lm\sqrt{1 + (L_x/L_y)^2} \tag{26c}$$

or:

$$L_x = Lm\sqrt{1/[1 - (Lm/L_y)^2]} \tag{26d}$$

If y is parellel to the axis of the doubly plunging fold, the upper limit to the wavelength in the orthogonal x-direction is where the fold is dome-shaped, in which case $L_x = L_y$ and*:

$$L_x = Lm\sqrt{2} \approx 1.41\ Lm$$

Figure 17 shows structural contour maps for doubly plunging folds ranging from cylindrical (Fig. 17A) to dome-shaped (Fig. 16D). The wavelengths in the x-direction are correctly proportioned according to eq. 26c, so that the wavelength, L_x, for the some-shaped fold shown in Fig. 17D is about 1.4 times that for the cylindrical fold shown in Fig. 17A.

The expression for the critical stress is also the same as that for cylindrical folding, and is derived by substituting eq. 26c into 25c:

$$\bar{\sigma}_{\text{crit}} = (3/2)\sqrt[3]{BB_0^2/6} \tag{26e}$$

* The referees pointed out that equations similar to these have been derived by Ghosh (1970) and by Dr. Siddans, at the University of Leeds, England. Dr. Siddans' results apparently are in an unpublished doctoral dissertation, and are not available to us. Therefore, we comment only upon the results of Ghosh. Ghosh (1970) derived equations similar to those in preceding paragraphs, for incompressible viscous materials. He also derived the reaction of an infinite elastic medium which is equivalent to that of eq. 23, if one lets Poisson's ratio be 0.5 (see Ghosh's eq. 8). Further, he derived results equivalent to eqs. 26a—26d, but for viscous materials (see Ghosh's eqs. 17 and 18). He did not derive equations analogous to our eqs. 26e—26g, but it is clear that one can perform these analyses for viscous materials if we start with Ghosh's eq. 14, putting it into the form of our eq. 25a. Thus, the results presented here and in following paragraphs are quite general, applying to elastic or viscous single layers or multilayers. Indeed, examination of the source of relations that provide eq. 26g and Fig. 18 indicates that these results could be derived for un unconfined plate or unconfined multilayer so that the results are independent of the relative elasticity or viscosity properties of the layer or layers and the medium.

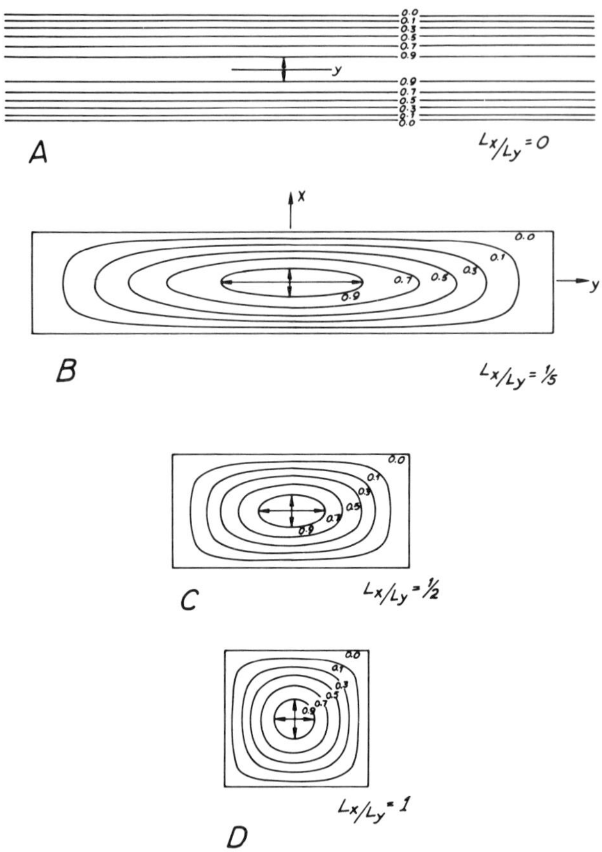

Fig. 17. Structural contour map of a single doubly plunging anticline or syncline, normalized with amplitude of fold. Complete fold pattern developed by reflecting contour map across rectangular boundaries. A. Cylindrical fold in which wavelength in y-direction is infinite. B. Doubly plunging fold with wavelength ratio of 1 : 5. C. Wavelength ratio of 1 : 2. D. Dome-shaped fold with wavelength ratio of 1 : 1.

Thus, as the driving stress, $\bar{\sigma}$, eq. 25b, approaches the critical stress defined by eq. 26e, doubly plunging folds with wavelengths defined by eqs. 26b—26d amplify without bound.

The plan shape of the doubly plunging fold which grows most rapidly, however, depends upon the stress ratio, $\bar{\sigma}^*$, eq. 24c. Examination of eq. 25b indicates that the driving stress, $\bar{\sigma}$, is maximized if the term in brackets in that equation:

$$[(l_x/\bar{l})^2 + \sigma^*(l_y/\bar{l})^2] \equiv [\sigma^*(L_x/L_y)^4 + (1 + \sigma^*)(L_x/L_y)^2 + 1]/[(L_x/L_y)^4 + (L_x/L_y)^2 + 1] \qquad (26f)$$

is maximized. The larger this term, the smaller the magnitude of the stress,

σ_{xx}, normal to the long axis of the fold, required for folding. We calculate the wavelength ratio that maximizes the driving stress in terms of the stress ratio by differentiating the term in eq. 26f with respect to wavelength ratio and setting the result equal to zero:

$$\sigma^* \equiv (\sigma_{yy}/\sigma_{xx}) = [(L_x/L_y)^2 + 2]/[(L_y/L_x)^2 + 2]; \qquad \sigma^* > 0 \tag{26g}$$

If $\sigma^* \leqslant 0$, eq. 26f is maximized if $l_y = 0$; that is, the folds are cylindrical.

Equation 26g is plotted in Fig. 18 for wavelength and stress ratios ranging from zero to one. According to the figure, and to eq. 26g, cylindrical folds ($L_y \approx \infty$) would be expected to develop if the stress ratio approaches zero ($\sigma_{yy} \to 0$). Dome-shaped folds ($L_y = L_x$) would be expected to develop if the stress ratio approaches unity ($\sigma_{yy} \to \sigma_{xx}$). However, only minor compression parallel to the long dimension of the doubly plunging fold can markedly change the wavelength ratio of the fold, producing ratios of 1 : 10 or even 1 : 15. Therefore, the critical stress is relatively insensitive to wavelength ratios less than 1 : 10, so that we would expect initial irregularities with wavelength ratios ranging from zero (cylindrical) to about 1 : 10 to amplify even though the state of stress were uniaxial, with maximum compression

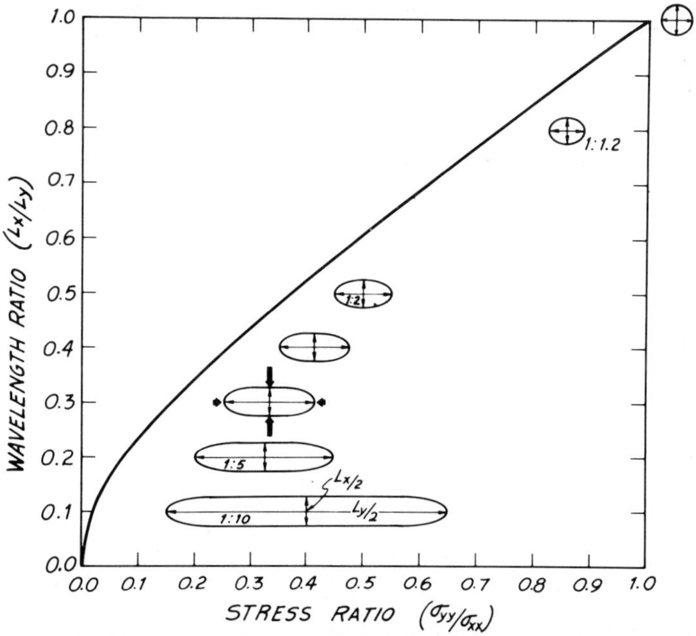

Fig. 18. Theoretical relation between wavelength ratio, L_x/L_y, and ratio of principal stress, σ_{yy}, parallel to long axis of fold and principal stress, σ_{xx}, normal to long axis of fold. Relation is valid for doubly plunging folds in single layer or multilayer, confined by media above and below or merely by a medium below. The relation should be useful for estimating ratios of principal horizontal stresses at the time of folding, for folds with wavelength ratios between about 1: 10 and 1 : 1.

normal, and zero stress parallel, to the long axis of the doubly plunging fold. The theory and Fig. 18 do suggest, however, that plan shapes of folds with ratios between 1 : 5 and 1 : 1 should be quite sensitive indicators of the state of stress at the time of folding.

Folding of multilayer with zero contact strength

In order to complete the analysis of doubly plunging folds, we will show that the solution for a multilayer is similar to that for a single layer and that the solution can be visualized in terms of results already discussed.

The differential equation describing equilibrium of an element of a multilayer with frictionless contacts is:

$$[B_1 t_1^3/(12t)]\nabla^4 w_1 - \sigma_{xx}[\{\partial^2(w_0 + w_1)/\partial x^2\} + \sigma^*\{\partial^2(w_0 + w_1)/\partial y^2\}] - B_v(\partial^2 w_1/\partial z^2) = 0 \quad (27a)$$

where ∇^4 is the fourth-order operator in brackets in eq. 24a, the initial deflection form is:

$$w_0 = \delta_0 \sin(l_x x) \sin(l_y y) \cos(\pi z/nT) \quad (27b)$$

and n is a factor which we will determine subsequently. It ranges from 1 for rigid media to infinity for no media. T is the thickness of the multilayer stack. We assume that the multilayer consists of N_P pairs of identical stiff and soft layers of combined thickness t between which there is zero contact strength.

The relevant solution to eqs. 27a and 27b is:

$$w = w_0 + w_1 = [\delta_0/(1 + \bar{\sigma}/\bar{\sigma}_{crit})] \sin(l_x x) \sin(l_y y) \cos(\pi z/nT) \quad (28a)$$

where the critical stress is:

$$\bar{\sigma}_{crit} = [B_1 t_1^3/(12t)](\bar{l})^2 + B_v(\pi/nT\bar{l})^2 \quad (28b)$$

The factor n is derived by equating the normal stresses σ_{zz} at the interface between the multilayer and the confining media. Thus:

$$n \operatorname{ctn}(\pi/2n) = (B_v/B_0)(\bar{L}/T) \quad (28c)$$

The equation for the factor n for a doubly plunging fold, eq. 28d is different from that for a cylindrical fold, eq. 3c only by the way the wavelength, $\bar{L}$, is defined:

$$\bar{L} = 2\pi/\bar{l} \equiv L_x L_y/\sqrt{L_x^2 + L_y^2} \quad (28d)$$

As before, the wavelength requiring the minimum critical stress is determined by differentiating eq. 28b with respect to wave number and setting the result equal to zero, bearing in mind that the factor n is a function of the wave number, as indicated in eqs. 28c and 28d. The resulting wavelength is:

$$\bar{L}_R = L_R \quad (29a)$$

where L_R is the *Ramberg wavelength* for cylindrical folding, eq. 3a, except that n is defined by eq. 28c. Further, the wavelength in the x-direction is:

$$L_x = L_R\sqrt{1 + (L_x/L_y)^2} \tag{29b}$$

or:

$$L_x = L_R\sqrt{1/[1 - (L_R/L_y)^2]} \tag{29c}$$

Equation 26g and Fig. 18 are valid for single- or multi-layers, so all the conclusions we drew concerning plan shapes of doubly plunging folds in a single layer are valid for multilayers confined by elastic or rigid media. Therefore, we simply reiterate the conclusions, that folds approaching symmetric dome shapes in multilayers should not form unless the horizontal stresses are nearly the same, but that doubly plunging folds in multilayers with wavelength ratios of L_x/L_y less than about 0.1 can grow markedly even though the horizontal stress parallel to the long dimension of the fold is zero.

We have not considered effects of principal stresses inclined to the principal, x- and y-axes of folds. Timoshenko and Gere (I, 1961, p. 379) explain some of the steps for such an analysis. However, we need not perform the analysis here because the doubly plunging folds that require the least principal stresses are those oriented with their major axes parallel to the axes of principal stresses. Suppose there is an initial or incipient doubly plunging fold with principal axes inclined to the principal stress directions. If amplitudes are small, the initial fold can be represented in terms of Fourier series of sinusoidal forms referred to the principal stress axes just as the initial form of an irregular cylindrical fold can be expressed in terms of Fourier series (e.g., Johnson, I, 1970, p. 119). Thus, we can analyze the amplification of the sinusoidal components of the initial doubly plunging fold by means of the theory presented in earlier pages. We could show, in this way, that the doubly plunging fold that amplifies most markedly will be the one which requires the minimum critical stress, and that the principal axes of the resulting fold will be parallel to the principal stress axes.

We suggest that this is one way to produce sigmoidal-shaped folds, such as the Huasna syncline. Perhaps the initial fold was oriented roughly north—south, as is the axis of the Huasna syncline at midlength, and the initial Huasna syncline grew in length in the northwest—southeast direction, roughly normal to the direction of maximum compression, and roughly parallel with the smaller folds within the Monterey Formation (Fig. 3).

The wavelengh ratio for the Huasna syncline is on the order of 1 : 5 so, according to Fig. 18, the horizontal stress parallel to the long axis the fold was about 0.075 as large as the stress normal to the long axis. The wavelength ratio for a doubly plunging fold at midlength immediately east of the Huasna syncline (Fig. 3) is about 1 : 4, indicating a stress ratio of about 0.12. Apparently, the regional stress state at the time of folding was one of maximum compression in the southwest—northeast direction, and intermediate compression, with a magnitude of about 0.1 that of the maximum, in the northwest—southeast direction. Further, the maximum compression was approximately normal to the traces of the large faults in this part of California.

ACKNOWLEDGEMENT

The research reported in this part was supported partly by the Shell Fund, Department of Geology and the Penrose Fund, Geological Society of America, but largely by the National Science Foundation Grant no. GA-36917. We thank these organizations for their generous support.

We wish to thank Graig McHendrie, who took many of the photographs of folds at The Eagle. To Perfecto Mary we are grateful for preparing the final drafts of the many complicated diagrams.

An early draft of the manuscript was reviewed by Dr. Raymond Fletcher of Stanford University and by Dr. David Pollard of the U.S. Geological Survey in Menlo Park. Their criticisms and editorial assistance greatly improved the manuscript. We are grateful to the referees, Professor Hans Ramberg and Dr. Peter Cobbold, for bringing to our attention the related work by Ghosh and Siddans.

REFERENCES NOT CITED IN CHAPTER 3

Atwater, T., 1970. Implications of plate tectonics for the Cenozoic tectonic evolution of western North America. Geol. Soc. Am. Bull., 81: 3513—3536.

Bailey, E.H., Irwin, W.P. and Jones, D.L., 1964. Franciscan and related rocks, and their significance in the geology of western California. Calif. Div. Mines Geol. Bull., 183, 177 p.

Berkland, J.O., Raymond, L.A., Kramer, J.C., Moores, E.M. and O'Day, M., 1972. What is Franciscan? Bull. Am. Assoc. Pet. Geol., 56: 2295—2302.

Dickinson, W.R., 1970. Relations of andesites, granites and derivative sandstones to arc-trench tectonics. Rev. Geophys. Space Phys., 8: 813—360.

Ghosh, S.K., 1970. A theoretical study of intersecting fold patterns. Tectonophysics, 9: 559—569.

Hall, C.A., Jr. and Corbató, C.E., 1967. Stratigraphy and structure of Mesozoic and Cenozoic rocks, Nipomo Quadrangle, southern Coast Ranges, California. Geol. Soc. Am. Bull., 78: 559—582.

Hamilton, W., 1969. Mesozoic California and the underflow of Pacific mantle. Geol. Soc. Am. Bull., 80: 2409—2430.

Handin, J. and Hager, R.V., Jr., 1957. Experimental deformation of sedimentary rocks under confining pressure: Tests at room temperature on dry samples. Bull. Am. Assoc. Pet. Geol., 41: 1—50.

Hill, C.E., 1976. Fold Forms in the Monterey and Santa Margarita Formations, San Luis Obispo County, California. M.S. Thesis, Department of Geology, Stanford University, California.

Honea, E., 1975. The Effect of Contact Strength on the Shape of Buckle Folds, Including Field Examples in South Darwin Canyon, California. Ph.D. dissertation, Department of Geology, Stanford University, California, 94 p.

Honea, E. and Johnson, A.M., 1976. A theory of concentric, kink and sinusoidal folding and of monoclinal flexuring of compressible, elastic multilayers. IV. Development of sinusoidal and kink folds in multilayers confined by rigid boundaries. Tectonophysics, 30: 197—239.

Jennings, C.W., 1958. Geologic Map of California. Olaf P. Jenkins Edition, San Luis Obispo sheet. California Division of Mines and Geology.

Jennings, C.W., 1973. State of California preliminary fault and geologic map. Calif. Div. Mines Geol., Prelim. Rep. 13.

Johnson, A.M. and Ellen, S.D., 1974. A theory of concentric, kink and sinusoidal folding and of monoclinal fluxuring of compressible, elastic multilayers. I. Introduction. Tectonophysics, 21: 301—339.

Johnson, A.M. and Honea, E., 1975a. A theory of concentric, kink and sinusoidal folding and of monoclinal flexuring of compressible, elastic multilayers. II. Initial stress and nonlinear equations of equilibrium. Tectonophysics, 25: 261—280.

Johnson, A.M. and Honea, Elmont, 1975b. A theory of concentric, kink and sinusoidal folding and of monoclinal flexuring of compressible, elastic multilayers. III. Transition from sinusoidal to concentric-like to chevron folds. Tectonophysics, 27: 1—38.

King, V.L., 1943. Huasna area development. In: Geologic Formations and Economic Development of the Oil and Gas Fields of California. Calif. Div. Mines Geol. Bull., 118: 448—449.

Page, B.M., 1970. Sur-Nacimiento fault zone of California: Continental margin tectonics. Geol. Soc. Am. Bull., 81: 667—690.

Ramberg, I.B. and Johnson, A.M., 1976. A theory of concentric, kink and sinusoidal folding and of monoclinal flexuring of compressible, elastic multilayers. V. Asymmetric folding in interbedded chert and shale of the Franciscan Complex, San Francisco Bay Area, California. Tectonophysics, 32: 295—320.

Reches, Z. and Johnson, A.M., 1976. A theory of concentric, kink and sinusoidal folding and of monoclinal flexuring of compressible, elastic multilayers. VI. Asymmetric folding and monoclinal kinking. Tectonophysics, 35: 295—334.

Scofield, D.H., 1975. Fold Forms in the Monterey and Santa Margarita Formations along Lopez Canyon, Tar Spring Ridge (7 1/2 ') Quadrangle, San Luis Obispo County, California, M.S. Thesis, Department of Geology, Stanford University, California, 61 p.

Taliaferro, N.L., 1943. Geology of the Huasna area. In: Geologic Formations and Economic Development of the Oil and Gas Fields of California. Calif. Div. Mines Geol. Bull., 118: 443—447.

Vedder, J.G. and Brown, R.D., 1968. Structural and stratigraphic relations along the Nacimiento fault in the southern Santa Lucia Range and San Rafael Mountains, California. In: W.R. Dickinson and A. Grantz (editors), Proceedings of Conference on Geologic Problems of San Andreas Fault System. Stanford Univ. Publ. Geol. Sci., 11: 242—259.

Vesić, A.S. and Clough, G.W., 1968. Behavior of granular materials under high stresses. J. Soil Mech. Found. Div., Am. Soc. Civ. Eng., Proc., SM3: 661—668.

Wilcox, R.E., Harding, J.P. and Seely, D.R., 1973. Basic wrench tectonics. Bull. Am. Assoc. Pet. Geol., 57: 74—96.

Chapter 10

FOLDING OF LINEAR ELASTIC AND POWER-LAW MATERIALS AND REVISION OF LINEARIZED THEORY

INTRODUCTION

This chapter has two primary purposes. It serves as a postscript or an appendix to the series of papers presented in earlier chapters, providing an opportunity to amplify or revise some of the analyses. Most important is the elimination of two special assumptions in the linearized theory presented in Chapter 4 and used in Chapters 5 and 6 to analyze single-layer and multi-layer folds. While attempting to eliminate the special assumptions, I was fortunate to read several papers by Rivlin * (e.g., 1948, 1949, 1956) which markedly clarified the entire theory of nonlinear elasticity, thereby allowing a fresh derivation of the theory of initial and incremental deformations. The postscript also allows comparison of predictions of wavelengths according to the linearized theory with those by Hans Ramberg for viscous materials, by Maurice Biot for elastic and viscous materials, and by Raymond Fletcher for viscous materials.

A second purpose of this chapter is to extend the analyses presented in the other chapters. Thus, we use an important technique introduced by Fletcher (1974) for analyzing effects on folding of initial perturbations of boundaries of layers. Also, we study the folding of a power-law material with

---

* See references at end of Chapter 1.

memory in the way Fletcher (1974) has, for a power-law material without memory. The final section of the chapter is an application of the power-law model, using constants estimated from experimental test results on rock reported by Handin and Hager (1957), in order to determine whether single-layer and multilayer folds are likely to develop in rock with such properties.

FOLDING OF LINEAR ELASTIC MATERIALS

We can analyze the folding of elastic materials under some conditions by using the linear theory of elasticity. We consider the amplification of initial perturbations of flat surfaces of layers, that is, we determine conditions under which the perturbations grow in amplitude. This is the approach Fletcher (1974, 1977) has introduced to study folding of viscous materials. According to this approach, we imagine that interfaces between layers are initially nonplanar and that the shapes of the interface surfaces can be expressed in terms of Fourier series. The initial shape, for example, might be a sine form:

$$v_{\mathrm{i}} = \delta_{\mathrm{i}} \sin(lx) \tag{1a}$$

where the i refers to initial deflection, δ_i is the initial amplitude of the sinusoidal wave and l is the wave number:

$$l = 2\pi/L \tag{1b}$$

where L is wavelength. Then, the layers are shortened and thickened uniformly, producing displacements U and V, and nonuniformly, producing displacements u_0 and v_0, as a result of uniform S_{xx}, S_{yy}, and nonuniform, s_{xx}, s_{yy}, s_{xy}, components of stress. In the linear theory we ignore interactions between the uniform and nonuniform states of stress and strain so they can be simply superimposed. Further, the boundary conditions are defined for the initial, undeformed boundaries. Thus the displacement gradients must be infinitesimal for the analysis to be valid. In these ways the linear theory is fundamentally different from the nonlinear and linearized theories we discussed in Chapter 4.

Equilibrium and rheological equations

Let us consider both compressible and incompressible elastic materials. For compressible materials, the stress—strain relations for the uniform state are:

$$S_{xx} = (2G + \lambda)(\partial U/\partial x) + \lambda(\partial V/\partial y) \tag{2a}$$

$$S_{yy} = (2G + \lambda)(\partial V/\partial y) + \lambda(\partial U/\partial x) \tag{2b}$$

and for the nonuniform stresses and strains:

$$s_{xx} = (2G + \lambda)(\partial u_0/\partial x) + \lambda(\partial v_0/\partial y) \tag{2c}$$

$$s_{yy} = (2G + \lambda)(\partial v_0/\partial y) + \lambda(\partial u_0/\partial x) \tag{2d}$$

$$s_{xy} = G(\partial v_0/\partial x + \partial u_0/\partial y) \tag{2e}$$

For incompressible materials:

$$S_{xx} - S_{yy} = 4G(\partial U/\partial x) \tag{3a}$$

$$s_{xx} = p + 2G(\partial u_0/\partial x) \tag{3b}$$

$$s_{yy} = p + 2G(\partial v_0/\partial y) \tag{3c}$$

$$s_{xy} = G(\partial v_0/\partial x + \partial u_0/\partial y) \tag{3d}$$

$$p = (s_{xx} + s_{yy})/2 \tag{3e}$$

We have assumed plane strain conditions. The constants, G and λ are Lamé's constants, where G is the shear modulus and:

$$\lambda = 2G\nu/(1 - 2\nu) \tag{4}$$

in which ν is Poisson's ratio. The equilibrium equations, where body forces are negligible, are:

$$\partial s_{xx}/\partial x + \partial s_{xy}/\partial y = 0 \tag{5a}$$

$$\partial s_{xy}/\partial x + \partial s_{yy}/\partial y = 0 \tag{5b}$$

The uniform stresses, S_{xx} and S_{yy}, automatically satisfy equilibrium, so the equilibrium equations involve only the nonuniform stresses.

Substituting the rheological equations 2c—2e and 3b—3d into equilibrium equations 5a and b, we derive, for *compressible materials*:

$$(2G + \lambda)(\partial^2 u_0/\partial x^2) + (G + \lambda)(\partial^2 v_0/\partial x \partial y) + G(\partial^2 u_0/\partial y^2) = 0 \tag{6a}$$

$$(2G + \lambda)(\partial^2 v_0/\partial y^2) + (G + \lambda)(\partial^2 u_0/\partial x \partial y) + G(\partial^2 v_0/\partial x^2) = 0 \tag{6b}$$

For *incompressible materials*, using the relation of incompressibility:

$$\partial u_0/\partial x = -\partial v_0/\partial y \tag{7a}$$

$$G[(\partial^2 u_0/\partial x^2) + (\partial^2 u_0/\partial y^2)] + \partial p/\partial x = 0 \tag{7b}$$

$$G[(\partial^2 v_0/\partial x^2) + (\partial^2 v_0/\partial y^2)] + \partial p/\partial y = 0 \tag{7c}$$

Boundary conditions

Before proceeding to the solution of the equilibrium equations, we will derive expressions for the boundary conditions. Figure 1 shows part of a boundary between two elastic materials or between an elastic material and a stress-free surface. The stresses acting on the surface can be expressed in

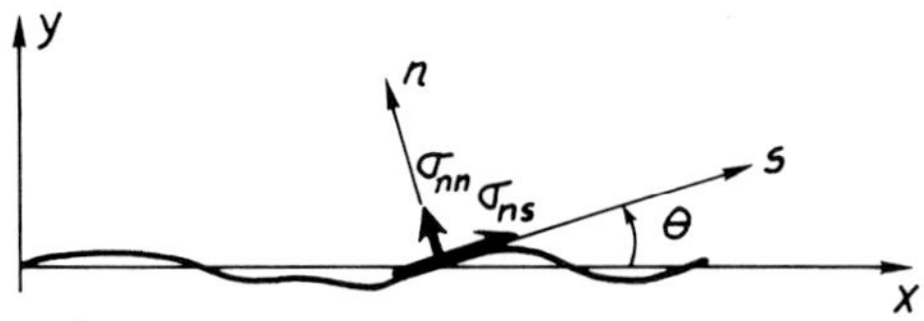

Fig. 1. Initial configuration and boundary conditions on an interface. Coordinates s and n are local coordinates, tangent and normal to the interface, which is nonplanar.

terms of a normal stress, σ_{nn}, and a shear stress, σ_{ns}, which are parallel to local coordinates n and s. The slope angle of the surface is locally θ, the angle between the x-direction and the s-direction (Fig. 1). The stresses acting in the x- and y-directions are σ_{xx}, σ_{yy} and σ_{xy}, where, according to definitions given earlier:

$$\sigma_{xx} = S_{xx} + s_{xx} \tag{8a}$$

$$\sigma_{yy} = S_{yy} + s_{yy} \tag{8b}$$

$$\sigma_{xy} = s_{xy} \tag{8c}$$

We can derive the relation between the stresses acting on the surface and the stress components in the x- and y-directions by means of Mohr's circle, and they are:

$$\sigma_{nn} = \sigma_{yy} \cos^2\theta + \sigma_{xx} \sin^2\theta - 2\sigma_{xy} \sin\theta \cos\theta \tag{8d}$$

$$\sigma_{ns} = (\sigma_{yy} - \sigma_{xx}) \cos\theta \sin\theta + \sigma_{xy}(\cos^2\theta - \sin^2\theta) \tag{8e}$$

Now we introduce another approximation. We assume that the slope angle θ is so small that the sine is approximately equal to the tangent:

$$\mathrm{d}v_\mathrm{i}/\mathrm{d}x = \tan\theta \simeq \sin\theta$$

and that the cosine is nearly equal to unity. Further, we assume that products of the initial slope and the nonuniform stress, such as $(s_{xx})(\mathrm{d}v_\mathrm{i}/\mathrm{d}x)$, are negligible, but that products of the initial slope and the uniform stresses are significant. (We already have assumed that products of the uniform stresses and the nonuniform strains and rotations are negligible.) With these assumptions, eqs. 8 provide:

$$\sigma_{nn} \simeq S_{yy} + s_{yy} = \sigma_{yy} \tag{8f}$$

$$\sigma_{ns} \simeq s_{xy} + (S_{yy} - S_{xx})(\mathrm{d}v_\mathrm{i}/\mathrm{d}x) \tag{8g}$$

Sinusoidal solutions

Incompressible material

We solve the equilibrium equations for incompressible material by eliminating the mean stress, p, between the two equations and introducing a displacement function ψ such that:

$$u_0 = \partial\psi/\partial y \qquad \text{and} \qquad v_0 = -\partial\psi/\partial x \tag{9a}$$

which satisfy the condition of incompressibility, eq. 7a. Thus, eqs. 7 and 9a provide the biharmonic equation in ψ:

$$\partial^4\psi/\partial x^4 + 2(\partial^4\psi/\partial x^2\partial y^2) + \partial^4\psi/\partial y^4 = 0 \tag{9b}$$

The sinusoidal solution to eq. 9b is:

$$\psi = (1/l)[\{a + b(ly - 1)\}\exp(ly) - \{c + d(ly + 1)\}\exp(-ly)\}]\cos(lx) \tag{10a}$$

where a, b, c and d are arbitrary constants. Substituting eq. 10a into eqs. 9a, we find:

$$u_0 = \partial\psi/\partial y = [(a + bly)\exp(ly) + (c + dly)\exp(-ly)]\cos(lx) \tag{10b}$$

$$v_0 = -\partial\psi/\partial x = [\{a + b(ly - 1)\}\exp(ly) - \{c + d(ly + 1)\}\exp(-ly)]\sin(lx) \tag{10c}$$

The form of ψ in eq. 10a was chosen so that eqs. 10b and c would be of the same form as Fletcher's (1977, eqs. 17) solution for a viscous material.

The mean stress is computed by substituting eq. 10b into eq. 7b and integrating:

$$p = -G(\partial u_0/\partial x) - G\int(\partial^2 u_0/\partial y^2)\,dx$$

$$p = -2Gl[b\exp(ly) - d\exp(-ly)]\sin(lx) + p_0 \tag{10d}$$

where p_0 is a constant.

Now we will consider the solution for the folding of a stiff (*1*) layer of thickness T confined above and below by soft (*2*) media (Fig. 2). Equation 10 apply to both the layer and the medium, but the constants, $a \ldots d$, are different. The origins of coordinates are different also, so that y_2 in the

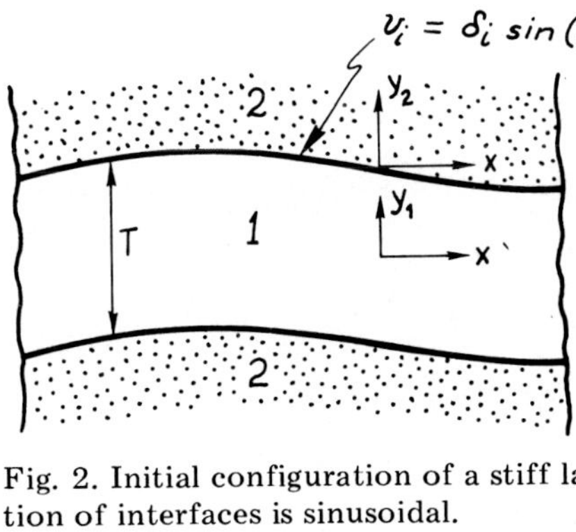

Fig. 2. Initial configuration of a stiff layer (*1*) between soft media (*2*). Initial configuration of interfaces is sinusoidal.

medium is zero at the upper interface and y_1 in the layer is zero at middepth (Fig. 2).

The nonuniform stresses and displacements vanish at infinity within the medium, so that:

$$a_2 = b_2 = 0 = p_0)_2$$

Thus, for the medium, eqs. 10 become:

$$u_0)_2 = (c_2 + d_2 ly_2) \exp(-ly_2) \cos(lx) \tag{11a}$$

$$v_0)_2 = -[c_2 + d_2(ly_2 + 1)] \exp(-ly_2) \sin(lx) \tag{11b}$$

$$p_2 = 2G_2 ld_2 \exp(-ly_2) \sin(lx) \tag{11c}$$

The mean stress and the displacement, u_0, vanish at middepth in the layer, and the mean stress is antisymmetric about middepth, so, from eq. 10d, we get:

$$p_0)_1 = 0; d_1 = b_2$$

From eq. 10b:

$$c_1 = -a_1$$

Thus, eqs. 10 for the layer become:

$$u_0)_1 = [(a_1 + b_1 ly_1) \exp(ly_1) - (a_1 - b_1 ly_1) \exp(-ly_1)] \cos(lx) \tag{12a}$$

$$v_0)_1 = [\{a_1 + b_1(ly_1 - 1)\} \exp(ly_1) + \{a_1 - b_1(ly + 1)\} \exp(-ly_1)] \sin(lx) \tag{12b}$$

$$p_1 = -2G_1 lb_1 [\exp(ly_1) - \exp(-ly_1)] \sin(lx) \tag{12c}$$

According to eqs. 8, 11 and 12, the boundary stresses at $y_2 = 0$ are:

$$\sigma_{nn})_2 = S_{yy} + l2G_2(c_2 + d_2) \sin(lx) \tag{13a}$$

$$\sigma_{ns})_2 = [-4G_2(\partial U/\partial x) l\delta_i - l2G_2 c_2] \cos(lx) \tag{13b}$$

because:

$$S_{yy} - S_{xx} = -4G(\partial U/\partial x)$$

$$\sigma_{nn})_1 = S_{yy} + l2G_1[\{a_1 + b_1(k - 1)\} \exp(k) - \{a_1 - b_1(k + 1)\} \exp(-k)] \sin(lx) \tag{13c}$$

$$\sigma_{ns})_1 = [-4G_1(\partial U/\partial x) l\delta_i + l2G_1\{(a_1 + b_1 k) \exp(k) + (a_1 - b_1 k) \exp(-k)\}] \cos(lx) \tag{13d}$$

where

$$k = lT/2 . \tag{13e}$$

Perfect slip. If the layer and medium are not bonded together, the displacements, u_0, are unspecified at the contacts but at:

$y_2 = 0$ and $y_1 = +T/2$

$$v_0)_2 - v_0)_1 = 0 \quad \text{and} \quad \sigma_{nn})_2 - \sigma_{nn})_1 = 0 \qquad (13f)$$

$\sigma_{ns})_2 = 0$ and $\sigma_{ns})_1 = 0$

Thus, the matrix of coefficients becomes:

c_2	d_2	a_1	b_1	
-1	-1	$-(e^k + e^{-k})$	$-[k(e^k - e^{-k}) - (e^k + e^{-k})]$	0
$\bar{G}$	$\bar{G}$	$-(e^k - e^{-k})$	$-k(e^k + e^{-k}) + e^k - e^{-k}$	0
-1	0	0	0	$2(\partial U/\partial x)\delta_i$
0	0	$e^k + e^{-k}$	$k(e^k - e^{-k})$	$2(\partial U/\partial x)\delta_i$

According to eq. 11b, the change in amplitude due to folding of the upper contact, where $y_2 = 0$, is equal to $-c_2 - d_2$. Therefore, we can solve the matrix for these two coefficients and determine the wavelength-to-thickness ratio, L/T, which is equal to π/k, for which the amplitude is maximized. This provides the ratio for folds which will amplify most for a given magnitude of axial strain, $\partial U/\partial x$, which is negative if compressive. The results of the calculations will be presented in Fig. 6B. According to Fletcher's analysis:

$$(c_2 + d_2)/[(\partial U/\partial x)\delta_i] = 2/[1 - \{e^{2k} - e^{-2k} + \bar{G}(e^{2k} + e^{-2k} + 2)\}/4k] \qquad (14a)$$

where δ_i is the initial amplitude of sinusoidal waves, and:

$\bar{G} = G_2/G_1$

$k = lT/2 = \pi T/L$

$l = 2\pi/L$

Here G_2 is the modulus of the medium and G_1 is the modulus of the layer, T is the thickness of the layer, and L is the wavelength. The waves also are amplified by the overall, uniform shortening, so the total amplitude, δ, is:

$$\delta = -c_2 - d_2 - (\partial U/\partial x)\delta_i + \delta_i \qquad (14b)$$

Bonded contacts. If the layer and medium are bonded together, both components of displacement must be continuous across contacts. Thus, at $y_2 = 0$ in the medium and at $y_1 = +T/2$, at the top of the layer:

$v_0)_2 - v_0)_1 = 0$

$u_0)_2 - u_0)_1 = 0$

$$\sigma_{nn})_2 - \sigma_{nn})_1 = 0$$
$$\sigma_{ns})_2 - \sigma_{ns})_1 = 0 \quad (14c)$$

The matrix of coefficients becomes:

c_2	d_2	a_1	b_1	
-1	-1	$-(e^k + e^{-k})$	$-[k(e^k - e^{-k}) - (e^k + e^{-k})]$	0
$\bar{G}$	$\bar{G}$	$-(e^k - e^{-k})$	$-k(e^k + e^{-k}) + e^k - e^{-k}$	0
$-\bar{G}$	0	$-(e^k + e^{-k})$	$-k(e^k - e^{-k})$	$(\bar{G} - 1)(\partial U/\partial x)\delta_i$
1	0	$-(e^k - e^{-k})$	$-k(e^k + e^{-k})$	0

According to Fletcher's solution (Fletcher, 1977, eq. 29), the growth of amplitude of the initial sinusoidal interfaces is:

$$(c_2 + d_2)/[(\partial U/\partial x)\delta_i] = 2(1 - \bar{G})/[1 - \bar{G}^2 - \{(1 + \bar{G})^2 e^{2k} - (1 - \bar{G})^2 e^{-2k}\}/4k] \quad (15)$$

According to Fletcher (1977), both eqs. 14a and 15 predict wavelength-to-thickness ratios equal to those for the elementary plate theory:

$$L = 2\pi T \sqrt[3]{G_1/6G_2}$$

for large wavelength-to-thickness ratios.

In Fig. 6A we shall compare wavelengths of folds that have maximal growths of amplitude, according to eq. 15, with analogous wavelengths predicted by theories of Ramberg (1970), Biot (1965a) and the linearized theory.

Compressible material

We solve equilibrium equations (6) for compressible materials by determining displacements, u_0 and v_0, that satisfy them. First we eliminate u_0 between the two equations and derive a biharmonic equation in v_0:

$$\partial^4 v_0/\partial x^4 + 2(\partial^4 v_0/\partial x^2 \partial y^2) + \partial^4 v_0/\partial y^4 = 0 \quad (16a)$$

Then we eliminate v_0 between the two equations to derive:

$$\partial^4 u_0/\partial x^4 + 2(\partial^4 u_0/\partial x^2 \partial y^2) + \partial^4 u_0/\partial y^4 = 0 \quad (16b)$$

The sinusoidal solution to eq. 16a is:

$$v_0 = [(a + bly)\exp(-ly) + (c + dly)\exp(ly)]\sin(lx) \quad (17a)$$

The differential equation for u_0, eq. 16b, is of the same form as that for v_0, eq. 16a, so the solutions for u_0 and v_0 should be similar. However, the solutions must satisfy equilibrium equations 6, so we select the cosine solution to eq. 16b:

$$u_0 = [(e + fly)\exp(-ly) + (g + hly)\exp(ly)]\cos(lx)$$

where $e \ldots h$ are constants. We can eliminate constants $e \ldots h$ by expressing them in terms of $a \ldots d$. We substitute the equations for the displacements into eq. 6a or 6b, and set the coefficients of terms in $ly\exp(-ly)$, $\exp(-ly)$, $ly\exp(ly)$, and $\exp(ly)$ equal to zero. These terms provide four equations with which we can express constants $e \ldots h$ in terms of constants $a \ldots d$. The result is:

$$u_0 = -[\{a + b(ly + 4\nu - 3)\}\exp(-ly) - \{c + d(ly + 3 - 4\nu)\}\exp(ly)]\cos(lx) \tag{17b}$$

where $3 - 4\nu = (\lambda + 3G)/(\lambda + G)$.

This completes the solution for the displacements. We evaluate the constants, as before, by using boundary conditions 13f or 14c and eqs. 8f and 8g. We shall not proceed further with the solution here.

ELIMINATION OF SPECIAL ASSUMPTIONS IN LINEARIZED THEORY

The solutions presented in earlier pages were based on the assumption that products of nonuniform deformations and uniform stresses are negligible, but that products of the initial slopes of boundaries of layers and uniform stresses are significant. Indeed, these latter products cause the lateral deflections which we associate with folding. The folding is stable according to this theory; the magnitudes of the deflections are determinant. In the linearized theory, developed in Chapter 4, we considered only interactions between the nonuniform, incremental deformations and uniform initial stresses, and the folding predicted by the theory is a result of instability; the magnitudes of the deflections are indeterminant. Thus the theories are fundamentally different. However, we can modify the linearized theory somewhat to account for effects of initial slopes of boundaries. We do so in the Appendix of this chapter.

In the discussion preceding the derivation of the linearized equations for a linear elastic material, in the paragraph below fig. 6 of Chapter 4, we indicated that we must *assume* the relationship between the final stress state $\sigma_{\bar{x}\bar{x}}$, $\sigma_{\bar{y}\bar{y}}$, $\sigma_{\bar{x}\bar{y}}$, and the initial and incremental stress states. Further, following eq. 23 there, we indicated that certain stresses, $S_{\bar{x}\bar{x}} \ldots$ acting on the edges of the finally deformed element *probably* are equal to the initial stresses, $S_{\bar{X}\bar{X}} \ldots$. Now we shall eliminate both of these special assumptions by deriving the relationships rather than assuming them. In the process of searching for means to eliminate these assumptions, we have found a way of simplifying the entire analysis, so we will present most of the analysis here. The following derivations are based on study of Rivlin * (1948, 1949 and 1956),

* See references at end of Chapter 1.

Green and Zerna (1968), Fung (1965), Green et al. (1952) and Mooney (1940). The papers by Rivlin are lucid. The reader is expected to be familiar with at least the first few pages of Chapter 4.

Sequence of plane-strain deformations

I will deviate somewhat from the notation used in Chapter 4, so I begin by describing the sequence of deformations and the notation. Figure 3A shows an element, $ABCD$, which becomes deformed, first by uniform strains in the X- and Y-directions into element $\tilde{A}\tilde{B}\tilde{C}\tilde{D}$ (Fig. 3B), and then by nonuniform strains and rotations into element $\hat{A}\hat{B}\hat{C}\hat{D}$ shown in Fig. 3C. Points within the nondeformed element are defined by coordinates X and Y (Fig. 3A). Displacements during the first deformation are designated by U and V. Thus, the position of a point (x,y) after the first deformation is defined by:

$x = X + U \qquad \text{and} \qquad y = Y + V$

and the first deformation is defined by:

$$\begin{bmatrix} dx \\ dy \end{bmatrix} = \begin{bmatrix} (1 + \partial U/\partial X) & 0 \\ 0 & (1 + \partial V/\partial Y) \end{bmatrix} \begin{bmatrix} dX \\ dY \end{bmatrix}$$

or:

$$\begin{aligned} dx &= \lambda_X \, dX \equiv (1 + \partial U/\partial X)\, dX \\ dy &= \lambda_Y \, dY \equiv (1 + \partial V/\partial Y)\, dY \end{aligned} \tag{18}$$

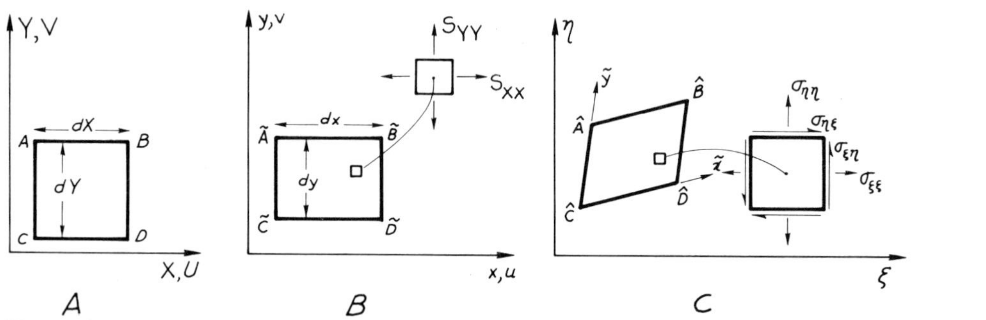

Fig. 3. States of an element of a body. A. Before deformation, the element is rectangular, with dimensions dY and dX. Coordinates X and Y refer to positions of points in the undeformed body. Displacements U and V refer to displacements of points from their original positions, during the *initial* deformation of the element, which leads to the state shown in B. In the initial state the element has dimensions dx and dy which are related to the dimensions dX and dY and the finite extension ratios. The state of initial stress is homogeneous and uniform. Principal stresses are parallel to x- and y-directions. If the initial stresses are sufficiently large, there is an *instability*, which allows an alternative equilibrium form for the element, shown in C. The deformation which leads from the uniform state shown in B to the nonuniform state shown in C is called the *incremental* deformation. The stresses acting within the nonuniformly deformed body are shown in C.

where λ is the finite extension. We assume that there is no shear in the x- and y-directions during the first deformation. This is the initial state of the body. Then, if the body becomes unstable, as discussed in some detail in Chapter 4, the element deforms again as shown in Fig. 3C. The displacements for the second deformation are defined by u and v *relative* to the second state. Thus:

$$d\xi = du + dx$$

$$d\eta = dv + dy$$

or:

$$\begin{bmatrix} d\xi \\ d\eta \end{bmatrix} = \begin{bmatrix} (1+u_x) & u_y \\ v_x & (1+v_y) \end{bmatrix} \begin{bmatrix} dx \\ dy \end{bmatrix} \tag{19}$$

where

$$u_x \equiv \partial u/\partial x;\ u_y \equiv \partial u/\partial y;\ v_x \equiv \partial v/\partial x;\ v_y \equiv \partial v/\partial y\ . \tag{20}$$

Substituting eqs. 18 into 19, we get:

$$d\xi = (1+u_x)\lambda_X\, dX + u_y \lambda_Y\, dY \tag{21a}$$

$$d\eta = v_x \lambda_X\, dX + (1+v_y)\lambda_Y\, dY \tag{21b}$$

Figure 4A shows a body which is circular in its undeformed state (Fig. 3A), and is elliptical in its finally deformed state (Fig. 4B). The long axis of the ellipse is proportional to λ_1 and the short one to λ_3. In particular:

$$\lambda_1 = ds_2'/ds_0' \qquad \text{and} \qquad \lambda_3 = ds_2''/ds_0'' \tag{22}$$

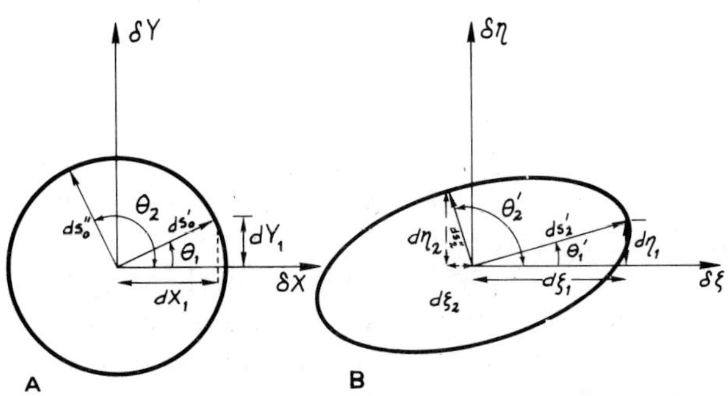

Fig. 4. Diagrams used to derive expressions for principal finite extension ratios for final state of body. A. A circle drawn on body before deformation, showing orientations of line elements, ds_0' and ds_0'', which become parallel to principal directions of finite extension ratios in final state, B. The circle has become an ellipse in the final state, with axes ds_2'' and ds_2'.

where ds'_2, for example, is half the long dimension of the ellipse, and ds'_0 is the length of the same line, before deformation. The direction of ds'_0 shown in Fig. 4A is the direction of the line element, which, when strained, is parallel to the long axis of the ellipse, Fig. 4B. Thus θ_1 is the angle between the X-direction and the line element ds'_0 in the undeformed body, and θ'_1 is the direction this line element assumes in the deformed body, when its length becomes ds'_2. According to Figs. 4 and to eq. 21a, the projection of the line element ds'_2, onto the ξ-axis is:

$$d\xi_1 = (1 + u_x)\lambda_X dX_1 + u_y \lambda_Y dY_1$$

and according to Figs. 4 this equation can be written:

$$ds'_2 \cos(\theta'_1) = (1 + u_x)\lambda_X \cos(\theta_1)\, ds'_0 + u_y \lambda_Y \sin(\theta_1) ds'_0$$

Using eq. (22), we find:

$$\lambda_1 \cos(\theta'_1) = (1 + u_x)\lambda_X \cos(\theta_1) + u_y \lambda_Y \sin(\theta_1) \quad (23a)$$

Similarly, considering the projection of the line element onto the η-axis, we derive:

$$\lambda_1 \sin(\theta'_1) = v_x \lambda_X \cos(\theta_1) + (1 + v_y)\lambda_Y \sin\theta_1 \quad (23b)$$

The projection of line element ds''_2, which is parallel to the short axis of the ellipse, Fig. 4B, onto the ξ-axis is, according to eq. 21a:

$$d\xi_2 = ds''_2 \cos(90 + \theta'_1) = -ds''_2 \sin(\theta'_1) = [-(1 + u_x)\lambda_X \sin(\theta_1) + u_y \lambda_Y \cos(\theta_1)]\, ds''_0$$

Thus:

$$\lambda_3 \sin(\theta'_1) = (1 + u_x)\lambda_X \sin(\theta_1) - u_y \lambda_Y \cos(\theta_1) \quad (23c)$$

Similarly:

$$\lambda_3 \cos(\theta'_1) = (1 + v_y)\lambda_Y \cos(\theta_1) - v_x \lambda_X \sin(\theta_1) \quad (23d)$$

Initial stress state

Now we have defined the two states of deformation of an element of a body. The next step is to derive relations between stresses and strains. Here we follow Rivlin (1948, 1956). We shall derive the relations for three-dimensional states of stress and strain and then simplify them for conditions of plane strain. Consider an element of a body subjected to principal normal stresses, S_{XX}, S_{YY} and S_{ZZ}. These stresses are defined as intensities of forces per unit of area of the *deformed* element, so that they are in units of force per unit of final area. These are the *initial stresses* (Fig. 3B). They act parallel to the X-, Y- or Z-axis which is parallel to the x-, y- or z-axis. Thus, we are not dealing with the peculiar stress components referred to curvilinear axes, as in Chapter 4, and the analysis presented here will be much simpler than that presented there.

Compressible materials

In order to derive the stress—strain relations, we consider the change in work, per unit of *original* volume of an element, done by slightly varying the principal extension ratios:

$$\delta W = S_{XX} \lambda_Y \lambda_Z \delta \lambda_X + S_{YY} \lambda_Z \lambda_X \delta \lambda_Y + S_{ZZ} \lambda_X \lambda_Y \delta \lambda_Z \tag{24a}$$

With the chain rule, we can derive another expression for the change in work per unit of original volume, that is, the change in strain energy, of the element:

$$\delta W = (\partial W/\partial \lambda_X)\delta \lambda_X + (\partial W/\partial \lambda_Y)\delta \lambda_Y + (\partial W/\partial \lambda_Z)\delta \lambda_Z \tag{24b}$$

Comparing eqs. (24a) and (24b):

$$S_{XX} = (1/\lambda_Y \lambda_Z)(\partial W/\partial \lambda_X) \tag{25a}$$

$$S_{YY} = (1/\lambda_X \lambda_Z)(\partial W/\partial \lambda_Y) \tag{25b}$$

$$S_{ZZ} = (1/\lambda_X \lambda_Y)(\partial W/\partial \lambda_Z) \tag{25c}$$

These equations can be written in more general form by using the three strain invariants (See Appendix):

$$I_1 = \lambda_X^2 + \lambda_Y^2 + \lambda_Z^2$$

$$I_2 = \lambda_X^2 \lambda_Y^2 + \lambda_X^2 \lambda_Z^2 + \lambda_Y^2 \lambda_Z^2 \tag{26a}$$

$$I_3 = \lambda_X^2 \lambda_Y^2 \lambda_X^2$$

Then, for example, with the chain rule, eq. 25a can be written in the form:

$$S_{XX} = (1/\lambda_Y \lambda_Z)[(\partial W/\partial I_1)(\partial I_1/\partial \lambda_X) + (\partial W/\partial I_2)(\partial I_2/\partial \lambda_X) + (\partial W/\partial I_3) \cdot (\partial I_3/\partial \lambda_X)] \tag{26b}$$

Further, using eqs. 26a:

$$(\partial I_1/\partial \lambda_X) = 2\lambda_X \tag{26c}$$

$$(\partial I_2/\partial \lambda_X) = 2\lambda_X(\lambda_Y^2 + \lambda_Z^2) \tag{26d}$$

$$(\partial I_3/\partial \lambda_X) = 2\lambda_X \lambda_Y^2 \lambda_Z^2 \tag{26e}$$

Substituting eqs. 26c—26e into 26b, we get:

$$S_{XX} = (2/\sqrt{I_3})[\lambda_X^2(\partial W/\partial I_1) + (I_2 - \lambda_Y^2 \lambda_Z^2)(\partial W/\partial I_2) + I_3(\partial W/\partial I_3)] \tag{27a}$$

Similar expressions are readily derived for the other two stress components. If the material is compressible but subjected to *plane strain*, in which case $\lambda_Z = 1$, eq. 27a becomes:

$$S_{XX} = (2/\sqrt{I_3})[\lambda_X^2\{(\partial W/\partial I_1) + (\partial W/\partial I_2)\} + I_3\{(\partial W/\partial I_2) + (\partial W/\partial I_3)\}] \quad (27b)$$

Similarly:

$$S_{YY} = (2/\sqrt{I_3})[\lambda_Y^2\{(\partial W/\partial I_1) + (\partial W/\partial I_2)\} + I_3\{(\partial W/\partial I_2) + (\partial W/\partial I_3)\}] \quad (27c)$$

$$S_{ZZ} = (2/\sqrt{I_3})[(\partial W/\partial I_1) + (\lambda_X^2 + \lambda_Y^2)(\partial W/\partial I_2) + I_3(\partial W/\partial I_3)] \quad (27d)$$

If the material is incompressible, the stresses are determined only to within an arbitrary constant, say Q, and the third invariant, I_3, is unity. In order to derive expressions for stresses in this case, we replace the normal stresses with $S_{XX} - Q$, $S_{YY} - Q$, and $S_{ZZ} - Q$ in eqs. 25. Thus, eqs. 27 are replaced by:

$$S_{XX} = 2\lambda_X^2[(\partial W/\partial I_1) + (\partial W/\partial I_2)] + 2(\partial W/\partial I_2) + Q \quad (28a)$$

$$S_{YY} = 2\lambda_Y^2[(\partial W/\partial I_1) + (\partial W/\partial I_2)] + 2(\partial W/\partial I_2) + Q \quad (28b)$$

$$S_{ZZ} = 2[(\partial W/\partial I_1) + (\lambda_X^2 + \lambda_Y^2)(\partial W/\partial I_2)] + Q \quad (28c)$$

The quantity Q is similar to a mean stress or pressure in linear theories of elasticity and viscosity. It can be replaced with a quantity that is somewhat more familiar. Let $Q = P + P_0$, where:

$$P_0 = -4(\partial W/\partial I_2) - 2(\partial W/\partial I_1)$$

Then eqs. 28a—28c become:

$$S_{XX} = 2(\lambda_X^2 - 1)[(\partial W/\partial I_1) + (\partial W/\partial I_2)] + P \quad (29a)$$

$$S_{YY} = 2(\lambda_Y^2 - 1)[(\partial W/\partial I_1) + (\partial W/\partial I_2)] + P \quad (29b)$$

$$S_{ZZ} = 2[(\lambda_X^2 + \lambda_Y^2 - 2)(\partial W/\partial I_2)] + P \quad (29c)$$

As in linear elasticity theory, knowledge of the state of strain specifies the stress components to within an arbitrary constant. In the linear theory the constant is the mean stress but in the nonlinear theory, eqs. 29, it is not, as can be verified by summing eqs. 29 and dividing by three. Further, the stress S_{ZZ} is equal to the mean stress for plane strain of a linear material, but normally is not for the nonlinear material, as indicated in eq. 29c. Rivlin (1948, p. 467—468) explains how the constant Q (or P) is determined.

In general the derivatives of the strain energy function, W, are variable, but in our analyses we will assume that they are constants. We will consider only two simple expressions for strain energy. One is the *neoHookean model* (Truesdell and Noll, 1965, p. 350):

$$W = (G/2)[(I_1 - 3)] \quad (30a)$$

for which eqs. 28 become the finite-strain analogs to stress—strain relations for the incompressible elastic material of linear elasticity theory. Here G is the modulus of rigidity of the material. Mooney (1940) proposed the model:

$$W = [(G + H)/4](I_1 - 3) + [(G - H)/4](I_2 - 3) \tag{30b}$$

which describes closely the deformation of rubber-like materials, even for very large strains corresponding to finite elongations of about 2.5 (150%) for a stiff rubber and about 6 (500%) for a soft rubber. The constant H has no analog in linear elasticity theory. For two types of rubber tested by Mooney, $H/G = 0.445$ for the stiff rubber and $H/G = 0.223$ for the soft rubber. The ratio must satisfy the limits: $-1 \leqslant (H/G) \leqslant 1$ (Truesdell and Noll, 1965, p. 350). The constant G is the modulus of rigidity or the shear modulus of the Mooney material, as in linear elasticity theory.

Thus we have defined the initial stress state in the deformed element shown in Fig. 3B. We could substitute either of eqs. 30 into eqs. 29a—29c, and calculate the stresses as functions of the finite elongations.

Incremental stresses for incompressible materials

The next step in the analysis is to consider further the second stage of deformation, represented in Fig. 3C, where there are both uniform initial stresses and nonuniform incremental stresses acting on an element. We shall continue to assume that the material is incompressible and that the body is in a state of plane strain. Let us designate σ_1, σ_{ZZ} and σ_3 as the principal stresses in the element. Then, according to eqs. 29:

$$\sigma_1 = 2(\lambda_1^2 - 1)[(\partial W/\partial I_1^*) + (\partial W/\partial I_2^*)] + P_0 \tag{31a}$$

$$\sigma_3 = 2(\lambda_3^2 - 1)[(\partial W/\partial I_1^*) + (\partial W/\partial I_2^*)] + P_0 \tag{31b}$$

$$\sigma_{ZZ} = 2[(\lambda_1^2 + \lambda_3^2 - 2)(\partial W/\partial I_2^*)] + P_0 \tag{31c}$$

Here I^* is an invariant for the total strains (Table I; also, see Appendix).

The maximum principal stress is acting at an angle θ_1' from the ξ-axis in the finally deformed body (Fig. 4B). Thus, using Mohr's circle, we can derive expressions for the stresses acting on sides of an element oriented parallel to the x- and y- or the X- and Y- or the ξ- and η-directions, which are equivalent orientations (Fig. 3C):

$$\sigma_{\xi\xi} = \sigma_1 \cos^2(\theta_1') + \sigma_3 \sin^2(\theta_1') \tag{32a}$$

$$\sigma_{\eta\eta} = \sigma_1 \sin^2(\theta_1') + \sigma_3 \cos^2(\theta_1') \tag{32b}$$

$$\sigma_{\eta\xi} = (\sigma_1 - \sigma_3) \sin(\theta_1') \cos(\theta_1') \tag{32c}$$

Substituting eqs. 31 into 32a, we find:

$$\sigma_{\xi\xi} = 2[(\partial W/\partial I_1^*) + (\partial W/\partial I_2^*)][(\lambda_1^2 - 1)\cos^2(\theta_1') + (\lambda_3^2 - 1)\sin^2(\theta_1')] + P_0$$

TABLE I

Expressions for invariants

$I_1 = g_{xx} + g_{yy} + g_{zz} = \lambda_1^2 + \lambda_2^2 + \lambda_3^2$

$I_2 = g_{xx}g_{yy} + g_{xx}g_{zz} + g_{yy}g_{zz} - g_{xy}^2 - g_{xz}^2 - g_{yz}^2 = \lambda_1^2\lambda_2^2 + \lambda_1^2\lambda_3^2 + \lambda_2^2\lambda_3^2$

$I_3 = g_{xx}g_{yy}g_{zz} + 2g_{xy}g_{xz}g_{yz} - g_{xx}g_{yz}^2 - g_{yy}g_{xz}^2 - g_{zz}g_{xy}^2 = \lambda_1^2\lambda_2^2\lambda_3^2$

For I^0 invariants:

$g_{xx} = \lambda_X^2; \quad g_{yy} = \lambda_Y^2; \quad g_{zz} = \lambda_Z^2; \quad g_{xy} = g_{xz} = g_{yz} = 0$

(For plane strain: $g_{zz} = 1 = \lambda_z$)

For I^* invariants:

$g_{xx} = [(1 + u_x)^2 + v_x^2 + w_x^2]\lambda_X^2; \quad g_{yy} = [(1 + v_y)^2 + u_y^2 + w_y^2]\lambda_Y^2$

$g_{zz} = [(1 + w_z)^2 + v_z^2 + u_z^2]\lambda_Z^2$

$g_{xy} = [w_x w_y + (1 + u_x)u_y + (1 + v_y)v_x]\lambda_X\lambda_Y$

$g_{xz} = [v_z v_x + (1 + w_z)w_x + (1 + u_x)u_z]\lambda_X\lambda_Z$

$g_{yz} = [u_y u_z + (1 + v_y)v_z + (1 + w_z)w_y]\lambda_Y\lambda_Z$

(For plane strain: $g_{zz} = 1; \quad g_{xy} = g_{xz} = w = u_z = v_z = 0$)

Finally, substituting eqs. 23 into this result and rearranging:

$$\sigma_{\xi\xi} = 2[(\partial W/\partial I_1^*) + (\partial W/\partial I_2^*)]\,[(1 + u_x)^2\lambda_X^2 - 1 + u_y^2\lambda_Y^2] + P_0 \tag{33a}$$

Similarly:

$$\sigma_{\eta\eta} = 2[(\partial W/\partial I_1^*) + (\partial W/\partial I_2^*)]\,[(1 + v_y)^2\lambda_Y^2 - 1 + v_x^2\lambda_X^2] + P_0 \tag{33b}$$

$$\sigma_{\eta\xi} = 2[(\partial W/\partial I_1^*) + (\partial W/\partial I_2^*)]\,[(1 + u_x)v_x\lambda_X^2 + (1 + v_y)u_y\lambda_Y^2] \tag{33c}$$

$$\sigma_{ZZ} = 2(\partial W/\partial I_2^*)[(1 + u_x)^2\lambda_X^2 + (1 + v_y)^2\lambda_Y^2 + u_y^2\lambda_Y^2 + v_x^2\lambda_X^2 - 2] + P_0 \tag{33d}$$

Thus far we have made few assumptions. Both the initial, uniform strains and the nonuniform strains may be finite. We have made no specific assumptions about the form of the strain-energy function, W. Those in eqs. 30 where merely presented as examples. In order to linearize the equations, however, we must make a further assumption. We shall assume that the displacement gradients, $u_x \ldots v_y$, referred to coordinates on the initially deformed body, are so small that products are negligible. Thus, we assume that products such as v_x^2, $u_xv_x \ldots$ are negligible. In this way, we derive expressions for the *incremental* stresses s_{xx}, s_{yy} and s_{xy}.

According to eqs. 29, the initial stresses are:

$$S_{XX} = 2(\lambda_X^2 - 1)[(\partial W/\partial I_1^0) + (\partial W/\partial I_2^0)] + P \tag{34a}$$

$$S_{YY} = 2(\lambda_Y^2 - 1)[(\partial W/\partial I_1^0) + (\partial W/\partial I_2^0)] + P \tag{34b}$$

$$S_{ZZ} = 2[(\lambda_X^2 + \lambda_Y^2 - 2)(\partial W/\partial I_2^0)] + P \tag{34c}$$

where I^0 are invariants for the initial strains (Table 1). Thus, we can write eqs. 33 as:

$$\sigma_{\xi\xi} = S_{XX} + s_{xx}$$

$$\sigma_{\eta\eta} = S_{YY} + s_{yy}$$

$$S_{ZZ} + s_{zz} = P + p + 2(\partial W/\partial I_2^*)[(1 + 2u_x)\lambda_X^2 + (1 + 2v_y)\lambda_Y^2 - 2]$$

$$\sigma_{\xi\eta} = s_{xy} \tag{34d}$$

where:

$$s_{xx} = 4[(\partial W/\partial I_1^*) + (\partial W/\partial I_2^*)]u_x\lambda_X^2 + p + 2(\lambda_X^2 - 1)\{(\partial W/\partial I_1^*) - (\partial W/\partial I_1^0) + (\partial W/\partial I_2^*) - (\partial W/\partial I_2^0)\} \tag{34e}$$

$$s_{yy} = 4[(\partial W/\partial I_1^*) + (\partial W/\partial I_2^*)]v_y\lambda_Y^2 + p + 2(\lambda_Y^2 - 1)\{(\partial W/\partial I_1^*) - (\partial W/\partial I_1^0) + (\partial W/\partial I_2^*) - (\partial W/\partial I_2^0)\} \tag{34f}$$

$$s_{xy} = 2[(\partial W/\partial I_1^*) + (\partial W/\partial I_2^*)]\,[v_x\lambda_X^2 + u_y\lambda_Y^2] \tag{34g}$$

The invariants, I^* and I^0 are defined in Table I, Equations 34e and 34f simplify somewhat if the material is simple; for example, those defined in eqs. 30, because for those materials:

$$(\partial W/\partial I_1^*) = (\partial W/\partial I_1^0) \quad \text{and} \quad (\partial W/\partial I_2^*) = (\partial W/\partial I_2^0)$$

and the terms in braces in eqs. 34e, f vanish.

For the Mooney material, eq. 30b, eqs. 34 become, for plane strain:

$$S_{XX} = G(\lambda_X^2 - 1) + P \tag{35a}$$

$$S_{YY} = G(\lambda_Y^2 - 1) + P \tag{35b}$$

$$S_{ZZ} = [(G - H)/2](\lambda_X^2 + \lambda_Y^2 - 2) + P \tag{35c}$$

$$s_{xx} = 2G\lambda_X^2(\partial u/\partial x) + p \tag{35d}$$

$$s_{yy} = 2G\lambda_Y^2(\partial v/\partial y) + p \tag{35e}$$

$$s_{xy} = G[(\partial v/\partial x)\lambda_X^2 + (\partial u/\partial y)\lambda_Y^2] \tag{35f}$$

$$s_{zz} = (G - H)(\partial u/\partial x)(\lambda_X^2 - \lambda_Y^2) + p \tag{35g}$$

Analogous expressions for compressible materials are derived in the Appendix.

Equilibrium equations

As explained in Chapter 4, the equilibrium equations refer to the final, deformed elements of the body. Thus:

$$\frac{\partial \sigma_{\xi\xi}}{\partial \xi} + \frac{\partial \sigma_{\xi\eta}}{\partial \eta} = 0 \tag{36a}$$

$$\frac{\partial \sigma_{\eta\eta}}{\partial \eta} + \frac{\partial \sigma_{\xi\eta}}{\partial \xi} = 0 \tag{36b}$$

The displacement gradients for the incremental stresses, eqs. 34, are in terms of coordinates x and y in the initially deformed body, so eqs. 36 must be transformed as in Chapter 4, eqs. 1—11. However, because the initial stresses, S_{XX} and S_{YY}, are constant, substitution of eqs. 34 into eqs. 11 of Chapter 4 provides:

$$\partial s_{xx}/\partial x + \partial s_{xy}/\partial y = 0 \tag{37a}$$

$$\partial s_{yy}/\partial y + \partial s_{xy}/\partial x = 0 \tag{37b}$$

Substituting eqs. 35 for a Mooney material into these equilibrium equations:

$$\lambda_X^2(\partial^2 u/\partial x^2) + \lambda_Y^2(\partial^2 u/\partial y^2) + \partial(p/G)/\partial x = 0 \tag{38a}$$

$$\lambda_X^2(\partial^2 v/\partial x^2) + \lambda_Y^2(\partial^2 v/\partial y^2) + \partial(p/G)/\partial y = 0 \tag{38b}$$

where the conditions of incompressibility are:

$$\partial u/\partial x = -\partial v/\partial y; \qquad \lambda_X \lambda_Y = 1 \tag{38c}$$

We eliminate p between eqs. 38 and derive the fourth-order equation by introducing the displacement potential:

$$v = -\partial\psi/\partial x; \qquad u = \partial\psi/\partial y \tag{39a}$$

so eqs. 38 combine to:

$$\partial^4\psi/\partial y^4 + 2\mathbf{A}(\partial^4\psi/\partial x^2\partial y^2) + \mathbf{B}(\partial^4\psi/\partial x^4) = 0 \tag{39b}$$

where:

$$\mathbf{A} = (\lambda_X^4 + 1)/2 \qquad \text{and} \qquad \mathbf{B} = \lambda_X^4 \tag{39c}$$

These final equations are of the same general form as, but much simpler than, those solved in Chapter 5.

Boundary conditions

In Chapter 4 we derived boundary conditions in terms of stresses acting on finally deformed sides of a body whose sides *originally were parallel to* the x- or y-direction (eqs. 20, Chapter 4). We will use the same approach here, but the boundary conditions will be different. The stress components with which we are dealing now, $\sigma_{\xi\xi}$, $\sigma_{\xi\eta}$, and $\sigma_{\eta\eta}$ (Fig. 3C), where:

$$\sigma_{\xi\xi} = S_{XX} + s_{xx} \tag{40a}$$

$$\sigma_{\eta\eta} = S_{YY} + s_{yy} \tag{40b}$$

$$\sigma_{\xi\eta} = s_{xy} = s_{yx} \tag{40c}$$

$$\sigma_{ZZ} = s_{zz} + S_{ZZ} \tag{40d}$$

act on elements in the finally deformed body and are components in the Cartesian ξ-, η- and Z-directions, which are respectively parallel to the X-, Y-, and Z-directions and the x-, y-, and z-directions (Fig. 3). In all the problems with which we will deal, the sides of bodies will remain parallel with X- and Y-directions during the initial, uniform deformation (Fig. 3B). However, during the incremental deformation, the sides of the bodies normally will not be parallel with these directions. For example, element $\tilde{A}\tilde{B}\tilde{C}\tilde{D}$ which is rectangular in the initial state, Fig. 3B, becomes diamond-shaped, $\hat{A}\hat{B}\hat{C}\hat{D}$, in the final state, Fig. 3C. The sides of the diamond-shape are parallel to the curvilinear axes, $\tilde{x}$ and $\tilde{y}$ (Fig. 3c). We must be able to specify the boundary stresses acting parallel and normal to these curvilinear coordinates in order to

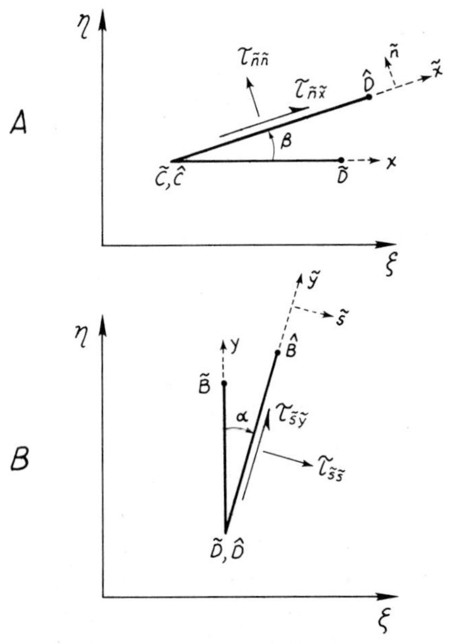

Fig. 5. Boundary stresses acting on interfaces originally parallel to x-direction in A and to y-direction in B. In finite-elasticity theory, boundary conditions must be specified on boundaries in their final positions, after deformation.

solve buckling problems. The boundary stresses are in units of force per unit of area of the *finally deformed body.*

We shall define the boundary stress components as shown in Fig. 5. For example, $\tilde{x}$ and $\tilde{n}$ are orthogonal curvilinear coordinates for a side originally parallel to the x-direction, and $\tilde{y}$ and $\tilde{s}$ are orthogonal curvilinear coordinates for a side originally parallel to the y-direction. Using the notation defined in Fig. 5A:

$$\begin{aligned} \tau_{\tilde{n}\tilde{n}} &= \sigma_{\eta\eta} \cos^2(\beta) + \sigma_{\xi\xi} \sin^2(\beta) - 2\sigma_{\xi\eta} \sin(\beta)\cos(\beta) \\ \tau_{\tilde{n}\tilde{x}} &= (\sigma_{\eta\eta} - \sigma_{\xi\xi}) \cos(\beta)\sin(\beta) + \sigma_{\xi\eta}[\cos^2(\beta) - \sin^2(\beta)] \end{aligned} \qquad (41)$$

However, to the *first-order* in incremental displacement gradients:

$$\sin(\beta) \simeq \partial v/\partial x \qquad (42)$$

so, using eqs. 40, and maintaining first-order accuracy:

$$\tau_{\tilde{n}\tilde{n}} = S_{YY} + s_{yy} \qquad (43a)$$

$$\tau_{\tilde{n}\tilde{x}} = (S_{YY} - S_{XX})(\partial v/\partial x) + s_{xy} \qquad (43b)$$

Similarly, considering the side $\hat{D}\hat{B}$ shown in Fig. 5b:

$$\tau_{\tilde{s}\tilde{s}} = S_{XX} + s_{xx} \qquad (43c)$$

$$\tau_{\tilde{s}\tilde{y}} = (S_{XX} - S_{YY})(\partial u/\partial y) + s_{xy} \qquad (43d)$$

COMPARISON OF PREDICTIONS OF WAVELENGTHS IN SINGLE LAYERS

Fletcher (1977) has derived an exact, first-order solution for folding of a linear viscous layer embedded in a linear viscous medium. The solution is different from that of Biot (e.g., 1960), and of Ramberg (e.g., 1970). It predicts wavelengths which are somewhat different from those predicted by the elementary, thin-plate analysis as well as from that of the Biot and Ramberg theories for ratios of viscosity of layer to medium of less than about 100.

Biot's theory is based on a theory of incremental deformations, introduced by Biezeno and Hencky (1928) for elastic materials. The theory was further developed for elastic materials and extended to viscous and viscoelastic materials by Biot (e.g., 1939, 1965a). The theory is based on what Biot (1965a, p. 26) calls "an important consideration in the whole procedure," which, in fact, is a special assumption. Briefly, the special assumption is that the principal stresses reorient in proportion to the incremental *rotation*, $(1/2)(\partial v_0/\partial x - \partial u_0/\partial y)$ (Biot, 1965a, p. 27). This special assumption is the most important difference between the theory of Biot, the theory introduced by Novozhilov (1953), the theory developed by Green et al.

(1952), and the theory presented in earlier pages. In the exact theory, a finite deformation can be expressed as a rotation followed by a pure strain (Rivlin, 1948, 1949). In our analysis, the rotation is expressed as the angle θ', which is a function of both the initial and incremental deformations, as it must be. The special assumption in Biot's analysis, therefore, is that the rotation is a function *only* of the incremental deformation. At least one other special assumption is apparent in his analysis (1965a, p. 61, eq. 2.12). Ironically, the linearized theory also has the virtue of being simpler than Biot's approximate theory; Biot's theory is very difficult to follow.

We shall briefly develop part of Biot's theory in order to contrast it with the exact, linearized theory. After we compare the basic equations of the two theories, we shall compare wavelengths predicted by various theories.

Comparison of Biot's theory with linearized theory of incompressible Mooney material

The special assumption in Biot's theory, stated above, results in several fundamental differences between the equations of his theory and those of the exact, first-order theory. The incremental stress—strain equations for the linearized theory are given in eqs. 35. Those for Biot's theory (1965a, p. 160) are:

$$s_{xx} = G(\lambda_X^2 + \lambda_Y^2)(\partial u/\partial x) + p \quad (44a)$$

$$s_{yy} = G(\lambda_X^2 + \lambda_Y^2)(\partial v/\partial y) + p \quad (44b)$$

$$s_{xy} = (G/2)(\lambda_X^2 + \lambda_Y^2)[(\partial v/\partial x) + (\partial u/\partial y)] \quad (44c)$$

Thus the symmetry of the stress—strain equations is fundamentally different. In Biot's theory, the initial state affects the incremental moduli for normal and shear stresses equally, so that the material behaves isotropically to incremental deformations. In the linearized theory, in contrast, the initial state affects each incremental modulus differently, eqs. 35, so that the material behaves *as though* it were anisotropic. The material is, in fact, isotropic for the total deformation, as indicated in eqs. 31.

The symmetry of the equilibrium equations also is different for the Biot and the linearized theories. Those for the linearized theory are given in eqs. 38a,b. According to Biot's theory (1965a, p. 161):

$$\lambda_Y^2(\partial^2 u/\partial x^2) + \lambda_Y^2(\partial^2 u/\partial y^2) + \partial(p/G)/\partial x = 0 \quad (45a)$$

$$\lambda_X^2(\partial^2 v/\partial x^2) + \lambda_X^2(\partial^2 v/\partial y^2) + \partial(p/G)/\partial y = 0 \quad (45b)$$

However, the fourth-order equations are identical according to the linearized and the Biot theories, eqs. 39b,c, as can be verified by combining eqs. 45, bearing in mind that $\lambda_X^2\lambda_Y^2 = 1$ for incompressible materials in plane strain.

Now, in solving problems, we solve the fourth-order equation, which is the same for the Biot and the linearized theories. The forms of these solutions, therefore, are the same. Any differences between solutions according to the two theories, therefore, must be results of differences between the ways the stress—strain relations, the boundary conditions, and the mean stress are specified.

Comparison of wavelengths predicted by Biot, Ramberg, Fletcher and linearized theories

The relevant sinusoidal solution of the fourth-order equation (39b) is:

$$\psi = (1/l)[a_2 \cosh(m_1 ly) + b_2 \cosh(m_2 ly)] \cos(lx) \tag{46}$$

for the layer, and:

$$\psi = (1/l)[a_1 \exp(-m_1 ly) + b_2 \exp(-m_2 ly)] \cos(lx) \tag{47a}$$

for the medium. Here:

$$\begin{aligned} m_1 &= +\sqrt{\mathbf{A} + \sqrt{\mathbf{A}^2 - \mathbf{B}}} = \sqrt{1/2}\sqrt{(\lambda_Y^4 + 1) + \sqrt{\lambda_Y^4 - 1}} \\ m_2 &= +\sqrt{\mathbf{A} - \sqrt{\mathbf{A}^2 - \mathbf{B}}} = \sqrt{1/2}\sqrt{(\lambda_Y^4 + 1) - \sqrt{\lambda_Y^4 - 1}} \end{aligned} \tag{47b}$$

and l is the wave number:

$$l = 2\pi/L$$

where L is wavelength. Eqs. 46 and 47 are valid for the Biot theory and the linearized theory.

The displacements for the layer are:

$$v = -\partial\psi/\partial x = [a_2 \cosh(m_1 ly) + b_2 \cosh(m_1 ly)] \sin(lx) \tag{48a}$$

$$u = \partial\psi/\partial y = [a_2 m_1 \sinh(m_1 ly) + b_2 m_2 \sinh(m_2 ly)] \cos(lx) \tag{48b}$$

and the mean stress is:

$$p = l[a_2 m_1 (A' - B' m_1^2) \sinh(m_1 ly) + b_2 m_2 (A' - B' m_2^2) \sinh(m_2 ly)] \sin(lx) \tag{49}$$

Here, for the linearized theory:

$$A' = G\lambda_X^2 \qquad \text{and} \qquad B' = G\lambda_Y^2 \tag{50a}$$

and for the Biot theory:

$$A'_{\mathrm{B}} = G\lambda_Y^2 = B'_{\mathrm{B}} \tag{50b}$$

For the medium:

$$v = [a_1 \exp(-m_1 ly) + b_1 \exp(-m_2 ly)] \sin(lx) \tag{51a}$$

$$u = -[a_1 m_1 \exp(-m_1 ly) + b_1 m_2 \exp(-m_2 ly)] \cos(lx) \tag{51b}$$

$$p = -l [a_1 m_1 (A' - B' m_1^2) \exp(-m_1 ly) + b_1 m_2 (A' - B' m_2^2) \exp(-m_2 ly)] \cdot$$
$$\cdot \sin(lx) \tag{52}$$

We eliminate the constants, a_1, b_1, a_2 and b_2 by means of the boundary conditions: Thus:

$$\text{at } y_L = +T/2; \; y_m = 0 \tag{53}$$

(where subscript L refers to the layer and m to the medium in this and following equations):

$$v_L - v_m = 0$$

$$a_2 \cosh(m_1 lT/2) + b_2 \cosh(m_2 lT/2) - a_1 - b_1 = 0 \tag{54a}$$

$$u_L - u_m = 0$$

$$a_2 m_1 (A'_L - B'_L m_1^2) \sinh(m_1 lT/2) + b_2 m_2 (A'_L - B'_L m_2^2) \sinh(m_2 lT/2)$$
$$+ a_1 m_1 + b_1 m_2 = 0 \tag{54b}$$

$$\tau_{\tilde{n}\tilde{n}})_L - \tau_{\tilde{n}\tilde{n}})_m = 0$$

(Using eqs. 43a, 35e, 50, 51 and 52):

$$l[a_2 m_1 (2G_L \lambda_Y^2 + A'_L - B'_L m_1^2) \sinh(m_1 lT/2) + b_2 m_2 (2G_L \lambda_Y^2 + A'_L - B'_L m_2^2)$$
$$\sinh(m_2 lT/2)] + l[a_1 m_1 (2G_m \lambda_Y^2 + A'_m - B'_m m_1^2)$$
$$+ b_1 m_2 (2G_m \lambda_Y^2 + A'_m - B'_m m_2^2)] = 0 \tag{54c}$$

Finally, using the boundary condition for the shear stresses:

$$\tau_{\tilde{n}\tilde{x}})_L - \tau_{\tilde{n}\tilde{x}})_m = 0$$

(Using eqs. 43b, 35f, 50, 51 and 52):

$$lG_L \lambda_Y^2 [a_2 (1 + m_1^2) \cosh(m_1 lT/2) + b_2 (1 + m_2^2) \cosh(m_2 lT/2)]$$
$$- lG_m \lambda_y^2 [a_1 (1 + m_1^2) + b_1 (1 + m_2^2)] = 0 \tag{54d}$$

These equations are for welded contacts between layer and medium and for the linearized theory.

Equations 54a and 54b are valid for Biot's theory, but eqs. 54c and 54d are not. The boundary conditions for Biot's theory are different from those for the linearized theory. According to Biot (1965a, p. 41), the boundary forces per unit of *initial* surface area, on surfaces originally parallel to the x-direction, are:

$$f_y = S_{22} + s_{22} + S_{22}(\partial u/\partial x) \tag{55a}$$

$$f_x = s_{12} - S_{11}[(\partial v/\partial x) + (\partial u/\partial y)]/2 - S_{22}[(\partial v/\partial x) - (\partial u/\partial y)]/2 \tag{55b}$$

in which we identify subscript 1 with x and 2 with y. Initial shear stresses are assumed to be zero. If we let S_{22} be zero and assume Mooney material, eqs. 55a and 55b become (Biot, 1965a, p. 160):

$$f_y = s_{22} = G(\lambda_X^2 + \lambda_Y^2)(\partial v/\partial y) + p \tag{55c}$$

$$f_x = G\lambda_X^2[(\partial v/\partial x) + (\partial u/\partial y)] \tag{55d}$$

using our notation for shear modulus and initial extension ratios. Therefore, eq. 54c is readily transformed for the boundary condition for Biot's theory by replacing the terms $2G_i\lambda_Y^2$ by $G_i(\lambda_Y^2 + \lambda_Y^2)$, where i is either L or m. Similarly, $G_i\lambda_Y^2$ in eq. 54d is replaced by $G_i\lambda_X^2$. Further A' and B' are defined in eq. 50a.

As explained in Chapter 4, eqs. 54 or similar equations for the Biot theory are solved by setting the determinant of the coefficients of the four constants, a_1, b_1, a_2 and b_2, equal to zero, and selecting the wavelength ratio which corresponds to the minimum λ_Y^2 (note that $\lambda_Y^2 = 1/\lambda_X^2$). We shall not discuss details of the solution here.

Figure 6A compares predictions of wavelength-to-thickness ratios according to theories of Biot (1965a), Ramberg (1970) and Fletcher (1977) and to the theories developed here.

Predictions of wavelength-to-thickness ratios as functions of modulus ratio for the linearized theory developed in Chapter 4 superimpose on those of the linearized theory developed here. The predicted ratios are consistently smaller than those of the elementary, thin-plate solution, also shown in Fig. 6A. For conditions where the confining stress (S_{22} of Biot) is zero, results according to Biot's theory superimpose upon those for the linearized theory. I can only conclude that Biot's series of special assumptions has compensated for one another for this problem. I have not compared solutions for other conditions. According to Fletcher's theory for viscous materials, wavelength-to-thickness ratios are consistently larger than those predicted by the elementary, thin-plate theory (Fig. 6A). Results of our analysis of a linear elastic material, presented in earlier pages, superimpose upon those of Fletcher. Ramberg's solution, based on analysis of folding of a thick plate, but incorporating a special assumption in the way that he introduces the effect of the uniform stresses, predicts wavelength ratios which are consistently less than those for any of the other theories. Thus, as indicated by Fletcher (1977), Ramberg's analysis does not improve upon that of the thin-plate theory developed by Biot and Ramberg.

Thus, we can conclude that exact, first-order solutions for a stiff layer embedded in soft media, with welded contacts, are significantly different according to the elasticity and the viscosity models. Folds in ideal viscous

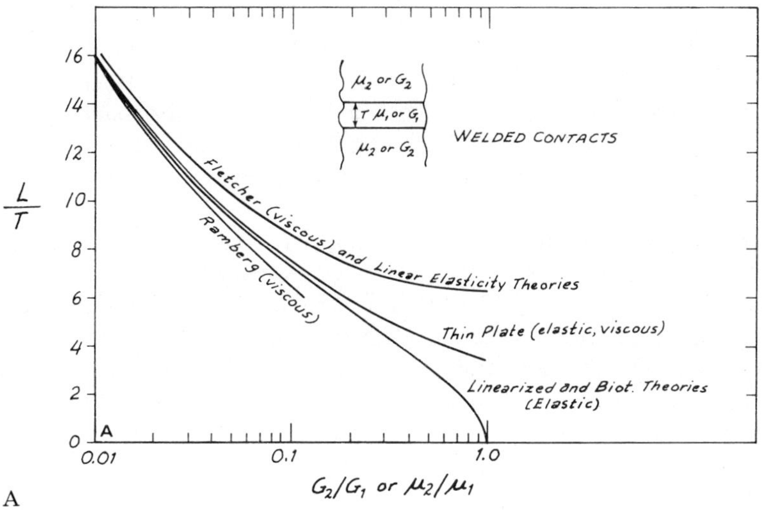

A

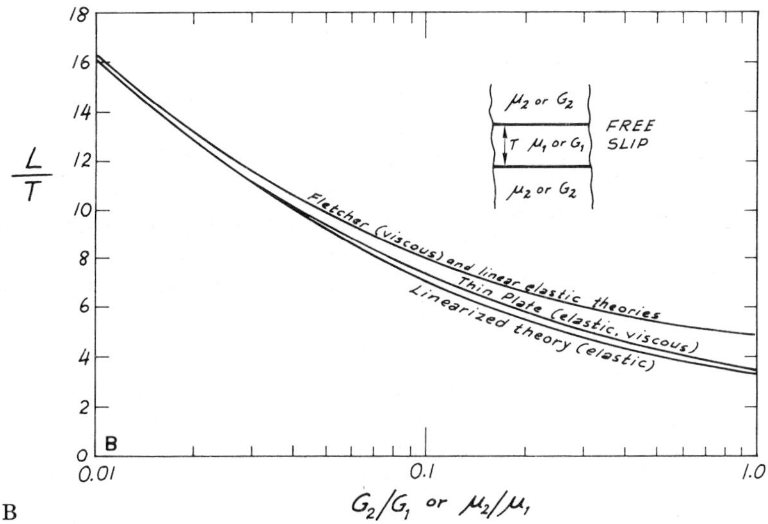

B

Fig. 6. Relations between dominant wavelength and viscosity or modulus ratio of layer to medium according to various theories. A. Contacts between stiff layer and media welded. Thin-plate or elementary theory, based on analysis of plates, is valid for elastic or viscous materials. Exact, first-order analysis of viscous folding, by Fletcher, predicts wavelengths that are somewhat longer than those predicted by elementary theory. Exact, first-order or linearized analysis of elastic folding, derived here, predicts wavelengths somewhat shorter than those of elementary theory. According to theory for viscous materials, wavelength-to-thickness ratio approaches limit of 2π as viscosity ratio approaches unity. According to theory for elastic materials, limit is zero. B. Contacts between stiff layer and media allow free slip. Elementary, thin-plate analysis rather closely predicts wavelengths for either elastic or viscous materials. Both exact theories predict that wavelengths approach finite values as viscosity or elasticity contrast between layer and medium vanishes.

materials tend to develop wavelengths longer than those predicted by the elementary, thin-plate theory, whereas folds in ideal (Mooney or neo-Hookean) elastic materials tend to develop wavelengths shorter than those of the elementary theory, approaching zero as the modulus ratio approaches unity. This is the fundamental difference between folding of elastic and viscous materials. The difference is important only if the modulus ratio or viscosity ratio is less than about 20 : 1.

The linearized and linear theories and the Fletcher theory also have been analyzed for conditions where there is frictionless contact between layer and medium. The wavelength ratios predicted by the linearized theory are nearly the same as those of the elementary theory, both approaching finite wavelength ratios as the modulus ratio approaches unity. Thus slippery contacts are sufficient to allow folding in such materials. Wavelength ratios predicted by Fletcher's theory for viscous material and by the linear elastic theory are only slightly larger than those of the elementary theory (Fig. 6B).

FOLDING OF POWER-LAW MATERIAL WITH MEMORY

The gross stress—strain relations of many rocks tested in the laboratory are nonlinear, especially as failure is approached (e.g., Handin and Hager, 1957). Strains at times of failure of many rocks are small, on the order of 10^{-4}—10^{-2}, so finite extension ratios are on the order of 0.99 to 1.01. Therefore, the nonlinearities cannot be explained in terms of those inherent to large deformations, as discussed in earlier paragraphs. Rather, the nonlinearities apparently reflect a lack of proportionality between measures of stress and measures of strain.

In order to analyze the folding of a nonlinear material we shall assume that some rock behaves according to the power-law model. In the one-dimensional form of this model, the shear strain, for example, is proportional to the shear stress raised to some power greater than unity. Further, we shall assume that strains are so small that geometric nonlinearities are insignificant. Thus, we essentially accept the assumptions of linear elasticity theory, that stresses are the same whether referred to deformed or to undeformed boundaries, and that strains and rotations are infinitesimal. We shall consider a power-law material with memory which is of the same form used by Fletcher (1974) to study single-layer folding of a power-law material without memory. The generalization of the one-dimensional power-law model is due to Calladine and Drucker (1962).

Our analysis of folding of power-law materials will be based on the growth of perturbations of interfaces between layers. The analysis, therefore, is not based on considerations of stability. We will consider both uniform and incremental stresses and strains, but the primary effects of the uniform stresses are to modify the effective elastic moduli for the incremental deformations and to induce shear stresses along perturbed boundaries.

Stress—strain relations to first order

The stress—strain relations for an incompressible, power-law material can be defined in terms of the second invariant, J_2, of the deviatoric stress tensor (e.g., Prager, 1961, p. 50), where, for plane strain:

$$J_2 = \sqrt{[(\sigma_{xx} - \sigma_{yy})/2]^2 + \sigma_{xy}^2} \tag{56}$$

Thus, following Fletcher (1974, eq. 1):

$$\sigma_{xx} = [2g/J_2^{(m-1)}](\partial u/\partial x) + p' \tag{57a}$$

$$\sigma_{yy} = [2g/J_2^{(m-1)}](\partial v/\partial y) + p' \tag{57b}$$

$$\sigma_{xy} = [g/J_2^{(m-1)}](\partial v/\partial x + \partial u/\partial y) \tag{57c}$$

where g is a constant with dimensions of stress to the mth power, m is a constant ($m \geqslant 1$), and p' is the mean stress:

$$p' = (\sigma_{xx} + \sigma_{yy})/2 \tag{57d}$$

The material is incompressible, so:

$$\partial u/\partial x = -\partial v/\partial y \tag{57e}$$

Here:

$$u = U + u_0; \quad v = V + v_0; \quad \sigma_{xx} = S_{xx} + s_{xx}; \quad \sigma_{yy} = S_{yy} + s_{yy};$$

$$\sigma_{xy} = s_{xy}; \quad p' = P + p \tag{57f}$$

where U and V are displacements associated with the uniform deformation, and u_0 and v_0 are associated with the nonuniform, incremental deformation. Similarly, the stresses S are the uniform stresses and the stresses s are the nonuniform stresses. P and p are the mean stresses for the uniform and nonuniform states, respectively.

Now, we will derive the stress—strain relations for the nonuniform, incremental stresses by substituting eqs. 57f into eqs. 57a—57c. To the first order of approximation in the nonuniform stresses:

$$J_2 = |(\sigma_{xx} - \sigma_{yy})/2| = |(S_{xx} - S_{yy} + s_{xx} - s_{yy})/2| \tag{58a}$$

Thus, the difference between eqs. 57a and 57b becomes:

$$[(S_{xx} - S_{yy} + s_{xx} - s_{yy})/2]^m = 2g(\partial U/\partial x + \partial u_0/\partial x) \tag{58b}$$

Using the binomial series expansion:

$$m[(S_{xx} - S_{yy})/2]^{(m-1)}[(s_{xx} - s_{yy})/2] = 2g(\partial u_0/\partial x) + 2g(\partial U/\partial x)$$

$$- [(S_{xx} - S_{yy})/2]^m \tag{58c}$$

to first order in nonuniform stresses.

The last two terms in eq. 58c sum to zero because they represent the stress-strain relation for the uniform state:

$$(S_{xx} - S_{yy})/2 = 2\hat{G}(\partial U/\partial x)$$

or:

$$S_{xx} = 2\hat{G}(\partial U/\partial x) + P \tag{59a}$$

$$S_{yy} = 2\hat{G}(\partial V/\partial y) + P \tag{59b}$$

where:

$$\hat{G} = g/|[(S_{xx} - S_{yy})/2]|^{(m-1)} \tag{59c}$$

$$P = (S_{xx} + S_{yy})/2 \tag{59d}$$

Accordingly, the remainder of eq. 58c is the stress—strain relation for the nonuniform state:

$$s_{xx} = (2\hat{G}/m)\,(\partial u_0/\partial x) + p \tag{60a}$$

$$s_{yy} = (2\hat{G}/m)\,(\partial v_0/\partial y) + p \tag{60b}$$

Similarly, we derive from eq. 43d that:

$$s_{xy} = \hat{G}\,[(\partial y_0/\partial x) + (\partial u_0/\partial y)] \tag{60c}$$

These are the stress—strain relations for the nonuniform stress state, valid to first order in the nonuniform stresses and displacement gradients. As indicated by Fletcher (1974), the effective normal modulus, $2\hat{G}/m$, can be smaller than the effective shear modulus, $\hat{G}$, if $m > 2$.

Equilibrium equations

We derive the equilibrium equations for the power-law material by substituting rheological equations (60) into the equilibrium equations:

$$[\hat{G}(2-m)/m](\partial^2 u_0/\partial x^2) + \hat{G}(\partial^2 u_0/\partial y^2) + \partial p/\partial x = 0 \tag{61a}$$

$$\hat{G}(\partial^2 v_0/\partial x^2) + [\hat{G}(2-m)/m](\partial^2 v_0/\partial y^2) + \partial p/\partial y = 0 \tag{61b}$$

where $\hat{G}$ is defined in eq. 59c. Note that the symmetry of these equations is fundamentally different from that of the equations for the linearized theory, eq. 38.

The equilibrium equations can be written as a single, fourth-order equation by introducing the displacement function, ψ, and eliminating p between equilibrium equations (61):

$$\partial^4\psi/\partial y^4 + 2\mathbf{A}(\partial^4\psi/\partial x^2\partial y^2) + (\partial^4\psi/\partial x^4) = 0 \tag{62a}$$

The mean stress, p, can be derived from eq. 61a or 61b:

$$p = -[\hat{G}(2-m)/m](\partial u_0/\partial x) - (\hat{G})\int(\partial^2 u_0/\partial y^2)\mathrm{d}x + p_0 \tag{62b}$$

where p_0 is a constant. Here:

$$u_0 = \partial\psi/\partial y \tag{62c}$$

$$v_0 = -\partial\psi/\partial x \tag{62d}$$

$$\mathbf{A} = (2-m)/m \tag{62e}$$

General sinusoidal solution

The relevant sinusoidal solution to eq. 62a can be derived from a trial solution of the form:

$$\psi = f(y)\cos(lx)$$

where $f(y)$ is some function of y. In this way we derive the solution:

$$\psi = (1/l)[\{C_1\cos(\alpha ly) + C_2\sin(\alpha ly)\}\exp(-\beta ly)$$
$$+ \{C_3\cos(\alpha ly) + C_4\sin(\alpha ly)\}\exp(\beta ly)]\cos(lx) \tag{63a}$$

where ψ is the displacement function, l is the wave number, equal to $2\pi/L$, where L is wavelength, and (e.g., Johnson, 1970, p. 95):

$$\beta = +[(1/2)(1+\mathbf{A})]^{1/2} \tag{63b}$$

$$\alpha = +[(1/2)(1-\mathbf{A})]^{1/2} \tag{63c}$$

in which $\mathbf{A}$ is defined in eq. 62e.

Substituting eq. 63a into eqs. 62c and 62d, we get:

$$u_0 = [\{-(C_1\alpha + C_2\beta)\sin(\alpha ly) + (C_2\alpha - C_1\beta)\cos(\alpha ly)\}\exp(-\beta ly)$$
$$+ \{-(C_3\alpha - C_4\beta)\sin(\alpha ly) + (C_4\alpha + C_3\beta)\cos(\alpha ly)\}\exp(\beta ly)]\cos(lx) \tag{64a}$$

$$v_0 = [\{C_1\cos(\alpha ly) + C_2\sin(\alpha ly)\}\exp(-\beta ly)$$
$$+ \{C_3\cos(\alpha ly) + C_4\sin(\alpha ly)\}\exp(\beta ly)]\sin(lx) \tag{64b}$$

Finally, substituting eq. 64a into eq. 62b, we derive an expression for the mean stress, p:

$$p = l[\hat{G}(2-m)/m][\{-(C_1\alpha + C_2\beta)\sin(\alpha ly) + (C_2\alpha - C_1\beta)\cos(\alpha ly)\}\exp(-\beta ly)$$
$$+ \{-(C_3\alpha - C_4\beta)\sin(\alpha ly) + (C_4\alpha + C_3\beta)\cos(\alpha ly)\}\exp(\beta ly)]\sin(lx)$$

$$-l[\hat{G}][\{\{C_1(\alpha^3-3\alpha\beta^2)+C_2(3\alpha^2\beta-\beta^3)\}\sin(\alpha ly)+\{C_1(3\alpha^2\beta-\beta^3)$$

$$+C_2(3\alpha\beta^2-\alpha^3)\}\cos(\alpha ly)\}\exp(-\beta ly)+\{\{C_3(\alpha^3-3\alpha\beta^2)+C_4(\beta^3-3\beta\alpha^2)\}$$

$$\sin(\alpha ly)+\{C_3(\beta^3-3\beta\alpha^2)+C_4(3\alpha\beta^2-\alpha^3)\}\cos(\alpha ly)\}\exp(\beta ly)]\sin(lx)+p_0^* \tag{64c}$$

Equations 64 along with boundary conditions (8) and rheological equations (60) are sufficient to solve the problem of the folding of an initially deflected layer in a soft medium or to solve problems involving multilayers. We shall not solve the problems here, however, because Fletcher (1974) already has done so for materials without memory. Instead, we will derive a solution based on thin-plate theory for the folding of a layer in a soft medium.

Approximate, thin-plate solution — single layer

The reaction of the medium to the sinusoidal deflection of one surface is derived as outlined in previous paragraphs. The relevant sinusoidal solution is:

$$\psi=(1/l)[C_1\cos(\alpha ly)+C_2\sin(\alpha ly)]\exp(-\beta ly)\cos(lx) \tag{65a}$$

where:

$$\alpha=\beta\sqrt{m-1} \tag{65b}$$

$$\beta=\sqrt{1/m} \tag{65c}$$

Using the boundary condition that the shear stress vanishes at the interface, $y=0$, between the layer and the medium, we derive:

$$u_0=-C_1[(1/\alpha)\sin(\alpha ly)]\exp(-\beta ly)\cos(lx) \tag{66a}$$

$$v_0=(C_1/\alpha)[\alpha\cos(\alpha ly)+\beta\sin(\alpha ly)]\exp(-\beta ly)\sin(lx) \tag{66b}$$

$$p=(C_1 2l\hat{G}/\alpha)[(\beta^2-\alpha^2)\sin(\alpha ly)+\alpha\beta\cos(\alpha ly)]\exp(-\beta ly)\sin(lx) \tag{66c}$$

where C_1 is an arbitrary constant.

At the interface, $y=0$, the normal stress is:

$$s_{yy}=2C_1 l\hat{G}\beta\sin(lx)$$

and the displacement, v_0, is:

$$v_0=C_1\sin(lx)$$

so that the stress—displacement relationship for the interface is:

$$s_{yy}=2\hat{G}l\sqrt{1/m}\,v_0;\qquad y=0 \tag{67}$$

The distributed load, q, per unit of length, used in plate theory, is proportional to twice the normal stress defined in eq. 67 because the medium is above and below the layer (e.g., see Johnson, 1970, p. 266). Thus:

$$q = 8\hat{G}_m \pi\sqrt{1/m_m}\, b v_0/L \tag{68}$$

where b is unit width of the plate, L is wavelength, and m_m and $\hat{G}_m$ are constants for the medium, defined as in eqs. 57 and 59c.

The differential equation for an initially deflected stiff layer in a soft medium is:

$$(4\hat{G}_l/m_l)I'(\mathrm{d}^4 v_0/\mathrm{d}x^4) - (S_{xx})_l(\mathrm{d}^2 v_0/\mathrm{d}x^2) + [8\hat{G}_m \pi v_0/(t\sqrt{m_m}\, L)]$$

$$= (S_{xx})_l(\mathrm{d}^2 v_i/\mathrm{d}x^2) \tag{69a}$$

where v_i is initial deflection, assumed to be sinusoidal, v_0 is the additional deflection, produced by the layer-parallel stress, $(S_{xx})_l$, within the stiff layer, and t is the thickness of the stiff layer.

$$v_i = \delta_i \sin(2\pi x/L)$$

$$I' = t^2/12 \tag{69b}$$

The solution to eq. 69a is:

$$v = v_0 + v_i = \delta_i \sin(2\pi x/L)/[1 + \{(S_{xx})_l/(S_{xx})_{\mathrm{crit}}\}] \tag{70a}$$

where:

$$(S_{xx})_{\mathrm{crit}} = [16\hat{G}_l \pi^2 I'/m_l L^2] + [2\hat{G}_m L/(t\pi\sqrt{m_m})] \tag{70b}$$

The wavelength requiring the minimum critical stress is:

$$(L/t) = 2\pi\sqrt[3]{\hat{G}_l\sqrt{m_m}/(6m_l\hat{G}_m)} \tag{71}$$

This is identical in form to the first approximation of Fletcher's (1974, eq. 9) solution for a power-law material without memory. It is of the same form as the expression for the wavelength of a stiff elastic layer, G, in a soft elastic medium, G_0:

$$(L/t) = 2\pi\sqrt[3]{G/6G_0}$$

because $\hat{G}_l/m_l$ is the effective modulus of the layer and $\hat{G}_m/\sqrt{m_m}$ is the effective modulus of the medium.

The wavelength is a function of the initial stress state because both $\hat{G}_m$ and $\hat{G}_l$ are functions of the initial stresses. We can express the wavelength, eq. 71, in terms of the initial, layer-parallel strain, $\partial U/\partial x$:

$$(L/t) = 2\pi\{(\sqrt{m_m}/6m_l)\{g_l^{(1/m_l)}/g_m^{(1/m_m)}\}(-\partial U/\partial x)^{\{(m_m - m_l)/m_m m_l\}}\}^{1/3} \tag{72}$$

where g and m are constants for the layer and medium.

Approximate, thin-plate solution — multilayer

The solution for the *Ramberg wavelength* of folds in a multilayer comprised of identical layers with frictionless contacts is basically the same, regardless of the rheological properties of the layers and media. The wavelength is (eqs. 5, Chapter 9):

$$L_{\mathrm{R}} = 2T\sqrt{\pi n/N}\sqrt[4]{1/[12(1-D)]} \qquad (73a)$$

where:

$$D = 1/[1 + (\pi/2n)\{\mathrm{ctn}(\pi/2n) + \tan(\pi/2n)\}] \qquad (73b)$$

$$n\,\mathrm{ctn}(\pi/2n) = [\hat{G}_l\sqrt{m_m}/\hat{G}_m\,m_l]\,(L_{\mathrm{R}}/T) \qquad (73c)$$

Here N is the number of layers, T is the total thickness of the multilayer stack and n is a factor determined by eq. 73c. If the medium is much stiffer than the multilayer, $n = 1$, $D = 0$, and the wavelength becomes the *Biot wavelength*:

$$L_{\mathrm{B}} \simeq 1.90\,T\sqrt{1/N} \qquad (74)$$

which is independent of the properties.

Thus, in order to determine the *Ramberg wavelength* we can use Figs. 12 of Chapter 9 by using the relation:

$$(B_0/B_1) \to (\hat{G}_m\,m_l/\hat{G}_l\sqrt{m_m}) \qquad (75)$$

because eqs. 73 are identical in form to eqs. 5 of Chapter 9.

PRELIMINARY EVALUATION OF SINGLE-LAYER AND MULTILAYER FOLDING OF ROCK WITH MEMORY

In Chapter 9 we derived tangent moduli for several rock types tested by Handin and Hager (1957) and showed that low tangent moduli favor the development of folds. Now we will determine the apparent constants of the same rocks, assuming that they behave as power-law materials with limited memory. We will be able to show in this way that single-layer folds in rocks with the properties reported by Handin and Hager (1957) are quite unlikely but that multilayer folds are possible.

We shall use experimental data on rocks reported by Handin and Hager (1957) and on Chatahoochee sand by Vesić and Clough (1968). The data were not presented for the purposes for which we will use them, which is always an exercise fraught with uncertainties. Further, we will assume that the materials are incompressible. This assumption is known to be incorrect for Chatahoochee sand for which volumetric strains are on the order of normal strains. We do not know the volumetric strains in rocks tested by Handin and Hager (1957), but they probably are significant for some of the rocks.

Thus, the use of these data to determine the constants corresponding to an incompressible, power-law material with memory is questionable. Nevertheless, the data probably are adequate to draw qualitative, preliminary conclusions, which is the sole aim here.

I have plotted data presented by Vesić and Clough (1968) and Handin and Hager (1957) on logarithmic graph paper in order to determine the constants, g and m, in the rheological equation:

$$[(S_{xx} - P_c)/2]^m = 2g\partial U/\partial x \quad (76)$$

where P_c is confining pressure, S_{xx} is axial stress and $\partial U/\partial x$ is axial strain. The data are roughly linear on logarithmic graph paper for certain ranges of strain, shown in the last two columns of Table II. The first column identifies the rock type and the source of test data. The third and fourth columns indicate apparent values of the rheological constants, m and g. The units of g are kilobars to the mth power. The values of the exponent, m, range from 1.4 to 4.6. The variability of m can be high for the same rocks tested under different confining pressures; viz. 1.7 for "shale" at 1 kbar and 4.6 at 2 kbar. It is high also for different orientations of test samples under the same confining pressure; viz. 2.8 for one orientation of Yule marble and 4.2 for

TABLE II

Apparent parameters according to power-law model

Rock	$-P_c$ (kbar)	m	$2g$ (kbar)^m	$-\partial U/\partial x$ min.	$-\partial U/\partial x$ max.
Marianna ls. (H & H *, fig. 4)	2	2.2	36	0.006	0.15
Marianna ls.	1	2.47	18.5	0.006	0.1
Barnes ss. (H & H, fig. 14)	2	4.74	351	0.015	0.08
Weeks Cr. ss. (H & H, fig. 24)	2	2.36	23	0.01	0.08
Green R. sh. (H & H, fig. 26)					
normal to layering	2	2.36	23	0.01	0.12
normal to layering	0.5	3.15	4	0.03	0.12
parallel to layering	1	4.0	0.2	0.02	0.04
parallel to layering	2	3.1	10	0.02	0.12
"shale" (H & H, fig. 25)	2	4.6	120	0.01	0.05
"shale" (H & H, fig. 25)	1	1.7	18	0.01	0.06
Repetto slt. (H & H, fig. 30)	2	1.54	18	0.02	0.08
Repetto slt. (H & H, fig. 30)	1	1.9	14	0.02	0.08
Chatahoochee sand (V & C **, fig. 9)	0.6	1.4	3.3	0.05	0.2
Chatahoochee sand (V & C, fig. 9)	0.4	1.5	2.6	0.05	0.2
Yule marble (H & H, figs. 17 & 18)					
"T cylinders"	1	2.8	1.11	~0	0.12
"L cylinders"	1	4.2	167	~0	0.2

* Handin and Hager, 1957.

** Vesić and Clough, 1968.

another; 2.4 for Green-River shale shortened parallel to layering and 3.1 for the same shale shortened normal to layering.

The values of g also are quite variable. g is equal to one-half the inverse of the strain corresponding to a stress difference, $S_{xx} - P_c$, equal to two, because in this case eq. 55a becomes:

$$2g = |-1/\partial U/\partial x|; \; |(S_{xx} - P_c)/2| = 1 \tag{77}$$

The larger is g, the smaller is the strain corresponding with a given stress difference and constant m. Values of $2g$ range from 0.2 $(\text{kbar})^4$ for Green River Shale to 351 $(\text{kbar})^{4.74}$ for Barnes sandstone (Table II). This parameter also is highly variable from rock type to rock type and even for the same rock type oriented differently, or subjected to different confining pressures. Thus it ranges from 1.1 $(\text{kbar})^{2.8}$ to 4.2 $(\text{kbar})^{4.2}$ for samples of Yule marble and from 10 $(\text{kbar})^{3.1}$ to 23 $(\text{kbar})^{2.4}$ for Green River Shale with different orientations. It is 120 $(\text{kbar})^{4.6}$ for the "shale".

Theoretical wavelengths of folds requiring the minimum axial shortening were computed for single layers of various rocks confined by media of other kinds of rock, using the elementary theory presented in eq. 53d. The constants presented in Table II were used in the computations and the values of axial strain, shown in column 3 of Table III, fall within the ranges indicated in the last two columns of Table II. The fourth column in Table III shows the theoretical ratios of wavelength to thickness of the layer. Where the theoretical wavelength ratio is less than about 3.46, the medium behaves as though it is stiffer than the layer, so the solution is invalid. This ratio depends upon the amount of axial strain, as shown for example for theoretical folding of a layer of Chatahoochee sand in a medium of Green River shale (Table III). The ratio is 2.9, which is less than the minimum allowable value of 3.46, for axial strains of -0.01, but is 4.6 for axial strains of -0.2.

The critical strain $(\partial U/\partial x)_{crit}$, the value of the axial strain required for very large amplification of folds, is an indication of the amount of strain required for significant folding. Its magnitude is invariably larger than the allowable strain indicated in the last column of Table II for each rock type and test condition. Thus, the only possible combination of rocks considered which might produce significant folding in single layers would be a layer with properties of Repetto siltstone within a medium with properties of Green River shale. The largest strain allowed for Repetto siltstone is on the order of -0.08 and the critical strain is -0.14.

The column labeled δ/δ_0 in Table III indicates the amplification of sinusoidal waves with an initial amplitude of δ_0. The amplification in all cases is very small, less than 2.2.

Thus, we must conclude that, if these data are representative of rocks under conditions where the rocks behave much as power-law materials with memory, single-layer folds cannot be developed. Single-layer folds, therefore, probably imply viscous-like behaviors of rock. This conclusion is necessarily tentative because we cannot be certain that the test data are representative.

TABLE III

Wavelength-to-thickness ratios for single-layer folds according to power-law model

Layer or layers	Medium	$-\partial U/\partial x$	L/t **	$-(\partial U/\partial x)_{crit}$	δ/δ_0	$\hat{G}_m m_l/(\hat{G}_l\sqrt{m_m})$
Marianna ls.	"shale"					
(2 kbar) *	(2 kbar)	0.01	2.9			1.7
		0.025	3.1			1.4
		0.05	3.3			1.2
Marianna ls.	Green R. sh.					
(2 kbar)	(Par. lay., 1 kbar)	0.05	5.4	0.1	2.2	0.6
		0.025	5.1			0.7
		0.01	4.8			0.8
Weeks Cr. ss.	Green R. sh.					
(2 kbar)	(Par. lay., 2 kbar)	0.01	3.3			1.1
		0.05	3.4			1.01
Weeks Cr. ss.	"shale"	0.01	2.7			2.1
(2 kbar)	(2 kbar)	0.05	3.0			2.3
Weeks Cr. ss.	Green R. sh.					
(2 kbar)	(Par. lay., 1 kbar)	0.01	4.5	0.22		0.5
		0.05	4.9	0.14	1.57	0.35
Repetto slt.						
(1 kbar)	(Par. lay., 1 kbar)	0.02	4.8			0.4
		0.08	5.8	0.28	1.4	0.2
Chatahoochee						
sand (0.6 kbar)	(Par. lay., 1 kbar)	0.01	2.9			1.7
		0.1	4.1			0.6
		0.2	4.6	0.95	1.3	0.4

* Confining pressure.

** If $L/t < 2\pi\sqrt[3]{1/6} \simeq 3.46$, the "medium" is stiffer than the "layer".

The conclusion is quite interesting, even if tentative.

The conclusion cannot be extended to multilayered rocks because folds in multilayers with frictionless contacts generally require much lower shortening than folds in single layers. The last column in Table III presents the effective modulus ratio for a multilayer confined by a medium. For example, for a multilayer of Marianna limestone confined by "shale", the effective modulus ratio is 0.59 for an axial strain of —0.01 and is 0.83 for an axial strain of —0.05. We could use these ratios in eq. 73c presented in earlier paragraphs to predict the ratio of wavelength to total thickness for the multilayer. Or, we could use Figs. 12 of Chapter 9 to compute wavelength ratios, merely by replacing B_0/B_1 in those figures with the values presented in the last column of Table III, that is, we merely use relation 75.

The minimum critical stress ratios presented in Figs. 12 of Chapter 9 are valid for a power-law material if we replace:

$$(\sigma_{crit})m_l/\hat{G}_l \rightarrow (\sigma_{crit})/B_1 \tag{78}$$

in Figs. 12. Thus, we multiply σ_{crit}/B_1 presented in Figs. 12 (Chapter 9) by values of G_l/m_l to compute the critical stress required for folding of the multilayer. Further, we divide the values of σ_{crit}/B_1 in those figures by 2 m_l to compute the critical shortening required for folding. Thus:

$$-(\partial U/\partial x)_{crit} = (\sigma_{crit}/B_1)/(2m_l) \tag{79}$$

in Figs. 12 of Chapter 9.

For example, suppose that a multilayer is comprised of frictionless layers of Marianna limestone in a medium of "shale". The amount of shortening for folding will be assumed to be the value for failure of the "shale" in laboratory tests, about —0.05 (Table II). The exponent, m, for the limestone is 2.2 (Table I). The effective modulus ratio, B_0/B_1, according to eq. 75 and Table III is 1.2. According to eq. 79:

$$(\sigma_{crit}/B_1) = -2m_l(\partial U/\partial x) \simeq -2(2.2)(-0.05) \simeq 0.22$$

Finally, according to Fig. 12A of Chapter 9, the number of layers of limestone can be less than ten to satisfy these conditions. If the strain were —0.01 rather than —0.05, there would have to be about 35 layers for folds to develop in the multilayer.

Therefore, although single-layer folds cannot develop in rocks with the properties reported here, multilayer folds can readily develop if there is a sufficient number of layers in the multilayer.

APPENDIX

I am placing a series of derivations in this Appendix which originally were in the text of Chapter 10. The derivations are peripheral relative to those presented in the text, yet they are necessary for an extension of analyses presented there. The derivation of invariants was disruptive to the flow of the text, so it is placed here also.

Invariants

We need the strain invariants for analysis of three-dimensional deformation in the text, so we will derive them here. The analysis is similar to that for two dimensions presented in the text. Let a line segment of length ds_0 be oriented arbitrarily within the undeformed body. Its length can be expressed in terms of its projections on the Cartesian axes so that:

$$(ds_0)^2 = dX^2 + dY^2 + dZ^2 \tag{80a}$$

After the two successive deformations, described above, the length of the line element is ds. Its length can be expressed as the product of the length ds_0 and the finite extension ratio λ. Further, its final length can be expressed in terms of its projection on the Cartesian axes for the finally deformed body (Fig. 3C), so:

$$(ds)^2 = (ds_0\lambda)^2 = (d\xi)^2 + (d\eta)^2 + (d\zeta)^2 \tag{80b}$$

where ζ is the axis parallel to the z-direction. We can write equations similar to eqs. 21 but for three-dimensional deformations. Thus, for example:

$$d\xi = (1 + u_x)\lambda_X dX + u_y \lambda_Y dY + u_z \lambda_Z dZ$$

and similar expressions can be derived for $d\eta$ and $d\zeta$. Doing so, and substituting the results into eq. 80b, we derive:

$$ds^2 = g_{xx}(dX)^2 + 2g_{xy}\, dX dY + g_{yy}(dY)^2 + g_{zz}(dZ)^2 + 2g_{xz}dX dZ + 2g_{yz}dY dZ \tag{80c}$$

where, for example:

$$g_{xx} = [(1 + u_x)^2 + v_x^2 + w_x^2]\lambda_X^2;\ g_{xy} = [w_x w_y + (1 + u_x)u_y + (1 + v_y)v_x]\lambda_X\lambda_Y$$

(the other g's are defined in Table I). However, according to eq. 80a:

$$ds_0^2 = (ds/\lambda)^2 = (dX^2 + dY^2 + dZ^2) \tag{80d}$$

Now, eq. 80c is an expression for an ellipsoid and eq. 80d is an expression for a sphere the radius of which is proportional to the square of the inverse of the finite extension ratio, λ. The ellipsoid is an inverse strain ellipsoid because its axes are proportional to the inverse of λ^2. By eliminating ds between eqs. 80c and 80d, we derive the condition for the diameter of the sphere to be equal to one principal axis of the ellipsoid:

$$dX[\lambda^2 dX] + dY[\lambda^2 dY] + dZ[\lambda^2 dZ] = dX[g_{xx}dX + g_{xy}dY + G_{xz}dZ]$$

$$+ dY[g_{yy}dY + g_{xy}\, dX + g_{yz}\, dZ] + dZ[g_{zz}dZ + g_{xz}\, dX + g_{yz}\, dY] \tag{80e}$$

This implies that coefficients of dX, dY and dZ within the brackets individually satisfy the equation so we have three equations of the form:

$$(g_{xx} - \lambda^2)dX + g_{xy}\,dY + g_{xz}\,dZ = 0$$

Or, writing eq. 80e in matrix form:

$$\begin{bmatrix} dX \\ dY \\ dZ \end{bmatrix} \begin{bmatrix} (g_{xx} - \lambda^2) & g_{xy} & g_{xz} \\ g_{xy} & (g_{yy} - \lambda^2) & g_{yz} \\ g_{xz} & g_{yz} & (g_{zz} - \lambda^2) \end{bmatrix} \begin{bmatrix} dX \\ dY \\ dZ \end{bmatrix} = 0 \tag{80f}$$

The increments dX, dY and dZ are arbitrary, so the determinant of the middle matrix must be zero. The principal extension ratios, therefore, are the three roots of the equation:

$$(\lambda^2)^3 - I_1(\lambda^2)^2 + I_2\lambda^2 - I_3 = 0 \tag{80g}$$

The quantities I_1, I_2 and I_3 are the three invariants of the finite extension ratios and are defined in Table I. Listed in Table I are the invariants for the sequence of two deformations described above and represented by I^* as well as the invariants for the initial deformations, represented by I^0.

Incremental stresses for compressible materials

We will derive expressions for the incremental stresses in a compressible material. The initial stresses are given by eqs. 27 if we use invariants I^0 (Table I). The principal stresses, resulting from the two stages of deformation are, for plane strain:

$$\sigma_1 = (2/\sqrt{I_3^*})[\lambda_1^2\{(\partial W/\partial I_1^*) + (\partial W/\partial I_2^*)\} + I_3^*\{(\partial W/\partial I_2^*) + (\partial W/\partial I_3^*)\}]; \quad \lambda_Z = 1$$

and the same expression for σ_3, except that λ_1^2 is replaced by λ_3^2. Further, the principal stress in the z-direction is given by eq. 27d if we replace I by I^*, λ_X^2 by λ_1^2 and λ_Y^2 by λ_3^2. Substituting these equations into eqs. 32, eliminating the principal extension ratios and the trigonometric functions of θ' by means of eqs. 23, and replacing I_3^* by its identity given in Table I, we derive, to first order in the incremental displacement gradients:

$$s_{xx} = 2[(\lambda_X/\lambda_Y)(W_1^* + W_2^*)(u_x - v_y) + \lambda_X\lambda_Y(W_2^* + W_3^*)(u_x + v_y)$$
$$+ (\lambda_X/\lambda_Y)(W_1^* + W_2^* - W_1^0 - W_2^0) + \lambda_X\lambda_Y(W_2^* + W_3^* - W_2^0 - W_3^0)] \tag{81a}$$

$$s_{yy} = 2[(\lambda_Y/\lambda_X)(W_1^* + W_2^*)(v_y - u_x) + \lambda_X\lambda_Y(W_2^* + W_3^*)(u_x + u_y)$$
$$+ (\lambda_Y/\lambda_X)(W_1^* + W_2^* - W_1^0 - W_2^0) + \lambda_X\lambda_Y(W_2^* + W_3^* - W_2^0 - W_3^0)] \tag{81b}$$

$$s_{xy} = 2(W_1^* + W_2^*)[v_x(\lambda_X/\lambda_Y) + u_y(\lambda_Y/\lambda_X)] \tag{81c}$$

where:

$$W_1^* = \partial W/\partial I_1^*; \quad W_2^* = \partial W/\partial I_2^*; \quad W_3^* = \partial W/\partial I_3^*$$

$$W_1^0 = \partial W/\partial I_1^0; \quad W_2^0 = \partial W/\partial I_2^0; \quad W_3^0 = \partial W/\partial I_3^0 \tag{81d}$$

and where I^* and I^0 are defined in Table I. Thus we have expressions for the incremental stresses for both incompressible, eqs. 35, and compressible, eqs. 81, substances. A relatively simple strain—energy function for compressible, rubber foams has been derived by Blatz (1969, eq. 55). The function involves the constants G and H of the Mooney material, eq. 30b, as well as a constant that is analogous to Poisson's ratio of linear elasticity.

Relations for initially perturbed boundaries

All the equations we have derived thus far assume that there is first a finite, initial deformation and then an infinitesimal deformation. Suppose, however, that the boundaries of a layer are initially perturbed, as Fletcher (1974, 1977) does for study of folding of viscous materials. Most of the equations change fundamentally if we assume initial perturbations. In this case we assume that the stress state is comprised *simultaneously* of uniform stresses, $S_{\xi\xi}$ and $S_{\eta\eta}$, plus *nonuniform* stresses, $s_{\xi\xi}$, $s_{\eta\eta}$, and $s_{\xi\eta}$, in the final state of the body. We will assume that the deformation is a result of *uniform* displacement gradients, U_X *and* V_Y ($U_Y = 0 = V_X$) plus nonuniform displacement gradients, u_X, u_Y, v_X, v_Y, where the subscripts imply differentiation with respect to coordinates X and Y in the undeformed body (Fig. 3A). The stress—strain relations can be derived by inspection from eqs. 33 by setting λ_X and λ_Y in *those* equations equal to unity and replacing u_x, for example, by $U_X + u_X$. To the first order in the nonuniform displacement gradients, eqs. 33 therefore become:

$$\sigma_{\xi\xi} = 2(W_1 + W_2)(\lambda_X^2 + 2u_X\lambda_X - 1) + P_0 \tag{82a}$$

$$\sigma_{\eta\eta} = 2(W_1 + W_2)(\lambda_Y^2 + 2v_Y\lambda_Y - 1) + P_0 \tag{82b}$$

$$\sigma_{\xi\eta} = 2(W_1 + W_2)(\lambda_X v_X + \lambda_Y u_Y) \tag{82c}$$

where W_1 and W_2 are defined as in eqs. 81d. If we assume a Mooney material, eq. 30b, these can be written as:

$$\sigma_{\xi\xi} = S_{\xi\xi} + s_{\xi\xi}; \quad \sigma_{\eta\eta} = S_{\eta\eta} + s_{\eta\eta}; \quad \sigma_{\xi\eta} = s_{\xi\eta}$$

$$S_{\xi\xi} = G(\lambda_X^2 - 1) + P \tag{83a}$$

$$S_{\eta\eta} = G(\lambda_Y^2 - 1) + P \tag{83b}$$

$$S_{ZZ} = (S_{\xi\xi} + S_{\eta\eta})/2 \tag{83c}$$

$$s_{\xi\xi} = 2Gu_X\lambda_X + p \tag{83d}$$

$$s_{\eta\eta} = 2Gv_Y\lambda_Y + p \tag{83e}$$

$$s_{\xi\eta} = G(v_X\lambda_X + u_Y\lambda_Y) \tag{83f}$$

Thus, the relations between nonuniform normal stresses and strains are significantly different from those for incremental values, eqs. 35. The conditions of incompressibility are different also. To first order in nonuniform rotation and shear strain, the requirement that the volume of an element be the same before and after deformation is (Chapter 4, eq. 8):

$$(\lambda_X + u_X)(\lambda_Y + v_Y) = 1$$

Thus, to first order in the nonuniform displacement gradients:

$$u_X\lambda_Y = -v_Y\lambda_X \tag{83g}$$

As before:

$$\lambda_X\lambda_Y = 1 \tag{83h}$$

The next step is to transform the equilibrium equations (36) into those for coordinates X and Y, where $d\xi = (dX + dU + du)$ and $d\eta = (dY + dV + dv)$. Skipping all intermediate steps, using eqs. 12 of Chapter 4, we derive, for example:

$$(\partial/\partial X)(2Gu_X + p\lambda_Y) + (\partial/\partial Y)(s_{\xi\eta}\lambda_X) = 0$$

Bearing in mind eqs. 83g and 83h, and using eq. 83f, this becomes:

$$(\partial^2 u/\partial X^2) + (\partial^2 u/\partial Y^2) + \lambda_Y\partial(p/G)/\partial X = 0 \tag{84a}$$

Similarly:

$$(\partial^2 v/\partial X^2) + (\partial^2 v/\partial Y^2) + \lambda_X\partial(p/G)/\partial Y = 0 \tag{84b}$$

These equations have very different symmetry from the corresponding equations for initial and incremental stress, eqs. 38a and 38b.

The coefficients of the fourth-order equations as well as the relations between the displacements and the displacement potention also are different. In light of eq. 83g, we introduce a displacement potential such that:

$$v\lambda_X = -\partial\psi/\partial X \tag{85a}$$

$$u\lambda_Y = \partial\psi/\partial Y \tag{85b}$$

Substituting these into eqs. 84 and eliminating p between the resulting equations, we derive the fourth-order eq. 39b, but the coefficients are:

$$\mathbf{A} = (1 + \lambda_Y^4)/2 \tag{85c}$$

$$\mathbf{B} = \lambda_Y^4 \tag{85d}$$

instead of those in eqs. 39c.

All that remains is to derive boundary conditions. Consider a surface which is approximately parallel to the X-axis in the body before loading. We assume that the surface is initially in the form of a sine wave and that its coordinates are:

$$y = \pm T/2 + v_i = \pm T/2 + \delta_i' \sin(l'X)$$

where δ_i' is the amplitude and l' is the wave number of the initial wave, and T is the initial thickness of the layer. The slope angle of the initial surface is β', where:

$$\tan(\beta') = dv_i/dX = l'\delta_i' \cos(l'X)$$

After the deformation, the boundary becomes parallel to the curvilinear coordinate $\tilde{x}$ and the local slope angle of the boundary is β (as in Fig. 5A). The tangent of the local slope angle can be derived from the transformation:

$$dY'/dX' = \tan(\beta)$$

where:

$$\begin{bmatrix} dX' \\ dY' \end{bmatrix} = \begin{bmatrix} (\lambda_X + u_X) & u_Y \\ v_X & (\lambda_Y + v_Y) \end{bmatrix} \begin{bmatrix} dX \\ dv_i \end{bmatrix}$$

so that:

$$\tan(\beta) = [v_X + (\lambda_Y + v_Y)(dv_i/dX)]/[(\lambda_X + u_X) + u_Y(dv_i/dX)]$$

Further:

$$\cos^2\beta = 1/(1 + \operatorname{ctn}^2\beta)$$

$$\sin^2\beta = 1/(1 + \tan^2\beta)$$

Therefore, to the first order in both the initial slope and the nonuniform displacement gradients:

$$\cos(\beta) \simeq 1$$

$$\sin(\beta) \simeq \lambda_Y(\partial v/\partial X) + (dv_i/dX)\lambda_Y^2; \ |\beta| \lesssim 15^\circ$$

Substituting these approximations into eqs. 41, we derive:

$$\tau_{\tilde{n}\tilde{n}} = S_{\eta\eta} + s_{\eta\eta} \tag{86a}$$

$$\tau_{\tilde{n}\tilde{x}} = (S_{\eta\eta} - S_{\xi\xi})[\lambda_Y(\partial v/\partial X) + (dv_i/dX)\lambda_Y^2] + s_{\eta\xi} \tag{86b}$$

Thus we have complete sets of equations to use either to evaluate instability or to evaluate growth of initial perturbations. Both sets of equations are exact to first order.

Chapter 11

SYNTHESIS

INTRODUCTION

The results of our research into processes of folding can be expressed in terms of mechanisms and mechanics of folding. During the course of our research, mechanisms of folding have been deduced by complex ineractions between perceptions of folds in the field and in experiments and theoretical deductions from mechanical analyses of idealizations of possible mechanisms. The mechanisms are in a constant state of flux; we continually change our ideas about developments in experiments and about what actually exists in the field. The mechanical analyses are more nearly permanent, being replaced or rejected only where logical errors are discovered or where special or unnecessary assumptions can be eliminated. However precise, the mechanical analyses become irrelevant as concepts of mechanisms change; however interesting, the mechanisms are speculative unless they are based on theoretical analysis.

Our current understanding of folds, based on both the mechanisms and mechanics of folding, will be presented in following pages. The discussion of mechanisms will be as complete as possible, including all that appear to be based on theory. The discussion here supersedes that presented in Chapter 1. However, the discussions of mechanics of folding will be restricted to analyses presented in Chapters 2 through 10 of this book and is therefore restricted to materials with memory. The discussion of mechanics presented here compliments that presented in Chapter 1, which summarized some of the more important mechanical analyses of folding of viscous, elastic, and nonlinear materials.

MECHANISMS OF FOLDING

I believe that one can explain sinusoidal, concentric-like, chevron, kink and asymmetric folds in terms of five basic mechanisms of folding: yielding between layers, yielding within layers, homogeneous strain, anisotropy due to interlayering of stiff and soft materials, and stress- or strain-induced anisotropy. These mechanisms will be discussed in general terms below but they must be specified in order to perform mechanical analyses.

Anisotropy due to interlayering of stiff and soft materials or to slip at contacts between layers

Multilayers comprised of stiff and soft materials behave in part as anisotropic materials, as shown by Biot (e.g., 1965a) *, but only in part. For example, multilayers with memory have different normal elasticity moduli for deformations parallel and normal to layering, as indicated in Chapter 2. These moduli are equal only if the materials are incompressible. Also, the multilayered materials shear in directions normal or parallel to layering much more easily than they shorten parallel to layering. An extreme example of an anisotropic body is where there are only stiff layers between which contact strength is zero, that is, between which slip is not resisted. The effective shear modulus of the multilayer is zero for such materials but the normal modulus is large, equal to the normal modulus of the stiff materials. In these senses, therefore, a multilayered body is anisotropic.

The deformation of multilayered materials, however, typically is inhomogeneous in detail, so that the multilayer does not behave as a *pervasively* anisotropic material. Rather, the multilayer behaves as an anisotropic material only for deformations which have large wavelengths relative to the scale of the thicknesses of layering. The inhomogeneous deformations can be represented, but only approximately, by considering bending of individual layers. Thus the normal and shear stresses must be equal across contacts between layers but they can be quite different within layers. To first order in the slope of layering, the inhomogeneous stress and strain distributions can be represented exactly, for viscous materials as explained by Fletcher (1977), and for elastic materials as explained in Chapter 10. In these cases we select fundamental elements which are small relative to thicknesses of layers and, following Hans Ramberg, match boundary conditions between layers.

Individual layers may have anisotropic but homogeneous rheological properties, for example due to preferred orientations of particles. Such

---

* References at end of Chapter 1.

materials are pervasively anisotropic, in contrast to multilayered materials, in terms of folds which have characteristic dimensions much larger than sizes of individual particles but of the same order as the thicknesses of layers with different properties.

Anisotropy due to initial stresses and strains

Another type of anisotropy might be as important in folding as is rheological anisotropy. This is "anisotropy" due to initial stresses or strains. Parrish (1973) and Fletcher (1974) showed that a power-law material without memory behaves essentially as an anisotropic viscous material. The effective viscosities for nonuniform normal strains can be much smaller than those for shear strains, thereby allowing bending to become relatively easy, even for thick layers. We showed the same behavior for a power-law material with memory (eqs. 60, Chapter 10). We have shown in Chapters 4 and 10 that simple elastic materials also behave as though they are anisotropic to incremental deformations. For example, an effect of highly compressive initial normal stresses along the axis of a layer can be considered to reduce the resistance of the layer to shear along planes normal to the axis. This is particularly evident in the expression for the incremental shear stress, eq. 35f of Chapter 10, where the coefficient of the slope, $\partial v/\partial x$, reduces as the finite extension ratio parallel to the axis, λ_X, reduces. Thus, the material behaves somewhat as though it were layered in the plane normal to the direction of compression (Fig. 7D, Chapter 4), but the "layers" have no thickness.

Yielding within layers

One consequence of an increase in compression in a homogeneous, elastic material is that the material may nearly "yield" elastically. We used the concept of elastic "yielding" in Chapter 5 to explain pinching at edges of deep beams of rubber and gelatin and to explain the formation of hinges in chevron and kink folds in these materials. The concept is based on the observation that the coefficient of the curvature term, $\partial^2 v/\partial x^2$, in differential equations 1 of Chapter 5, can become quite arbitrarily small. As it does, the curvature itself can become large, and a tight bend or pinch can form. Therefore, the concept of pinch formation advanced in Chapter 5 appears to be qualitatively sound. The resistance to bending and to increase in curvature will be least on the concave sides of layers, so the elastic "yielding" will start there and progress inward into the layer, reducing the resistance of the layer and allowing a tight hinge form. In this way we would explain tight hinges in elastic materials, especially rubber-like materials.

Another source of yielding could be nonlinear elastic behavior of the material itself. For example, Elmont Honea used elementary plate theory to show that a power-law model of material with memory can describe chevron-like forms in isolated layers (Johnson, 1970, fig. 8.10).

Of course, the yielding need not be elastic at all, rather it might be permanent, plastic deformation (e.g., Johnson, 1970, figs. 8.5 and 8.6). Thus there is a variety of ways to produce yielding in materials and different materials presumably yield for different reasons. Nevertheless, I would interpret sharp hinges of layers in kink bands or chevron folds to be the result of some mechanism of yielding, whether the yielding is due to high initial stresses or behaviors such as nonlinear elastic, brittle, elastic-plastic, nonlinear viscous, or visco-plastic.

Yielding between layers

One of the fundamental mechanisms of folding, according to our analyses of kink folding, is yielding *between* layers. The yielding might be along contacts between layers, as we assumed in analyses of kink folding of rubber strips, paper cards, phyllite and slate, in Chapters 6 and 8. The yielding might be plastic, described in terms of Coulomb's or Tresca's yield criterion. On the other hand, the yielding might be within a plastic material interbedded with viscous or elastic materials, as in experiments by Stephenson Ellen (e.g., Johnson, 1970, Chapter 8).

The yielding also might be a result of nonlinear behavior such as that of a power-law material. For example if a multilayered material was comprised of a power-law material interbedded with a linear material, the fold patterns could approach kink forms. Another model would be two power-law materials with different properties. Indeed, I would suggest these models for future investigations of mechanisms of kink folding. A problem with the methods we used is that the solutions are necessarily discontinuous so that one must use one set of equations to describe yielded parts of multilayers and another set of equations to describe nonyielded parts of multilayers. If one, instead, used a power-law material for the soft interbeds, one could derive solutions using one set of equations for the entire body. The analysis probably would have to be done with a computer, as in Parrish's (1973) analysis of folding of a single layer with power-law properties. This would be an interesting way, I believe, to investigate conditions of kink folding as opposed to conditions of symmetric sinusoidal and concentric-like folding and thus arrive at more general criteria of fold types than those deduced in Chapter 6.

Whatever the nature of the yielding, however, I am quite sure that kink folding involves the mechanism of yielding between layers in the general sense described here.

Superimposed strain

An important mechanism of folding is superimposed strain of passive markers. Many investigators have recognized the importance of homogeneous strain in changing fold patterns (e.g., Campbell, 1951; Ramberg,

1964b; Ramsay, 1967; Hudleston, 1973a). This mechanism is critical to the development of some asymmetric folds, as we showed in Chapters 7 and 8. Relatively homogeneous strain following initial growth of folds by buckling is probably the origin for most similar folds, in which thickness of layers is different in crests and on limbs of folds (e.g., Hudleston and Stephansson, 1973).

Instabilities

Various types of instabilities have formed the basis for all theories of folding developed thus far and most folding probably is a result of some type of instability. The types of flow responsible for similar folding, as visualized by Warren Carey (1962), for example, and shown in Fig. 4 of Chapter 1, are difficult to understand unless there is instability. I can think of no other reason for cyclical velocity or displacement distributions to develop as shown in fig. 4 of Chapter 1. Indeed, I think that a fruitful area of research in folding theory will be the search for different mechanisms which produce instabilities that can occur in different environments where we find folds.

The types of instabilities which already have been recognized in folding theory area:

(1) Gravity instabilities, where interlayered high- and low-density materials tend to rearrange so that materials with higher densities are below and material with lower densities are above (e.g., Odé, 1966; Fletcher, 1967; Ramberg, 1967, 1968).

(2) Instabilities arising from differences in layer-parallel and layer-normal normal stresses at contacts between layers with different viscosities. This is the basis of the Fletcher theory of folding (Fletcher, 1974, 1977; Smith, 1975). This type of instability is discussed in Chapter 10.

(3) Instabilities arising from changes in effective elasticity moduli and from rotation of initial normal stresses at contacts between layers (Chapter 4; Novozhilov, 1953; Toupin and Bernstein, 1961; Biot, 1965a; Green and Zerna, 1968). For an elastic material with planar contacts, a condition of instability exists if there are two or more states of deformation, both of which are compatible with the same stress boundary conditions, as explained in Chapters 4 and 10. This type of instability is restricted to materials with memory.

(4) Bending instabilities, where a layer first shortens and then bends unstably when axial stresses reach sufficiently high magnitudes, (e.g., Goodier, 1946; Currie et al., 1962). This is the elementary theory which has been the basis for most folding theory, including much of that reported in this book. The bending theory of instability, unlike the others, is approximate and based upon special assumptions. Analysis based on this instability generally is a valid predictor of wavelengths only where wavelengths are greater than about ten times the thicknesses of the stiff layers.

(5) Yielding instabilities — bending. If an elastic-plastic layer begins to

yield at hinges of folds, plastic zones grow and, when they have reached a certain size, the resistance to bending increases less rapidly than the bending moment produced by the buckling load, and thus the elastic-plastic layer becomes unstable (e.g., Prager and Hodge, 1951; Timoshenko and Gere, 1961; Johnson, 1970, p. 304). Many other rheological behaviors will produce similar bending instabilities, as discussed in previous paragraphs.

(6) Yielding instabilities — slippage. This type of instability was discussed at some length in Chapter 8, in connection with the analysis of kink folding. It involves the mechanism of yielding between layers, discussed above. In order for the instability to exist, the planes of slippage must be inclined at angles less than 45 degrees to the direction of maximum compression.

MECHANICS OF FOLDING OF MATERIALS WITH MEMORY

The primary purpose of the remainder of this chapter is to summarize what I consider to be the valid theoretical conclusions reached during the course of the research for this series of papers. The series of papers is much like a first draft of a manuscript for a book on some aspects of the origin of different fold forms. In order to present a coherent and logical picture, the manuscript should be completely reorganized and rewritten, eliminating misconceptions and concepts of fleeting interest. This is not possible, of course, with the reprinting of a collection of papers. Therefore, I will attempt in following pages to write an abstract of the book that might have developed as a result of reorganization and rewriting. I will attempt to provide a guide to the papers. This abstract should provide an indication of the mechanical analyses which have lasting value, in my opinion. It will not discuss the field or experimental observations, which are at least as valuable as the theoretical observations. These will have to be consulted in the original papers, especially those in Chapters 2, 3, 5, 6, 7, 8 and 9.

Folding of incompressible elastic material

In order to describe the deformation of a material with memory, the fundamental equations must be written in terms of strains because it is the strain that the material "remembers". Components of strain are expressed in terms of displacements from a reference state, normally the state of the undeformed body. The equilibrium equations, however, are expressed in terms of a reference state within the deformed body. For these reasons the basic equilibrium equations for a material with memory are somewhat different from those for a material without memory, as for example, a linear viscous material.

We shall review the essential steps in the derivation of equilibrium equations for finite deformation of a simple elastic substance. Analysis of the equilibrium equations provides explanations for many phenomena observed experimentally, including the formation of pinches at edges of deep beams, the formation of hinges in chevron folds, the transition from sinusoidal to concentric-like to chevron folds in multilayers, and the tendency of deep beams to deform by simple shear, much as a deck of slippery cards. We shall follow the derivation presented in Chapter 10 rather than that in Chapter 4, because the former is simpler.

Let us consider a material with memory which has undergone uniform finite initial compression or extension in the x- and y-directions. The initial normal stresses are designated S_{XX} and S_{YY}. They deform an element *abcd* (Fig. 1A) into the shape $a'b'c'd'$ (Fig. 1B). The stresses are in terms of force intensity per unit of *final* area of sides of the element. The displacements from the initial state have components U and V, and the components of initial strain are expressed in terms of derivatives of U and V with respect to the original coordinates, X and Y, or equivalently, in terms of finite extension ratios:

$$\lambda_X = (1 + \partial U/\partial X) \quad \text{and} \quad \lambda_Y = (1 + \partial V/\partial Y) . \qquad (1)$$

The finite extension ratios are required for the stress—strain relations for the initial state.

Now, if the uniform initial stresses and strains are sufficiently large, there are two or more possible states of equilibrium for the body. One is the uniformly deformed body, an element of which is shown in Fig. 1B. The others are nonuniformly deformed bodies, an element, $a''b''c''d''$, of which is shown in Fig. 1C. The components of stresses acting on the sides of the deformed body are referred to the final areas of elements. The strains in the element, Fig. 1C, are a result of first uniform strains and then nonuniform strains. The former are expressed by eqs. 1 and the latter are expressed in terms of gradients of displacements, u and v, relative to the initial state of the body (Fig. 1B). These latter, *incremental* displacement gradients are superimposed upon the initial strains not by addition but by matrix multiplication, as indicated in Chapter 10. The final stress state, σ, can be expressed

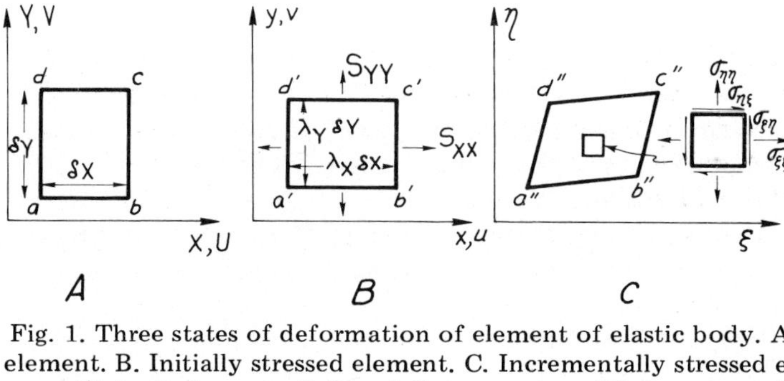

Fig. 1. Three states of deformation of element of elastic body. A. Undeformed, unloaded element. B. Initially stressed element. C. Incrementally stressed element. If initial stresses are sufficiently large, both B and C represent equilibrium states for the body, so the body is unstable.

exactly as the sum of initial stresses, S, and incremental stresses, s, which become defined in the course of the analysis:

$$\sigma_{\xi\xi} = S_{XX} + s_{xx}$$

$$\sigma_{\xi\eta} = s_{xy}$$

$$\sigma_{\eta\eta} = s_{YY} + s_{yy}$$

For a relatively simple elastic material, called a *Mooney* material, the incremental stresses are:

$$s_{xx} = 2G\lambda_X^2(\partial u/\partial x) + p \tag{2a}$$

$$s_{yy} = 2G\lambda_Y^2(\partial v/\partial y) + p \tag{2b}$$

$$s_{xy} = G[\lambda_X^2(\partial v/\partial x) + \lambda_Y^2(\partial u/\partial y)] \tag{2c}$$

in which p is a constant, similar to a mean stress, and G is the shear modulus. These incremental stresses are a result of the incremental displacement gradients but they also are functions of the state of initial strain. In effect, as shown in eqs. 2, the moduli for the incremental deformations are functions of the initial strains. Most important is the relationship for incremental shear strain. As a body is shortened in the X-direction, the finite extension ratio λ_X decreases and the finite extension ratio λ_Y increases. Indeed, for plane strain of an incompressible material, $\lambda_X\lambda_Y = 1$. Therefore, the coefficient of the slope $\partial v/\partial x$ of lines originally parallel to the x-direction decreases and the coefficient of the slope $\partial u/\partial y$ of lines originally parallel to the y-direction increases. I believe that this is the primary reason that deep beams theoretically tend to buckle in simple shear along lines parallel to the y-direction, as shown in Fig. 2A. In contrast, eqs. 2 indicate that normal strains in the x- and y-directions become more difficult with increasing initial shortening. Thus, subtracting eqs. 2a and 2b and using the condition of incompressibility of incremental strains, $\partial u/\partial x = -\partial v/\partial y$:

$$s_{xx} - s_{yy} = 2G(\lambda_X^2 + \lambda_Y^2)(\partial u/\partial x) \tag{3}$$

The coefficient of the normal strain always increases with increasing initial strain because, if there is shortening in the X-direction, λ_Y^2 increases more rapidly than λ_X^2 decreases, and the sum is always greater than or equal to two. These, then are the primary effects of the initial state of strain on the buckling of a simple elastic material.

A related effect can be observed by substituting eqs. 2 into the equilibrium equations. As shown in Chapter 10, the equilibrium equations become:

$$G[\lambda_X^2(\partial^2 u/\partial x^2) + \lambda_Y^2(\partial^2 u/\partial y^2)] + \partial p/\partial x = 0 \tag{4a}$$

$$G[\lambda_X^2(\partial^2 v/\partial x^2) + \lambda_Y^2(\partial^2 v/\partial y^2)] + \partial p/\partial y = 0 \tag{4b}$$

I believe that equilibrium equation 4b explains the tendency for pinches to develop at the core of a buckle in a deep beam of elastic material. As the axial shortening increases, that is, as the finite extension ratio, λ_X, decreases, the coefficient of the curvature term, $\partial^2 v/\partial x^2$, in eq. 4b decreases, thereby allowing large curvature to develop at the core of the buckle.

These are the primary contributions of the linearized theory to our understanding of folding of materials with memory.

Sinusoidal solutions to eqs. 4, using boundary conditions derived in Chapter 10, have been derived for several problems. Solutions to analogous equations, derived in Chapter 4, are presented in Chapters 5 and 6. One of the problems solved in Chapters 5 and 10 is the buckling of a deep beam, with a small ratio of length to thickness. Figure 2A shows the theoretical pattern of deformed grid lines on a deep beam with an original length-to-depth ratio of 2 : 1. Figure 2B shows deformed grid lines on a deep beam of photoelastic rubber, with an original length-to-depth ratio of 4.6 : 1 and Fig. 2C shows the corresponding theoretical pattern of grid lines. The patterns are quite similar. Other comparisons were presented in Chapter 5.

Comparison of solutions for wavelengths of sinusoidal folding derived with the linearized theory and solutions derived with the elementary theory indicates that the solutions are essentially the same for a single layer in a soft medium if the ratios of elasticity moduli are greater than about 10 : 1 (Fig. 8, Chapter 5). Comparison of solutions using the two theories for multilayers confined by rigid boundaries (Fig. 5, Chapter 6) indicates that, if there are at least ten layers in the multilayer, the elementary theory closely estimates wavelengths of folds. Therefore, the elementary theory probably is sufficiently accurate to estimate wavelengths of most folds in material with memory.

Elementary theory of sinusoidal folding of multilayers

The wavelengths of folds in multilayers composed of identical stiff layers and identical soft layers are defined according to the elementary theory by

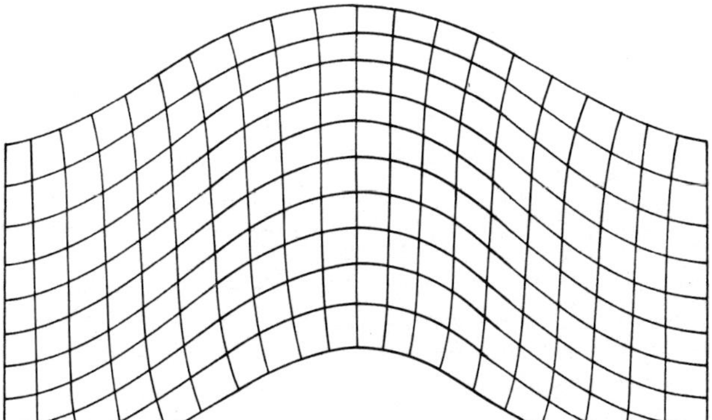

Fig. 2A.

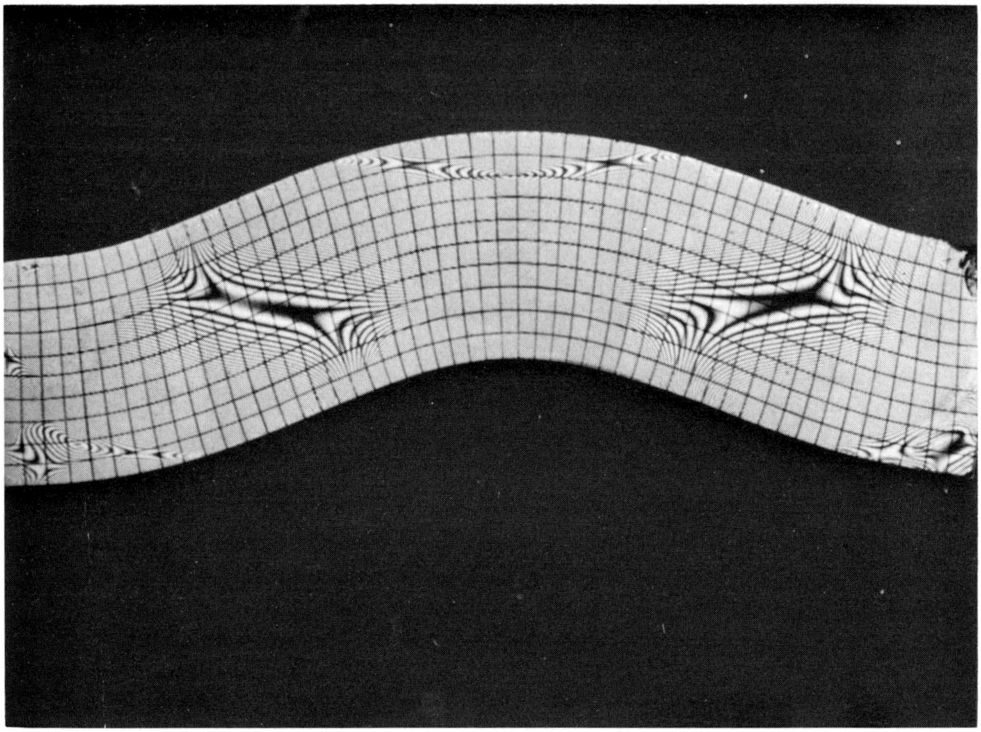

B

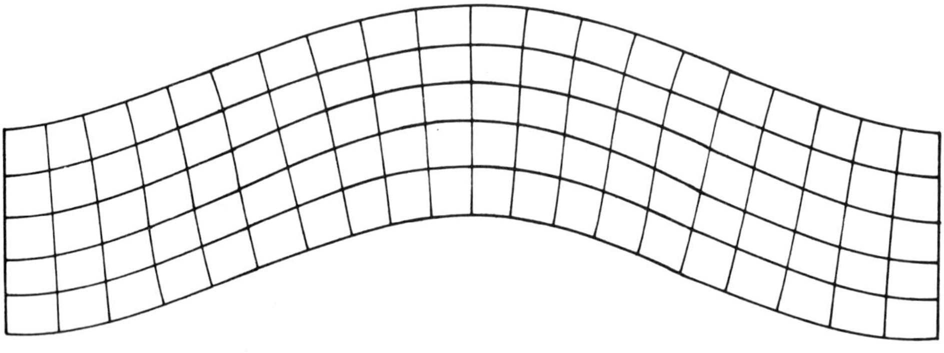

C

Fig. 2. Deformed grid lines on elastic deep beam. A. Theoretical pattern of deformation of rectangular grid on deep beam with small length-to-thickness ratio. Grid lines originally vertical near midheight in beam remain essentially vertical during the deformation, indicating essentially simple shear in the vertical direction, as indicated by theory. B. Pattern of grid lines on deep beam of photoelastic rubber. C. Theoretical pattern of deformation of grid lines for deep beam with dimensions nearly the same as those of the deep beam shown in B.

the *Ramberg wavelength* (eq. 3, Chapter 9):

$$L_R = 2T\sqrt{\pi n/N_p}\sqrt[4]{(B_1/12B_v')/(1-D)}\,; \quad N_p \geqslant 5 \tag{5a}$$

where n is a factor defined by the equation:

$$\sqrt{n}\,\mathrm{ctn}\,(\pi/2n) = (B_v/B_0)\,2\sqrt{\pi/N_p}\sqrt[4]{(B_1/12B_v')/(1-D)} \tag{5b}$$

$$D = 1/\{1 + (\pi/2n)[\mathrm{ctn}\,(\pi/2n) + \tan\,(\pi/2n)]\} \tag{5c}$$

$$B_v = [(t_1 + t_2)B_1 B_2/(B_1 t_2 + B_2 t_1)] \tag{5d}$$

and:

$$B_v' = B_v(1 + t_2/t_1)^3 \tag{5e}$$

Here t_i and B_i refer to stiff ($i = 1$) and soft ($i = 2$) layers and to the confining medium ($i = 0$).

$$B_i = E_i/(1 - \nu_i^2) \tag{5f}$$

where E is Young's modulus and ν is Poisson's ratio. T is the total thickness of the multilayer:

$$T = N_p(t_1 + t_2)$$

where N_p is the number of pairs of layers, if the multilayer is confined above and below by media. If the multilayer is confined below by a medium but has a free upper surface, we use $2T$ instead of T in the equations. That is, the wavelength for a multilayer with a free upper surface is twice that of the same multilayer with media above and below.

The factor n must be determined iteratively because it enters eq. 5b in a complex manner. The factor is a measure of the amplitude of folds at top and bottom of the multilayer as compared to the amplitude of folds at middepth, as shown in eq. 3d of Chapter 6. After n is determined, the wavelength can be determined from eq. 5a.

We presented some graphical solutions to these equations in Chapter 9, for the special case where there are no interbeds and where contacts between layers are frictionless. Here I will present another series of graphs for more general conditions. Figure 3 shows relations between modulus ratio B_v/B_1 and thickness ratio, t_2/t_1, for various ratios of moduli of soft to stiff layers B_2/B_1. These curves will be useful for estimating modulus ratios without having to solve eq. 5d. Further, Fig. 3 indicates conditions where only multilayer folds are possible and where both single-layer and multilayer folds can occur together.

If the layers in the multilayer are widely separated by soft media and the modulus ratio of layer to medium is greater than 10 : 1 (that is, if $L/t \gtrsim 8$), we use the familiar equation for the wavelength of a single layer in an infinite

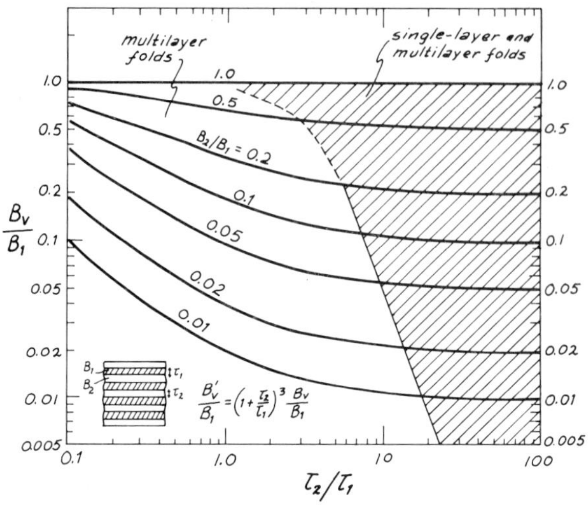

Fig. 3. Relations among averaged modulus ratio, B_v/B_1, thickness ratio, t_2/t_1 and modulus ratio, B_2/B_1, of layers for a multilayer. Here 1 refers to stiff and 2 to soft layer. Figure also shows thickness ratios for which multilayer can fold either as individual layers or as a unit. Given thickness and modulus ratios, one can use Fig. 4 to estimate wavelengths for multilayer.

medium:

$$L = 2\pi t \sqrt[3]{B_1/6B_0};\ L/t_{\text{med}} \lesssim 1;\ L/t \gtrsim 8;\ B_1/B_0 \gtrsim 10 \tag{6}$$

where t is the thickness of the layer and t_{med} is the thickness of the interbeds or of the confining media.

Figure 4 shows relations between the Ramberg wavelength, the number of layers and the ratio of the average modulus of the multilayer and the modulus of the medium, B_v/B_0. This modulus can be calculated from the relation:

$$B_v/B_0 = (B_v/B_1)(B_1/B_0)$$

where B_v/B_1 is given in Fig. 3.

The series of graphs in Fig. 4A is for wavelengths of multilayers solely comprised of stiff layers with frictionless contacts. The equations for the wavelengths in this case can readily be derived from eqs. 5 by replacing N_p with N, the number of layers, and B_v with B_1, the modulus of the stiff layers. Thus:

$$L_R = 2T\sqrt{\pi n/N}\sqrt[4]{1/[12(1-D)]};\ N \geqslant 10 \tag{7}$$

where n is defined by eq. 5b, with $\bar{B}_v \equiv B_1$, and D is defined by eq. 4c. The graphs are for confining media the properties of which range from soft, $B_0 = B_1/10$ to rigid, $B_0 \to \infty$. If the media are rigid, n in eqs. 5a and 6 is unity and D is zero, so the *Ramberg wavelength* reduces to the *Biot wavelength*, derived in Chapter 2, eq. 52.

Further, if the multilayer is comprised of layers of unequal thickness, we

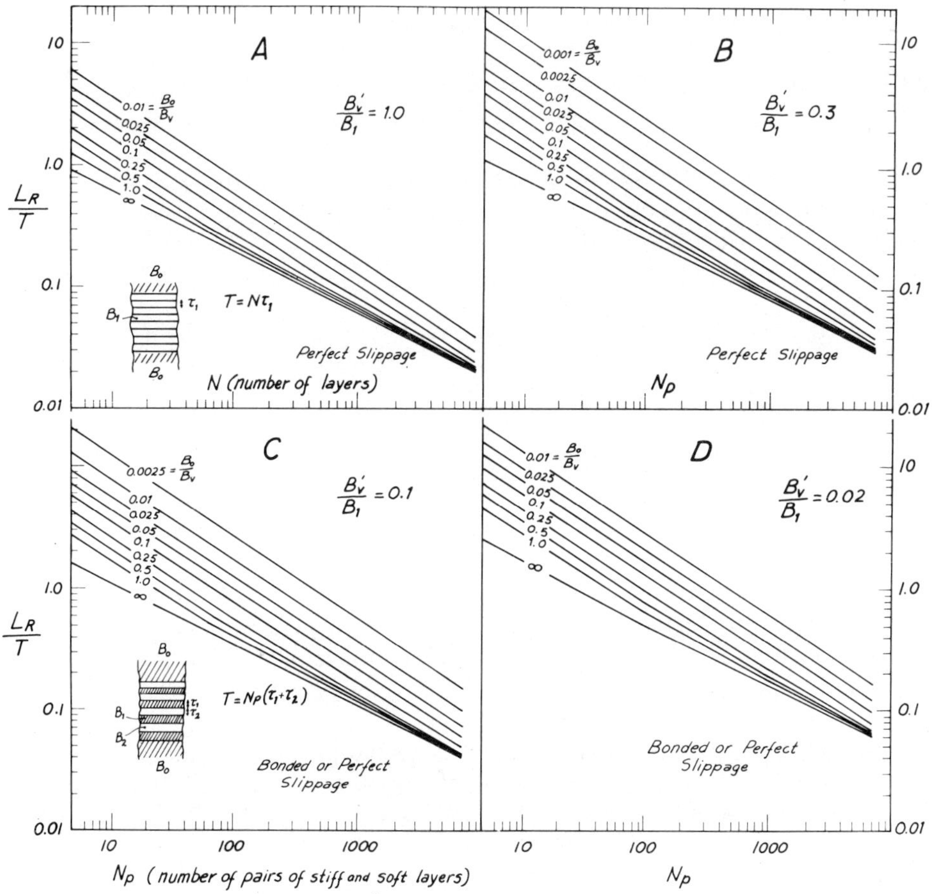

Fig. 4. Relations among Ramberg wavelength, number of layers and modulus ratio of multilayer to confining media for four averaged modulus ratios for multilayer. Relations are based on approximate theory.

use eq. 6 to compute approximately the *Ramberg wavelength*, using the *effective number* $\bar{N}$, of layers instead of the actual number, N where:

$$\bar{N} = T/\sqrt{(\sum_N t_i^2)/N} \qquad (8)$$

and

$$T = \sum_N t_i \,.$$

Folding of power-law material

We considered folding of a power-law material with memory in Chapter 10, following Fletcher's (1974) analysis of folding of a power-law material without memory. According to the one-dimensional form of the power-law model, the shear strain, ϵ_s, is proportional to the shear stress, τ, raised to some power, m:

$$\epsilon_s = (1/g)\tau^m$$

where g is a constant, analogous to a shear modulus. The analysis of a stiff layer with properties m_l and g_l in a medium with properties m_m and g_m indicated that the wavelength is:

$$L = 2\pi t \sqrt[3]{G_e/6} \tag{9a}$$

where

$$G_e = \hat{G}_l \sqrt{m_m}/m_l \hat{G}_m \tag{9b}$$

$$G_i = g_i/|S_{xx} - S_{yy}|^{(m_i - 1)}; \text{ where } i = l \text{ or } m\ . \tag{9c}$$

Thus, the effective shear modulus is a function of the initial stress state, $S_{xx} - S_{yy}$, of the body. Also, the expression for the wavelength for folding of the power-law material is directly analogous to that for a simple elastic material, eq. 6. These equations are derived in Chapter 10 and I refer the reader to that chapter for details of the derivations as well as attempts to apply the analysis to single-layer folding of several types of sedimentary rock.

Analysis of folding of a multilayer of power-law material indicates that the wavelength is defined by eq. 7 but the constant, n, is a function of G_e, eq. 9b, rather than B_0/B_1, as shown in eqs. 73 and 75 of Chapter 10. Again, one should consult Chapter 10 for details.

Transformation from sinusoidal to concentric-like to chevron folds

Although the elementary theory seems to be adequate to predict wavelengths of folds for most multilayers, it provides few insights into the transition from sinusoidal to concentric-like to chevron folds which we recognize in experiments with multilayers of elastic material. The linearized theory seems to provide the necessary insights.

Let us consider multilayers comprised of interbedded soft and stiff layers firmly bonded together, or multilayers in which there is perfect slippage between layers. The confining media can be stiff or soft. If the confining media are very stiff, the contact between the multilayer and the media will be assumed to be one of free slip. Otherwise, the contact can be bonded. These kinds of multilayers show a transition from low-amplitude, sinusoidal forms to concentric-like forms to chevron forms (Chapter 3, Figs. 19 and 22; Chapter 5, Figs. 10).

The multilayer is shortened parallel to layering and thickened normal to layering. As a result, *initial* stresses are developed within the multilayer, and the stiff layers carry higher layer-parallel compressions than the soft layers. This is the initial state of stress, which is assumed to be homogeneous within each stiff or soft layer. The only possible equilibrium configuration for the multilayer is uniform shortening and thickening. As the initial stresses approach certain critical values, however, there are two or more equilibrium configurations. The configuration of uniform shortening and thickening is one possibility, but it is *unstable.* The slightest perturbation will cause the configuration to change. Another equilibrium configuration is a pattern of sinusoidal waves, the sinusoidal fold form, so it is a possible alternative configuration. The wavelength of folds that require the minimum shortening and thichening depends upon a host of factors, including the rheological and dimensional properties of individual layers, the total thickness of the multilayer, and the properties of the confining media. The wavelength is approximately related to these factors as shown in eqs. 5 through 9.

The wavelengths of subsequent folds are largely determined by these early, sinusoidal waves, but the sinusoidal pattern becomes modified by higher-order effects. Just as the straight, shortened form of the multilayer becomes unstable when the shortening reaches a critical value, so does the sinusoidal pattern become unstable as amplitudes of the folds grow. The incremental stresses are related to the incremental, sinusoidal displacements u and v, as shown in eqs. 2. For a first, rough approximation, we can determine effects of a first increment of sinusoidal deformation on the second increment of deformation by adding the incremental normal strains associated with the first sinusoidal deformation to the initial strains. Thus, the first incremental strains behave much as initial strains for the second incremental strains. This procedure is nonrigorous, but rigor greatly complicates the analysis.

We can understand qualitatively the transition from sinusoidal folds to concentric-like folds in elastic multilayers if we accept the approximate procedure of adding incremental and initial stresses outlined above and if we assume that concentric-like folds are essentially sinusoidal folds in which the wavelengths change from place to place throughout the folds. Thus, wavelengths are assumed to be smaller near cores and larger near crests of concentric-like folds according to this nonrigorous way of visualizing folds.

We will consider only two higher-order effects. One is a change in axial compression in stiff layers near upper and lower boundaries of a multilayer as a result of different moduli of the media and the interbeds. The media and interbeds are assumed to be bonded to the stiff layers for this part of the analysis. If the media are *softer* than the interbeds, the wavelengths tend to decrease near cores of folds and increase in crests, tending to produce a concentric-like form in the multilayer. The reason is that a gradient of shear stress is set up across the stiff layer, as explained in Chapter 5, and the gradient in shear stress sets up a gradient of layer-parallel normal stresses along the layer, increasing compression in cores and decreasing compression in crests of folds. In contrast, if the media are *stiffer* than the interbeds, the wavelengths *increase* in cores of folds and decrease in crests, tending to produce a form different from a concentric-like form. An example was shown in Chapter 5, Fig. 10C3. This effect depends directly on relative shear moduli of interbeds and media, so if contacts between layers are frictionless, the effect will be absent. Also, the effect operates only near edges of multilayers. The second effect is more nearly pervasive. It is a change in confining stress normal to layers as fold amplitudes increase. It is a different result of the interaction between multilayer and media and can be expressed, approximately, in terms of the normal, incremental stress derived in Chapter 2, following eq. 26, and used in Chapter 9, following eq. 3b. The normal stress varies with the sine of y:

$$\sigma_n \simeq -B_v(\delta_0\pi/nT)\sin(\pi y/Tn)\sin(2\pi x/L)$$

where δ_0 is amplitude and $y = 0$ at middepth in the multilayer of thickness T. B_v is a modulus defined in eq. 5d, and n is a function of the stiffness of the confining media, defined in eq. 5b. Factor n ranges from infinity for no medium to unity for rigid media. As indicated in the equation above, the incremental normal stresses in anticlines, e.g., $0 \leq x \leq L/2$, are compressive (negative) for $y > 0$ and tensile for $y < 0$. The stresses are zero at middepth. Further, we showed in Chapter 5, Fig. 2, that tensile confining stresses tend to decrease wavelengths and compressive confining stresses tend to increase wavelengths in isolated layers subject to constant axial stress. Thus, we suggest that it is for these reasons that wavelengths tend to become shorter in cores of anticlines and synclines, and larger in crests, but remain unchanged at middepth; thus the similar, sinusoidal form trends to transform into a concentric-like form. I would feel more comfortable with these explanations if they were derived from a rigorous analysis of higher-order effects. The explanations are consistent with experimental observations, however. The transition from sinusoidal to concentric-like folds requires further rigorous analysis. The analysis presented here merely suggests that the rigorous analysis is worth the effort.

I believe that chevron folds in cores of concentric-like folds can be explained in terms of yielding in crests, as discussed at some length in Chapter 5 and in connection with mechanisms of folding in preceding paragraphs. Thus, I visualize chevron folding as a result of higher-order processes also.

The evolution of typical folds in multilayered materials with memory is shown in Fig. 5A. The folds begin as low-amplitude, sinusoidal waves (Fig. 5A*1*). With further shortening and growth of amplitude, a concentric-like pattern begins to develop, characterized by shorter wavelengths in cores of folds and longer wavelengths in crests (Fig. 5A*2*). Then the layers with the shortest wavelength, at cores of folds, yield elastically or permanently to produce a chevron form (Fig. 5A*3*). As shortening continues, other layers yield and the chevron form can largely replace the concentric-like form (Fig.

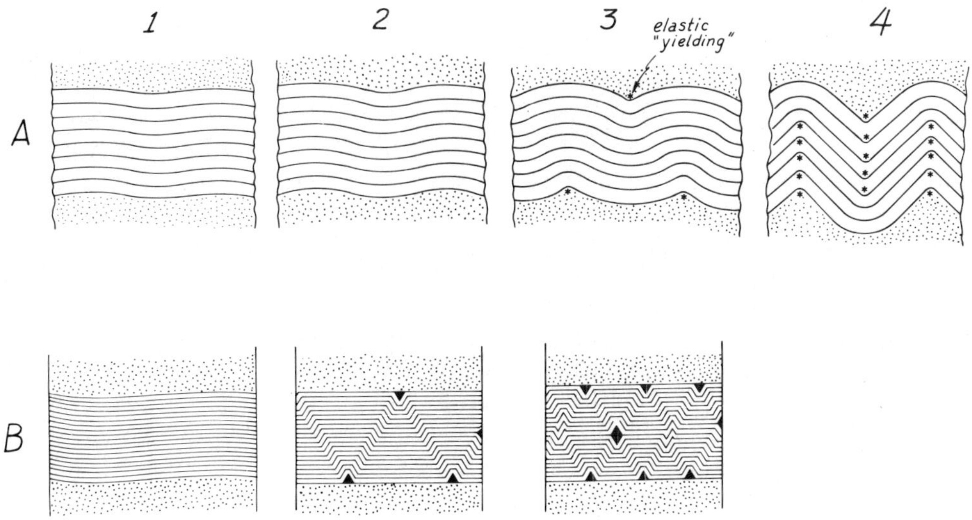

Fig. 5. Sequence of development of fold forms. A. Sequence from sinusoidal, to concentric-like to chevron forms. B. Sequence from low, sinusoidal forms in multilayer as a whole, to kink forms, with local chevron forms.

5A*4*). These chevron folds ideally can be distinguished from those produced by intersection of kink bands (e.g., Paterson and Weiss, 1966) because associated with the chevron folds are vestiges of the earlier concentric-like pattern (Fig. 5A*4*). The wavelengths of the final chevron folds, measured along layers, should be roughly equal to wavelengths of the initial sinusoidal folds. Furthermore, the combination of concentric-like and chevron form creates no "room problem". The fold pattern can extend indefinitely, because chevron folds are of the similar variety.

Theory of kink folding

Our theory of sinusoidal folding, summarized in previous pages, assumes either that contact strength between layers is so high that slippage never occurs or that contact strength is so low that slippage occurs everywhere. Our theory of kink folding, on the other hand, is based on the concept of finite strength of contacts between layers (Chapters 6 and 8). The concept can be generalized to include materials with nonlinear properties, I presume, but we have not done this. Rather, we will assume that contact strength is expressed in terms of the Terzaghi-Coulomb law of strength which postulates that strength is a function of cohesion and effective normal stress (eq. 6, Chapter 6).

Kink folding, in this view, is a result of two types of instability, bending instability and yielding instability. Also, kink folding can be initiated by perturbations in the layering, by low-amplitude sinusoidal folding, or by triggers such as faults which provide local initial slopes to the layers.

We will begin by considering kinking of multilayered materials with initial irregularities. Each layer will be assumed to have the same properties so that this part of the analysis should apply approximately to experimental kink folds produced in rubber, paper cards, phyllite, slate and probably other foliated materials. The multilayer is confined normal to layers so that contact strength is relatively high. If the multilayer is bounded rigidly, there can be no sinusoidal folding, but if the media are soft, there can be overall folding of the multilayer. The multilayer can fold sinusoidally only as a whole because the multilayer behaves as a homogeneous, nonlayered body until slip occurs locally. Let us suppose that the multilayer is bounded by rigid media and that it is subjected to maximum compression parallel to layers and minimum compression normal to layers. As the maximum compression increases relative to the minimum compression, the multilayer yields locally where layers have slight initial slopes. The magnitude of the stress ratio at the time of slippage depends upon the frictional and cohesive strengths of the contacts (and pore pressure) as well as the initial slope θ, of layers locally (Fig. 6A and eqs. 5 and 13, Chapter 8). If the strength is purely frictional, with a friction angle of ϕ, the stress ratio required to initiate slippage is:

$$(\sigma_3/\sigma_1) = \tan(\theta \pm \phi)/\tan\theta \tag{10}$$

where the sign of ϕ is selected to be the same as the sign of θ. Further, the slippage will be *unstable* because an increase in θ requires a decrease in stress ratio, as shown in Fig. 14 of Chapter 8. The unstable slippage can be the beginning of kink folding; if the material is foliated, a kink band will develop by buckling of individual layers.

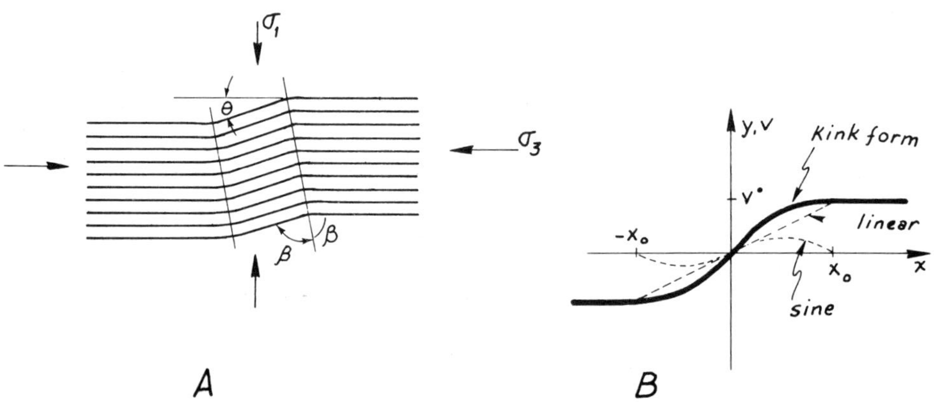

Fig. 6. Geometry of kink band and kink form. A. Idealized kink band, showing slope angle of layers in kink band, θ, and angle between layering and boundary of kink band, β. No dilitation in kink band. σ_3 is maximum compression and σ_1 is minimum compression. B. Ideal kink form, which is comprised of a sinusoidal form superimposed on a linear form. Width of kink band is $2x_0$.

Now let us consider the buckling aspect of the process, using elementary plate theory. According to eq. 22, Chapter 6, the critical axial load, P, for each layer required to initiate kinking is:

$$P = \tau t b/(dv/dx)_0 \tag{11a}$$

where τ is shear strength and $(dv/dx)_0$ is the slope angle of layers locally. For a frictional material:

$$\tau = -\sigma_n \tan\phi$$

and for small slope angles, $(dv/dx) = \tan\theta$. Also, $\sigma_1 = \sigma_n$ and $\sigma_3 = -P/tb$, where t is thickness and b is unit width of individual layers. Thus, eq. 11a can be rewritten as:

$$(\sigma_3/\sigma_1) \approx \tan\phi/\tan\theta \tag{11b}$$

which is a close approximation of the exact expression, eq. 10.

Consideration of buckling with plate theory allows us to do more than approximate eq. 10, however. The buckling theory developed in Chapter 6 indicates that equilibrium can be maintained by bending of the layer in the kink band. Analysis of the bending leads to the solution for the *kink form* (Fig. 6B), in individual layers (eq. 16, Chapter 6):

$$v = v^0[(1/\pi)\sin(\pi x/x_0) + (x/x_0)] \tag{12}$$

which is the sum of a sinusoidal term and a linear term. Here v^0 is the half height of the step of the kink form and x_0 is the half width (Fig. 6B). The properties of the kink form are discussed at length in Chapter 6. We can place limits on the width of kink bands but the width is difficult to predict for a variety of reasons. The width $2x_0$, is (eq. 24, Chapter 6):

$$2x_0 \leqslant 2\pi\sqrt{(dv/dx)_0 BI/\tau bt}$$

If the stress state is maintained at the critical stress required for initiation of slippage, we can replace the inequality with an equality. Also, we can rewrite the equation using variables used to derive eq. 11b, above:

$$(x_0/t) < \pi\sqrt{(\tan\theta/\tan\phi)(-B/12\sigma_n)}; \quad \sigma_n < 0 \tag{13}$$

Here σ_n is the confinement normal to layers and is negative if compressive. Thus, the width of the kink form, $2x_0$, is proportional to the thickness t, of layers. It is inversely related to the confinement, which is consistent with data reported by Weiss (1968) and Donath (1970), who experimented with paper cards and slate. Kink bands, however, can grow in width after they have locked, so it would be most difficult to determine properties of multilayers by measuring widths of kink bands. Indeed, the primary understanding we gain from considering kinking as a buckling phenomenon is that kink bands should have finite widths. If kink bands were simply a result of yielding instabilities, they should have zero widths.

Bending resistance may play only a fleeting role in the development of many kink bands. If the layers in kink bands yield within hinges (Fig. 10, Chapter 6), the bending resistance will approach zero and the further development of the kink bands will be related to yielding *between* layers.

By considering this yielding between layers we can predict the locking angle of kink bands with eq. 10 because for each stress ratio there are two angles, θ. The larger θ is the locking angle and is shown in Fig. 14, Chapter 8. The locking angle is a function of the stress ratio and the angle of internal friction, ϕ, according to this theory. The ultimate locking angle θ_L, is:

$$\theta_L = \pm(90^\circ - \phi/2) \tag{14}$$

If there is no dilitation within kink bands, the inclination (Fig. 6A) angle, β, of the kink band is related to the slopes of layers, θ:

$$\beta = \pm(90^\circ - \theta/2) \tag{15}$$

Thus, using eq. 14, we derive the minimum angle for the kink band relative to nonfolded layering:

$$\beta_{min} = \pm(45^\circ + \phi/4) \tag{16}$$

These relations are valid only if the principal stresses are parallel and normal to nonfolded layering. Other possibilities are considered in Chapter 8.

Thus, the theory of kinking provides estimates of conditions of initiation, initial widths, locking angles, and inclinations of kink bands relative to nonfolded layering. It also describes, approximately, the propagation of kink bands, as discussed in Chapter 8.

Criteria of kink folding

Thus far we have considered kink folding of multilayers comprised of identical materials. Sinusoidal folding was not possible unless the multilayer buckled as a unit within a soft medium or unless the contact strength between layers was zero. If we consider multilayers comprised of interbedded stiff and soft layers, the contacts between which can yield, we can derive criteria to distinguish conditions that favor kink folding on the one hand from conditions that favor sinusoidal and concentric-like folding on the other. The criteria were derived in Chapter 6. We will consider kink folds which initiate at irregularities in layering or at inflection points in a pattern of sinusoidal folds.

The first criterion, K_1, compares the strength, τ, of contacts between layers and the average shear modulus, G_a, of the multilayer:

$$K_1 = (\partial v_0/\partial x)(G_a/\tau) \tag{17a}$$

where $\partial v_0/\partial x$ is the maximum slope of initial irregularities in the layers. Here τ is the Terzaghi-Coulomb strength:

$$\tau = C - \sigma_n' \tan\phi \tag{17b}$$

in which C is cohesion, σ'_n is effective normal stress between layers and ϕ is angle of internal friction. The parameter G_a is the average shear modulus:

$$G_a = (t_1 + t_2) G_1 G_2/(t_1 G_2 + t_2 G_1) \tag{17c}$$

where t_i is thickness and G_i is shear modulus for stiff (i = 1) and soft (i = 2) layers. The average shear modulus is zero if contact strength is zero. The average shear modulus is essentially equal to the shear modulus of the stiff layers if the thickness of soft layers is very small, and is essentially equal to the shear modulus of the soft layers if the soft layers are extremely soft.

Now, according to the analysis in Chapter 6, if:

$$\tau = 0$$

which requires G_a to be zero, or more generally, if:

$$K_1 \approx 0$$

sinusoidal and concentric-like folding should develop. Therefore, if contact strength is large relative to the shear modulus, yielding will not occur. Or, if the shear strength is zero, yielding occurs everywhere. These conditions favor sinusoidal folding. On the other hand, if:

$$K_1 > 0$$

kink folding is favored. That is, if G_a is large relative to the contact strength, the slopes will be small at the time yielding of contact occurs locally and kink bands are expected. It follows that very low contact strengths, which may be due to high pore-water pressures, and very soft interbeds with high contact strengths, favor sinusoidal and concentric-like folding. Multilayers with finite contact strength and comprised of layers with nearly the same properties favor kink folding.

These predictions are in accord with our experiments, described in Chapters 6 and 8 and with experiments by Weiss (1968), who used decks of paper cards.

The second index, K_2, of kink folding that we derived in Chapter 6 is an attempt to compare the shear resistance to folding of a multilayer to the bending and lateral resistances to folding. The second criterion, unlike the first, requires that the confining media be very stiff relative to the multilayer. It is based on the assumption that the conditions required for yielding of contacts all along the layer at middepth in multilayers with contact strength are the conditions required for initiation of kink folding. Thus, if the average modulus, G_a, is relatively large compared to resistances to lateral deflection and bending, the axial load, P, will be low relative to the critical load for Biot waves, P_B, and the amplitudes of Biot waves will be small at the time yielding occurs at inflection points. This, we presume, describes conditions favorable to kink folding. The criterion is (eq. 45a, Chapter 6):

$$K_2 = 1 + G'N_p[(1 + t^*)^2\sqrt{3}/\pi]\,[G^*/(t^* + G^*)]\,[\sqrt{(B^* + t^*)/B^*}] \tag{18}$$

where $t^* = t_2/t_1$, $G^* = G_2/G_1$, $G' = G_1/B_1$, $B^* = B_2/B_1$, and N_p is the number of pairs of stiff (1) and soft (2) layers in the multilayer. According to our interpretation, if $K_2 >> 1$, kink folding is expected, whereas if $K_2 \approx 1$, sinusoidal and concentric-like folding is expected.

The criterion and the interpretation indicate that large ratios of G_2/G_1, that is, stiff interbeds, and large numbers of layers, favor kink folding. Thus, the primary information added by the second criterion is that thick multilayer stacks in which stiff and soft layers have comparable moduli will tend to kink rather than to fold sinusoidally and concentrically if there is contact strength. These predictions also are in accord with observations.

Transition from sinusoidal to conjugate kink to chevron folds

An idealization of the development of kink and chevron folds in elastic multilayers is shown in Fig. 5B. The multilayer contains a variety of initial waves. As it is shortened, initial waves with the *Ramberg wavelength* become amplified selectively (Fig. 5B*1*). When the largest slope at an inflection point of the amplified waves reaches a certain value, which depends upon the average shear modulus and the contact strength, the contacts of the layer at middepth yield locally. The yielding is unstable and the layer buckles, releasing confinement on parts of layers above synclinal hinges and below anticlinal hinges, thereby inducing contacts of adjacent layers to yield unstably. The kink band thus propagates spontaneously through the multilayer (Fig. 5B*2*). The stiff layers yield at hinges and rotate until they lock against adjacent layers as a result of contact strength. The locking angle fixes the direction of propagation of the kink band. As the kink band propagates, it meets stiff boundaries and reflects, producing a conjugate set of kink bands (Fig. 5B*3*). If the kink bands widen, they can produce a pattern of chevron folds, as shown by Paterson and Weiss (1966) *.

Asymmetric folding

Theoretical and experimental analyses of asymmetric folding reported in Chapter 8 indicate that the sense of asymmetry of folds in a multilayer depends upon the properties of the multilayer, the confinement normal to layering, the sense of layer-parallel shear, and the sequence of deformations.

There is a transition from symmetric, conjugate kink bands (Fig. 7A), where initial, layer-parallel shear stresses are zero, to monoclinal kink bands (Fig. 7C), where initial, layer-parallel shear stresses are high. The transition form is an asymmetric, conjugate fold, in which one set of kink bands has a lower slope than the other (Fig. 7B).

The tendency for kink bands to be monoclinal if principal stresses are inclined to nonfolded layering was explained as follows in Chapter 8.

---

* References at end of Chapter 1.

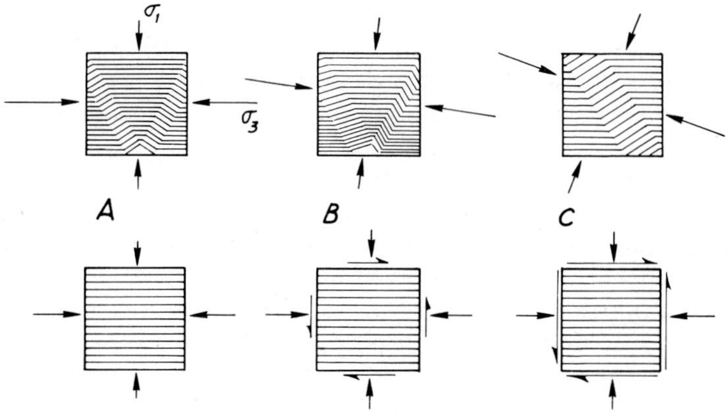

Fig. 7. Symmetric and asymmetric kink bands. A. Ideal, symmetric, conjugate kink bands formed where principal stresses act normal and parallel to layering. B. Ideal, asymmetric, conjugate kink bands formed where principal stress directions are inclined slightly clockwise relative to layering or to normal to layering. Sense of asymmetry would be as deduced from rule of thumb of "drag folds". C. Ideal asymmetric or monoclinal kink band, formed where principal stresses are inclined in a clockwise sense from the layering. Sense of asymmetry is opposite to that predicted by rule of thumb.

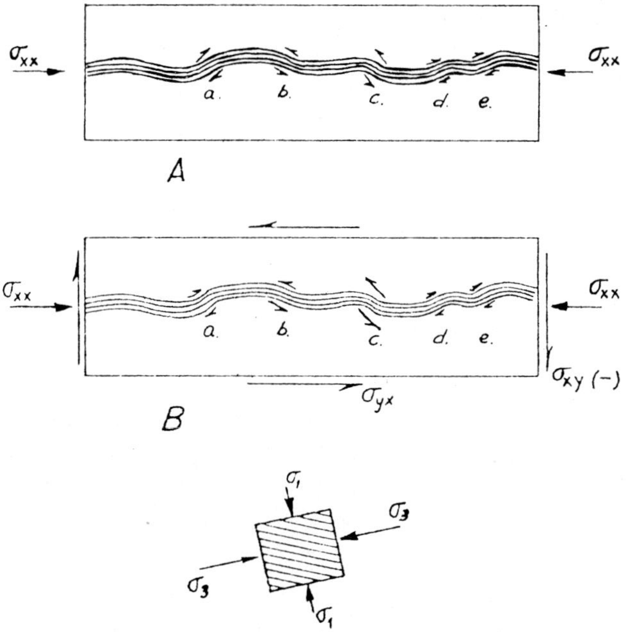

Fig. 8. Senses and magnitudes of layer-parallel shear induced by local slopes of layers within irregular multilayer. A. Initial stress state uniaxial, parallel to general layering. B. Initial stress state biaxial, with principal stresses oriented as shown in C.

Suppose that a multilayer contains two initial dips at *a* and *b* (Fig. 8A). The initial dip at *a* is left-lateral and that at *b* is right-lateral. The senses of layer-parallel shear stress induced at *a* and *b* by normal stresses σ_{xx} and σ_{yy}. acting normal or parallel to nontilted layering, are opposite as indicated in Fig. 8A. If the multilayer is also subjected to shear in the sense indicated in Fig. 8B, the layer-parallel shear stress on the initial dip *a* is reduced but that on *b* is increased. Therefore, slippage will tend to occur first at *b* rather than at *a*, and right-lateral kink bands will be formed for this state of stress, which is also shown in Fig. 6C. Further, the kink band at *b* will not reflect at a rigid boundary to produce the conjugate band if the principal stresses are sufficiently highly inclined to nonfolded layering, as explained in Chapter 8. In this way monoclinal kink bands are initiated.

The sense of layer-parallel shear can be deduced from the sense of asymmetry of the folds, for either conjugate (Figs. 9A*4*, 9B*4*) or monoclinal (Figs. 9A*3*, 9B*3*) kink folds. Left-lateral monoclinal kink bands are associated with a positive state of shear stress and right-lateral monoclinal kink bands are associated with a negative state of shear stress. I use the following rule of thumb to remember the sense of monoclinal kinking associated with a certain sense of shear. Consider the component of shear stress normal to layering, shown by dashed arrows in Figs. 9A2 and 9B2. The sense of shear normal to layering is reversed if the direction of maximum compression is changed as shown in those figures. Then, the sense of displacement of opposite sides of the monoclinal kink band are compatible with the sense of shear stress normal to layering.

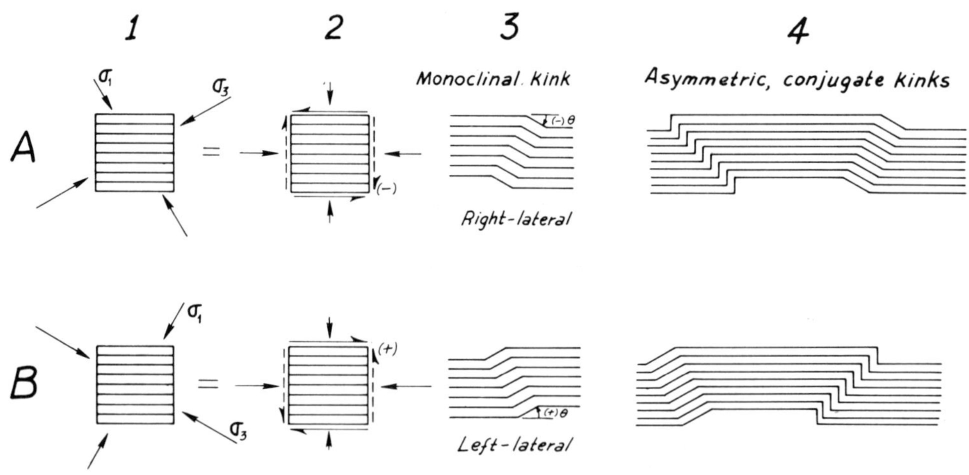

Fig. 9. Relation between orientation of maximum compression, σ_3, and senses of asymmetric folding developed in multilayers in which contact strength is overcome. A. Negative state of shear relative to layering produces either monoclinal kink band or asymmetric, conjugate kinks. Monoclinal kink is right-lateral. B. Positive state of shear produces either lef-lateral, monoclinal kink band or an asymmetric set of kinks.

For conjugate folds, in contrast, Figs. 9A*4* and 9B*4*, one uses the standard rule of thumb for drag flexures. The asymmetry indicates the sense of *layer-parallel* shear stress. Thus, it is only the monoclinal kink band which requires the special rule of thumb.

Figures 10 show ideal folds developed during different sequences of loading:

Simultaneous layer-parallel shearing and shortening. If smooth, sinusoidal folds or conjugate kink bands are developed, the sense of asymmetry is compatible with the rule of thumb of drag folds (Figs. 10A*4* and 10A*6*). If monoclinal kink bands are produced, the sense of shear is deduced as shown in Figs. 9. The smooth folds begin as symmetric forms and become asymmetric as their amplitudes increase.

First layer-parallel shortening, then shear. The kink folds are first conjugate and symmetric. Smooth, sinusoidal folds also begin as symmetric forms. Then both types become asymmetric in the sense of drag folds. If the layers are elastic, however, the conjugate kink folds may suddenly flip into a monoclinal mode (Chapter 7, Fig. 10C*2*).

First layer-parallel shear, then shortening. For conditions that favor

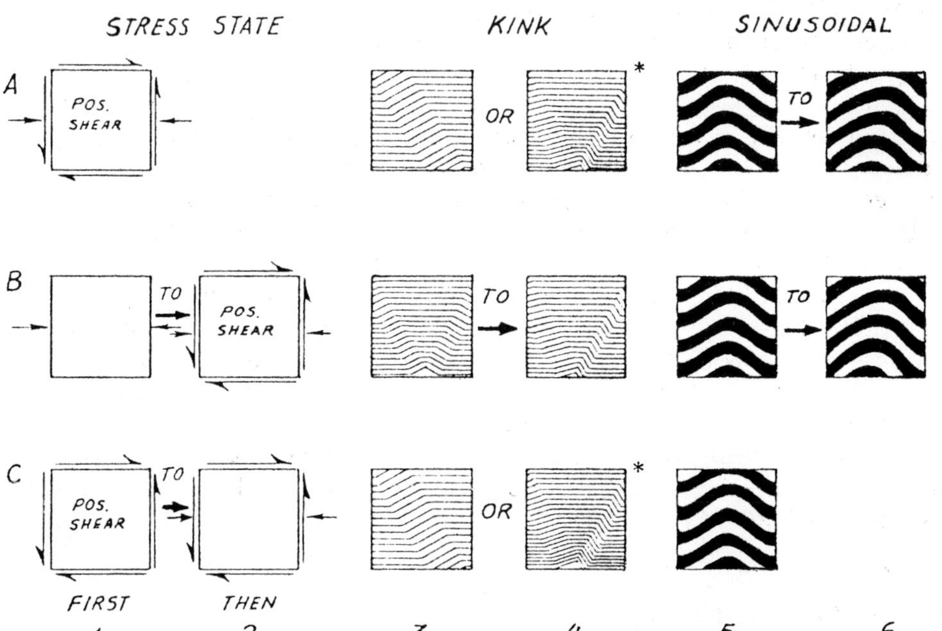

Fig. 10. Ideal fold patterns developed in different kinds of multilayers subjected to different states of stress and different deformation histories. Patterns deduced from experiments and theory. A. Simultaneous layer-parallel shear and shortening. B. First layer-parallel shortening and then shear. C. First layer-parallel shear, then shortening. Folds patterns marked (*) are theoretical; they have not been observed experimentally.

sinusoidal folding there is no asymmetry, at least for low-amplitude folds. For conditions that favor kink folding in materials with memory, the sense of asymmetry is the same as that for simultaneous shear and shortening. The kink bands may be monoclinal.

The effect of layer-parallel shear is twofold. If there is easy slip between layers, the shear stress between layers is negligible and asymmetry can be considered to be a result of superimposed uniform shear strain on a pattern of passive markers (Chapter 7). If properties favor kink folding, however, layer-parallel shear induces shear stresses which affect the slippage locally and thereby *mechanically* affect the fold pattern (Fig. 8). As the kink folds lock, at higher amplitudes, further layer-parallel shear deforms the fold pattern much as passive markers (Chapter 7). Thus, the layer-parallel shear *stress* is unimportant in sinusoidal folding, at least for low amplitude folds, but the layer-parallel shear strain produces asymmetry. Both layer-parallel shear stress and shear strain produce asymmetry in kink folding.

Doubly plunging folds

Nearly all theories of folding with which I am familiar consider folding under conditions of plane strain or plane stress. Thus, folds are imagined to be cylindroidal. Folds in fact terminate and most folds terminate at both ends unless they form with axes at steep angles to the earth's surface. The folds that terminate at both ends we have termed *doubly plunging* (Chapter 9).

We have developed theory for doubly plunging folds in Chapter 9, using elementary plate theory. The theory is based on the usual assumptions of plate theory as well as the assumption that the ideal shape of the simple doubly plunging fold is comprised of sinusoidal waves in two orthogonal directions. The wavelengths are L_x and L_y in the x- and y-direction, respectively (inset figure, Fig. 11A). In cylindrical folding theory we assume that L_y is infinite. Although we have developed the theory for elastic materials, it can readily be extended to viscous materials through the elementary theory. Ghosh (1970) has partly analyzed doubly plunging folds in viscous material, as indicated in a footnote in Chapter 9.

The theory indicates that the ratio of compressive normal stresses parallel to the x- and y-directions of the doubly plunging fold is related to the wavelength ratio (eq. 26g, Chapter 9):

$$(\sigma_{yy}/\sigma_{xx}) = [(L_x/L_y)^2 + 2]/[(L_y/L_x)^2 + 2] \qquad (19a)$$

Equation 19a is plotted in fig. 11A.

The wavelength in the x-direction is given by the equation:

$$L_x = L_0\sqrt{1 + (L_x/L_y)^2}$$

or, using eq. 19a:

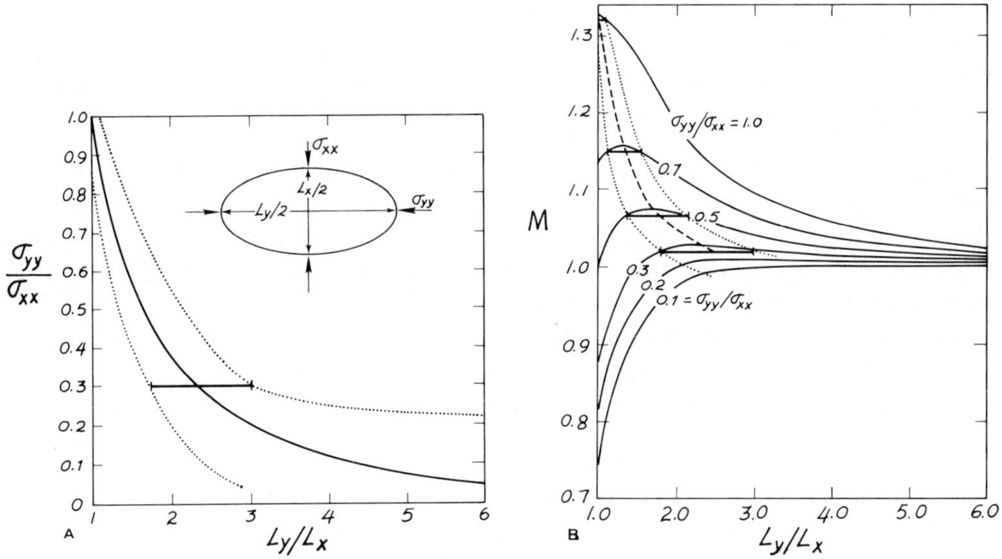

Fig. 11. Theoretical plan shapes of doubly plunging folds. A. Relation between ratio of horizontal stresses and ratio of wavelengths of doubly plunging folds (see inset figure). The relation is valid for single-layer or multilayer folds. B. Relation between parameter M and wavelength ratio for various values of stress ratio. The larger is M, the smaller is the critical stress in the x-direction required for marked amplification of folds. The maxima of these relations define the solid curve in A. The dotted lines indicate approximately the selectivity of the folding. The more widely spaced the dotted lines, the poorer is the selectivity.

$$L_x = L_0\sqrt{(\sigma_{yy}/\sigma_{xx}) \pm \sqrt{[(\sigma_{yy}/\sigma_{xx}) - 1]^2 + (\sigma_{yy}/\sigma_{xx})}} \tag{19b}$$

where L_0 assumes different values, depending upon the nature of the multilayer. If there is a single layer in a thick medium, L_0 is the L defined in eq. 6. For a multilayer confined by elastic media, L_0 is the *Ramberg wavelength*, L_R, defined in eq. 5a. If the confining media are very stiff, L_0 is the *Biot wavelength*, L_B, defined in eq. 52 of Chapter 2. Finally, if the multilayer is comprised of layers with frictionless contacts, with identical elasticity moduli but various thicknesses, L_0 in eq. 19b is defined by:

$$L_0 = L_R = 2T\sqrt{\pi n/\bar{N}}\sqrt[4]{1/[12(1-D)]}$$

where $\bar{N}$ is the *effective* number of layers, defined by eq. 8, n is defined in eq. 5b and D is defined in eq. 5c. T is the total thickness of the multilayer.

Therefore, we readily can compute the plan shapes of ideal doubly plunging folds if we know the stress ratio, σ_{yy}/σ_{xx}, of principal horizontal stresses, and the properties required to compute the wavelengths for cylindrical folds. Regardless of the elastic and dimensional properties of the multilayer, however, we can compute the stress ratio if we know the plan shape of the fold, by using eq. 19a or Fig. 11A. As mentioned in Chapter 9, however, the plan shapes of folds should be relatively insensitive to stress ratios if the stress ratios are less than about 0.2. This can be shown with Fig. 11B, which indicates the relation between the wavelength ratio and a parameter M. Now, M is the coefficient of the driving stress in expressions for amplification of waves, presented in Chapter 9. All we need to know here is that the larger is M the smaller is the critical stress required for folding. Thus, the maxima of the curves shown in Fig. 11B define the solid curve in Fig. 11A. The maxima are poorly defined for stress ratios less than about 0.3 and well defined for higher stress ratios. Therefore, if the stress ratio is low, the selectivity of the folding will be low; we would expect to see folds with a wide range of plan shapes if the stress ratio is less than about 0.3. The degree of selectivity is indicated approximately by dotted lines in Figs. 11. The band width was arbitrarily selected, as shown in Fig. 11B, to correspond with a factor of M of 0.01 less than the maximum value of M. According to Fig. 11A, if the stress ratio were 0.2, doubly plunging folds in which the long axes ranged from about 2 to nearly infinity times the short axes would be amplified about the same amount. It the stress ratio were 0.5, however, doubly plunging folds with long axes ranging from about 1.3 to 2.2 times the short axes would be amplified about the same amount. Clearly, the selectivity markedly increases with the stress ratio. Accordingly, where stress ratios are relatively low, we would expect to see a wide range of plan shapes of doubly plunging folds, and where stress ratios are relatively high, we would expect the doubly plunging folds to be dome-like and there should be a narrow range of plan shapes.

This is the state of our understanding of the mechanics of folding of multilayers with memory.

AUTHOR INDEX

SUBJECT INDEX